The Neurobiology of Aggression and Rage

Allan Siegel

CRC PRESS

Boca Raton London New York Washington, D.C.

Library of Congress Cataloging-in-Publication Data

Siegel, Allan.
The neurobiology of aggression and rage / Allan Siegel.
p. cm.
Includes bibliographical references and index.
ISBN 0-415-30834-8 (alk. paper)
1. Aggressiveness--Physiological aspects. 2. Anger--Physiological aspects. 3. Neuropsychology. I. Title.
[DNLM: 1. Neurobiology. 2. Aggression--physiology. 3. Rage--physiology. WL 102 S571n 2004]
QP401.S555 2004
155.2′32—dc22 2004057043

Direct all inquiries to CRC Press, 2000 N.W. Corporate Blvd., Boca Raton, Florida 33431.

Visit the CRC Press Web site at www.crcpress.com

International Standard Book Number 0-415-30834-8
Library of Congress Card Number 2004057043
Printed in the United States of America 1 2 3 4 5 6 7 8 9 0
Printed on acid-free paper

Dedication

To the memory of Dr. John P. Flynn, whose dedication to scientific truth, scholarship, and ethical conduct represented the gold standard for those who were privileged to study under his guidance.

"Who is strong? He who conquers his evil inclination, as it is said: 'Better is one slow to anger than a strong man, and one who rules over his spirit than a conqueror of a city' "
[Ethics of the Fathers, Chapter 4]

Preface

The basic goal of this book is to provide an up-to-date review and analysis of the biological factors and processes that are involved in the expression and control of rage and aggressive behavior. The major focus of the book is to provide an understanding of the anatomical substrates of the major forms of aggression as well as to examine their basic underlying physiological, neurochemical and genetic mechanisms. This analysis is determined mainly from animal research in order to illustrate how such knowledge helps us to understand the neurology of human aggression. A final chapter provides the author's views on how the neurobiology of aggression and rage can be utilized to understand and control our own aggressive tendencies.

The Author

Allan Siegel, Ph.D., is professor of neurosciences and psychiatry in the Departments of Neurology & Neuroscience and Psychiatry at the New Jersey Medical School in Newark. For the past 39 years, Dr. Siegel's research has been dedicated to the neurobiology of aggression and rage, including studies concerning neurophysiological and neurochemical mechanisms and neuroanatomical substrates for these processes. He has published over 260 papers and abstracts, including the Pretest Series in Neurosciences, and is co-authoring a textbook in neurosciences that is near completion.

Acknowledgment

Portions of the research presented in this book were supported by NIH Grant NS 07941 and by grants from the Harry Frank Guggenheim Foundation to the author.

Table of Contents

1 What Is Aggression?

Violence and rage have become major public health and social problems in the United States and many other countries around the world. In particular, during the past few years there have been more than 3 million violent crimes committed in the United States annually (Reiss et al., 1994), resulting in costs of billions of dollars to society. In addition, the homicide rate among adolescents increased 238% between 1985 and 1994 (Hennes, 1998). In the year 2000, an estimated 1.4 million violent crimes were reported (FBI Press release, Oct. 22, 2001).

Violence is influenced by cultural, environmental, and social forces that shape the manner in which it is expressed (Eron, 1987). Nevertheless, as supported by a significant body of data described below and in subsequent chapters, the neural basis of aggressive behavior in humans resembles that of animals, and the forms of aggression seen in humans parallel those observed in animals. Thus, some important questions are:

1. What are the forms of aggression that are studied in animal research?
2. How are they related to each other?
3. How do they relate to human forms of aggression?
4. What are the similarities, as well as differences, in the neural mechanisms that govern each of these forms of aggression studied in animals?

In this chapter, I attempt to address the first three questions, and consider the fourth in subsequent chapters.

In attempting to identify models of aggression, it is constructive to first try to define what is meant by "aggressive" behavior. Perhaps, the most common definition of aggressive behavior used is a variant of one that was described by Moyer (1968) who referred to aggression as "a behavior that causes (or leads to) harm, damage or destruction of another organism." While this definition would seem to include a wide variety of conditions normally associated with aggressive behavior, it does not necessarily include the classes of behavior that are associated with "threat" or related forms of "affect" that are intimidating in nature, and that are designed to control the behavior of others. The notion of threat can also be expanded to include categories of behavior linked under the general rubric of "hostility." Here, we have to distinguish between two levels of definition — one based on principles of social psychology and the other based on likely underlying neuronal mechanisms. In terms of social psychology, it is possible and even reasonable to postulate distinct behavioral or social mechanisms that identify "hostility" as a unique aggressive process relative to "threat" and "rage" (Kingsbury et al., 1997). However, in terms of the neural mechanisms that may underlie these behaviors, it is conceivable and perhaps likely that they are highly similar. In terms of the major themes of this book, wherever possible there will be attempts to collapse these into categories that are likely subserved by common behavioral and neuronal substrates.

At the animal level, aggressive responses may include threatening postures and vocalizations such as hissing and growling. At the human level, the outward expression of aggressive behavior can reach much higher levels of complexity and, accordingly, may include a much wider spectrum of responses. In particular, aggressive responses can be manifest in varieties of ways, extending from physical assault through verbal and postural expressions. Aggression can even be of a passive form in which the aggressive display is in the form of a deliberate failure of an individual to perform a given act or task that was requested of him.

ANIMAL MODELS OF AGGRESSION

From the discussion above, it would seem natural to assume that aggressive behavior is not a unitary phenomenon, but instead may reflect various processes captured under a single common theme referred to as "aggressive behavior." Complicating this matter further is the fact that different forms of aggression are identified operationally on the basis of experimental methodologies applied by various investigators to the study of aggression. In 1968, Moyer attempted to define types of aggression largely on the basis of these methodologies. In effect, what Moyer established were "operational" definitions of aggression based on the arrangement of various environmental conditions. Because of the common use of several methodologies and types of aggression, these categories of aggression described by Moyer are summarized below as a basis for further discussion.

1. *Fear-induced aggression.* In this form of aggression, the animal is placed in a position where it would like to escape, either because of the degree of confinement or the presence of a threatening animal in its environment. In this situation, however, escape is denied to the animal, and the animal turns, instead, to attack the attacking animal following its attempt to escape.
2. *Maternal aggression.* This form of aggression is one that occurs in most forms of mammalian species, but is used infrequently in aggression research. In this model, the critical stimulus is the presence of an organism in proximity to the young. Here, the closer the mother is situated to both the young and threatening organism, the more likely an attack.
3. *Inter-male aggression.* In this form of aggression, the presence of a male triggers aggressive reactions on the part of another male. The frequency of bouts that occur between a male and another male greatly outweigh those that occur between a male and a female or between two females. Inter-male aggression shares some similarities with the resident–intruder model (see below), but differs from it in that the interactions do not take place within the subject's territory.
4. *Irritable aggression.* This form of aggression occurs in response to a threat, intimidation, or an environmental condition that is annoying or

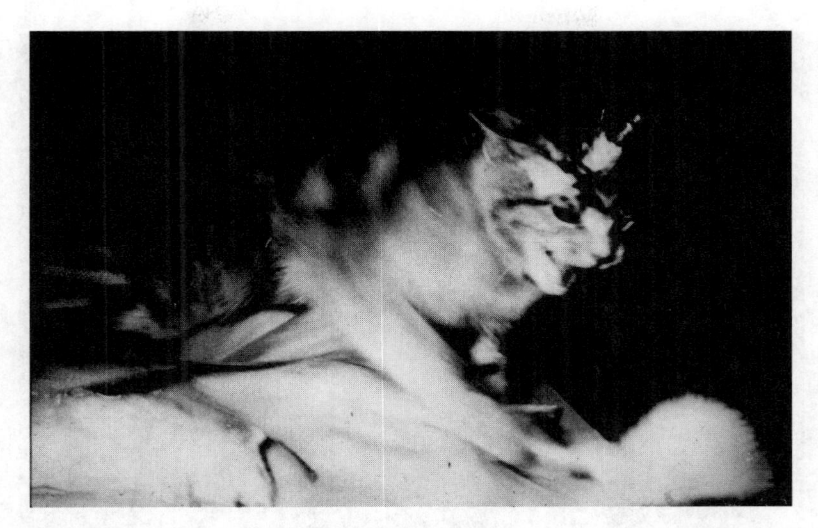

FIGURE 1.1 Defensive rage behavior in the cat. This response is elicited by electrical stimulation of the medial hypothalamus of the cat. The response induced closely mimics that seen under natural conditions when the cat is threatened by another species. This response includes piloerection, pupillary dilatation, arching of the back, striking at the animal that it perceives to be threatening, marked hissing, and retraction of the ears.

irritating. The response can be directed to a wide variety of subjects, and is generally not associated with escape or flight behavior. A related form of aggression is found in cats (Figure 1.1), and is referred to as "affective defense" (also called "defensive rage"). Considerable research has been conducted with this model of aggression in cats, and is considered in detail in subsequent chapters.

5. *Sex-related aggression*. This type of aggression is more commonly observed in humans and to a lesser extent in animals. In humans, this type of aggression may manifest itself in a variety of ways, some of which are strange or bizarre. Here, sexual arousal is frequently associated with increased levels of hostility or hostile fantasies. In animals, components of aggressive behavior are sometimes associated with sexual acts. In such instances, the aggressive and sexual aspects appear as components of the same behavioral act, thus creating difficulties in classifying these behaviors. Overall, few studies involving this form of aggression have been conducted, especially in animals.
6. *Predatory aggression*. This form of aggression is basically unrelated to the other forms of aggression described above and can be distinguished from them because (1) it is specifically triggered by the presence of a prey object within the visual field of the predator, and (2) a predator will display predation regardless of the environment in which it is placed. Predatory attack behavior has been studied most extensively in cats, where this response can be elicited by electrical stimulation of the lateral

FIGURE 1.2 Predatory attack behavior in the cat. This response is elicited by electrical stimulation of the lateral hypothalamus of the cat. The response induced closely mimics that seen under natural conditions when the cat is attacking a prey object. This response lacks the pronounced sympathetic components seen with defensive rage, but instead, is characterized by minor pupillary dilatation, an absence of vocalization, initial stalking of the prey object, and an attack aimed for the back of the neck, which is typically deadly to the prey.

hypothalamus (Figure 1.2). However, experimentation using a rodent model of predatory attack has also been described (Sandnabba, 1995). The use of the feline model of predatory attack in ethological and experimental settings has enabled investigators to obtain considerable information concerning the neural (Flynn, 1970, 1976; Siegel and Pott, 1988; Siegel and Brutus, 1990) and behavioral (Leyhausen, 1979) properties of this form of aggression. However, it should be noted that because it is different from other forms of aggression, predatory attack may be viewed by some ethologists not as a form of aggression, but rather as behavioral strategy in association with feeding behavior (Adams, 1980; Palanza et al., 1994). Nevertheless, most authors disagree with this notion, and its role in normal and pathological states is considered later in this chapter.

7. *Territorial aggression.* This form of aggression occurs when an intruder enters into an area that an animal has determined for itself to be its own domain. This is commonly known as a resident–intruder model. Tests using the resident–intruder model generally use animals of the same species, although a resident animal might also attack an intruder of a different species. Both environment and test stimuli are important in this type of aggression. For example, as the intruder moves toward the

periphery of the territorial demarcation, the likelihood of attack from the resident decreases. Several different types of intruders have been used such as an unfamiliar male, an olfactory bulbectomized male, or a lactating female. Each type of intruder generates a different probability of attack related to the sex of the resident. For example, a bulbectomized male mouse intruder will likely elicit attack from the resident male mouse, while a lactating female mouse intruder will induce attack more readily from a female resident than a male resident mouse (Whalen and Johnson, 1988).

OTHER MODELS OF AGGRESSION

Other models and operational definitions of aggression have been proposed that do not necessarily fit into the classification system proposed by Moyer. For example, one such model includes competition for a limited resource (Palanza et al., 1994). In this model, the subjects are placed into a situation in which they must interact to obtain basic commodities such as food and water, or to obtain access to a better territory or a mate. While this model may be viewed as a variation of the resident–intruder model, it also differs from it in several ways. First, the resident–intruder model involves interactions that take place in the resident's territory, whereas in the competition model, the interaction takes place in an area that is common to both animals. Second, there is no need for competition to take place since resources such as food and water are not limited. Another model is *infanticide*. Here, the subject is given access to pups and the variable measured is the death of the pups.

IS AGGRESSIVE BEHAVIOR BIMODAL?

Over the past few decades, a number of investigators have attempted to characterize aggressive behavior in either animals or humans in a bimodal manner. In the animal literature, the use of the terms *offense* and *defense* have been applied to describe what appear to be opposing ends of a continuum (Adams, 1979). A second bimodal classification scheme defines aggression along a continuum ranging from predatory to affective defense behavior (Flynn, 1970; Siegel and Pott, 1988; Siegel and Brutus, 1990). A third bimodal classification scheme was recently proposed by Kingsbury et al. (1997) for the study of human aggression. It suggests that aggressive behavior can be classified into two categories, which the author refers to as "factors." These include *instrumental aggression* and *hostile aggression*. The following discussion summarizes these classification schemes and approaches in the study of aggressive behavior.

OFFENSIVE AND DEFENSIVE BEHAVIOR

Considerable effort has been expended, mainly through studies in rodents, in attempting to characterize the terms "offensive" and "defensive" behaviors. In addition, several investigators have made attempts to identify the neuronal regions associated with each of these functions (see Adams, 1979; Blanchard et al., 1977a to 1977e; Blanchard and Blanchard, 1984; Mos et al., 1984). The neuronal concomitants of aggression in rats will be considered in Chapter 3.

Offense

Offensive behavior in rats can typically be demonstrated with the use of the resident–intruder model in which the alpha male (i.e., the resident) elicits the offensive behavior (see Figure 1.3). Typically, offensive behavior is characterized primarily as a form of threat together with a ritualized form of attack consisting of piloerection, a (lateral) bite and kick in which wounds are made mainly upon the back of the opponent. Since the wounds are basically superficial, it was suggested by Adams (1979) that offensive behavior provides a communicative function, namely, to serve notice of that animal's dominance over the territory that it considers its own. However, it has also been demonstrated that the site of the wounds is dependent on the environment of the animal. Most wounds involving largely the upper part of the back occur from rats housed individually in small cages, while rats housed in larger cages with a female showed a smaller percentage of bites on the back with a greater percentage of bites directed at the head and belly (Mos et al., 1984).

FIGURE 1.3 Resident–intruder model in the rat. This model, which is commonly used to study aggression in rodents, involves the placement of a rat, called the "intruder," in the cage of another rat, called the "resident." Typically, the resident attacks the intruder in a characteristic manner. Illustrations of these responses are shown in (A) in which a male resident rat attacks the intruder while leaping with a neck bite, and then the intruder escapes; and (B) in which one rat displays a defensive upright posture in response to a sideways threat by the other. (Courtesy of Dr. Klaus Miczek.)

FIGURE 1.3 (Continued)

Experimental studies involving offensive behavior have not been conducted in cats. However, a detailed ethological analysis of offensive behavior in felines has been carried out (Leyhausen, 1979). According to Leyhausen, as the cat begins to approach a conspecific, it draws itself high on its legs, stretching out its back and its head is pulled forward. The pupils are not dilated, but rather are slightly constricted, and the ears are turned outward in a sharp angle. As the offensive animal continues to approach the conspecific, it may elicit mewing and howling, which most likely constitutes a threat. It then raises its head, tilts it sideways while still focusing on the opponent, then turns the head from side to side, and as it passes the conspecific, attacks by attempting to bite the nape of the neck.

Defense

Most studies focused on aggressive behavior have involved the analysis of the behavioral, anatomical, physiological, neurochemical, and neuropharmacological properties of defensive behavior conducted principally in detail in both rodents and cats. In the rat, defense occurs in the presence of a predator, attacking conspecific, or any other stimulus that is perceived to be dangerous to the animal (Blanchard and Blanchard, 1989), and is sometimes alternatively referred to as "protective aggression" (Palanza et al., 1994). The behavioral characteristics of this response may include several of the following features: a lunge and bite attack, usually directed at the face or parts of the body, an upright posture, squealing, flight, freezing, submissive posture, and ultrasonic vocalization (Adams, 1979; Blanchard and Blanchard, 1989). Several components of natural defense in the rodent are shown in Figure 1.3. It has also been reported that defensive behavior can be elicited by electrical stimulation of the rat's hypothalamus (Panksepp and Trowill, 1969). However, brain stimulation–induced defense in the rat differs somewhat from that which is induced under natural conditions. In particular, brain stimulation–induced defense in the rat includes a sudden retreat followed by attempts to escape by jumping out of the cage.

In the cat, defensive behavior occurs in the presence of another animal (either a conspecific or an animal of another species) that is perceived to be a threat. The basic elements of this response include a flattening of the ears, a shrinking or lowering of the body, drawing in of the head, piloerection, hissing, pupillary dilatation, and a stiffening of the tail that becomes motionless (Leyhausen, 1979). Quite often, such a response pattern is sufficient to cause the opponent to retreat, which is the main objective of the defensive response. If the cat decides to attack, it does so by striking with its paw. The paw strike is the primary means by which it signifies its defense. If the opponent begins to attack, the cat eliciting defensive behavior will then roll over on its back with its underside facing the opponent (Leyhausen, 1979). Such a response can interrupt the advance of the opponent. The continued use of its forepaw may also be effective in warding off the opponent, especially if the forepaw strike lands on the head region of the opponent. If the opponent begins to attack, the cat that initially elicits defensive behavior may now change its strategy to that of offense as it attempts to scratch and bite the neck of the opponent. A similar pattern of responses is elicited by electrical stimulation of the cat's medial hypothalamus or midbrain periaqueductal gray (PAG) (Hess and Brugger, 1943; Flynn, 1970; Chi and Flynn, 1971; Fuchs et al., 1985a, 1985b; Siegel and Pott, 1988; Siegel and Brutus, 1990).

Aggression Induced by Brain Stimulation in Rat: Limitations of Offense–Defense Dichotomy

In contrast to cats, it is apparent that a variety of forms of aggression can be elicited by electrical stimulation of several areas of the hypothalamus in rats (Kruk et al., 1979, 1980, 1983; Panksepp, 1971; Panksepp and Trowill, 1969; Woodworth, 1971). As noted above, elements of defensive behavior (Panksepp and Trowill, 1969) as well as biting attack behavior (Woodworth, 1971) have been reported following stimulation of the medial and lateral hypothalamus of the cat, respectively. In addition, attack elicited from the intermediate hypothalamus (i.e., the region extending from the lateral aspect of the ventromedial hypothalamus to the medial aspect of the lateral hypothalamus) produces a complex attack pattern that may include biting of the head of the opponent, biting of the back accompanied by hind paw kicking of the flank, or a combination of jump attacks and clinch flights (Kruk et al., 1979, 1980, 1983, 1990). The type of attack varies with the strain of rat. For example, head bites are dominant with the inbred strain, CPB-WEzob, while bites directed to the back are common with random-bred albino Wistar CPB-WE rats.

At the time of biting, the attacking rat may place its front paws around the neck or back of the opponent, while the hind limbs are placed in a position to kick the flank and other parts of the opponent's body. If the opponent assumes an upright position, the attacking rat will modify its attacking pattern by eliciting jump attacks. In such an attack pattern, the attacking animal jumps a considerable distance to reach the opponent, at which time, it bites the head and kicks the body of the opponent with its hind paws. If the animal loses its balance after jumping, it may then alter its attack to that of a clinch fight.

It is of interest to note that the attacking rat initially bites the head, then the upper and lower parts of the back, but not the ventral surface of the opponent (if it assumes a submissive position after it is attacked) (Kruk et al., 1983). This form of hypothalamic attack may be directed against various targets. These include dominant and subordinate rats, dead and anesthetized rats, and mice. A male rat may attack another male or a female, especially one that is not receptive (Kruk et al., 1984; Mos et al., 1987). In terms of the form of attack initiated, two additional facts should be noted. The first is that the specific attack pattern is dependent on the behavior of the opponent, in which case the attacking animal may exhibit components of offense and defense in response to the opponent. The second is that the pathways and neural systems mediating offense and defense, although not well delineated in the rat, presumably lie within overlapping regions of the forebrain and brainstem. *Accordingly, stimulation of an attack site will likely activate regions that mediate offense as well as defense.*

In summary, the form of the attack is dependent on the (1) position of the opponent, including primarily the extent to which the head and back are exposed; (2) strategies adopted by the opponent; and (3) specific strain of the attacking animal. Consequently, the utility of applying the terms "offense" and "defense" in the context of hypothalamically elicited aggressive behavior in the rat becomes questionable. Thus, it would appear to be difficult if not illogical to attempt to apply these motivational categories to define and classify hypothalamically elicited aggression in the rat.

Defensive Rage (Affective Defense) Versus Predatory Attack

One school of thought would suggest that aggressive forms of behavior could be reduced to two categories — affective defense and predatory attack. If this were indeed the case, it should also be possible to reduce the seven types of aggression described by Moyer into these two categories. Indeed, categories of aggression that include fear-induced, maternal, inter-male, sex-related, irritable, and territorial aggression may all share a similar common feature, namely, an aggressive response based on the presence of elements of fear and/or threat, which may be real or perceived. Thus, it is reasonable to re-label these categories of aggression under the rubric "affective defense."

Attention should be drawn to one possible source of confusion relating to the nomenclature that I am attempting to establish here. I just noted above that mixed components of offense and defense may be present in the overall repertoire of the animal in expressing aggressive behavior. Therefore, in the present analysis, while

I may characterize the overall response patterns of six of seven Moyer categories as affective defense, it is possible and indeed likely that the component of offense may also be represented here as well. In particular, in the resident–intruder model, most authors agree that elements of offense are clearly present in the behavior of the resident rat on the intruder even though the motivation underlying the aggressive act may be one of perceived fear or threat to it by the intruder. Nevertheless, that offense may be present as a component of the behavioral repertoire of the animal is not inconsistent with the classification of the overall behavioral response as affective defense. In contrast to affective defense, predatory attack is limited to a single category of aggressive behavior, and therefore is far more restricted in its occurrence in nature.

Do Equivalents to Affective Defense and Predatory Attack Exist in Humans?

Affective Defense

To summarize our knowledge of affective defense behavior in animals, we may conclude that affective defense occurs in response to a real or perceived threat. It is usually immediate following the sighting of the threat stimulus within the environment, as thought and planning of the response appears to play little or no role in its organization. The response is also accompanied by marked sympathetic signs such as marked increases in heart rate and blood pressure, piloerection, vocalization, lowering of the back, and retraction of the ears.

It is instructive to attempt to compare affective defense in animals with what some authors refer to as equivalent response forms in humans (Meloy, 1988, 1997; Vitiello et al., 1990). These authors characterize affective defense behavior as impulsive, destructive to the object of the aggressor, and typically aversive to the aggressor. In addition, they further indicate that this response pattern includes poor modulation of behavior and high autonomic arousal. Questionnaire test items closely related to affective defense encompass the following: aggression that was unplanned, an individual who was totally out of control during the aggressive act, aggression that had no purpose, the person is exposed to physical harm during the time when he is aggressive, and an individual who damages his own property during the act of aggression (Vitiello et al., 1990).

The characteristics associated with affective defense behavior in humans have been independently described in detail by Meloy (1988). Several of these characteristics clearly overlap with those described by Vitiello. Included are an intense sympathetic arousal, an affective attack based on a real or perceived (which could be delusional) threat to the person, and an immediate (i.e., impulsive) response to the threat stimulus. Meloy extends his analysis by adding other properties of this form of aggression. Specifically, the goal object of affective defense is to reduce or eliminate the threat object from the environment, and thus presumably reduce the level of tension. A second feature is that the person or animal displaying affective aggression can also easily show displacement of the perceived threat from one object to another. For example, during the time period when affective defense is expressed, the aggressor may easily attack a third person who accidentally entered the room instead of his original target. A third characteristic is that as a result of sympathetic arousal, the behavioral repertoire is typically limited in time to events of short duration (usually lasting no longer than a few seconds or a minute). A fourth characteristic is that the aggressor elicits a ritualized or stereotyped posture displaying defense and attack prior to the initiation of the actual attack. Such posturing could take the form of clenching of the fists, other gestures, and the use of obscene language. Ostensibly, the goal of such behavior is to intimidate the existing threat carrier/object. Finally, the individual is able to subjectively experience the emotional state such as fear or anger that occurs during the time at which affective defense is elicited. In animals as well as humans, it is generally agreed that such states are basically aversive.

These descriptions of affective defense behavior in humans bear a striking resemblance to the term "episodic dyscontrol" that has been used by a variety of investigators (see Monroe, 1978). According to Monroe, episodic dyscontrol is a term that characterizes an "explosive personality," reflecting the absence of impulse control. This behavioral condition has often been associated with paranoia and altered perceptual states, and presumably occurs in response to stimuli that evoke fear, anger, or rage. (This response is currently referred to as an "intermittent explosive disorder" in the psychiatric diagnostic classification scheme.) Monroe proposes that it results from an imbalance of urge control mechanisms in which one of such mechanisms is overwhelmed by an intense drive or a poorly developed mechanism is overwhelmed by a normal urge. Episodic dyscontrol assumes the presence of excessive neuronal discharges from limbic structures to subcortical regions such as the hypothalamus and brainstem. In fact, this represents a central thesis by Monroe, who provided extensive evidence in support of this view. This notion as well as the likely neuronal mechanisms governing the expression and regulation of aggressive behavior will be more fully considered in a later chapter.

Predatory Attack

As we indicated above, predatory attack, which is common among a wide variety of species, has as its major objective the procurement of food for the aggressor, and for this reason, occurs across species. We also pointed out that few autonomic signs are present in this form of aggression aside from some mild pupillary dilatation. The response pattern is not associated with aversive properties, but is, instead, positively reinforcing. The question of importance here is whether a comparable form of behavior exists in humans. This question is addressed directly in the following discussion.

A review of the literature reveals that overwhelming numbers of studies conducted in humans have concerned forms of aggressive behavior most closely linked with affective defense. In fact, a perusal of the test batteries constructed over the years leads one to draw a similar conclusion. Almost all items taken from tests such as the Buss–Durkee hostility scale (1957) and Bussand Perry (1992) relate to affective forms of aggression rather than to predatory-like behavior. Nevertheless, a number of papers have been published over the past few years that have specifically identified the presence of predatory forms of violence in humans.

Perhaps, the most extensive description of human predatory aggression was provided by Meloy (1988, 1997). Of central importance in this description is that the characteristics of human predatory aggression contrast with those associated with affective defense. In particular, Meloy points out the virtual absence of sympathetic signs that are characteristic of affective defense. Because of the absence of these signs, it is often difficult to detect any response patterns that could be used to predict the onset of predatory aggression. It should also be noted that individuals displaying predatory violence might shift to affective aggression when the victim is in physical contact with the aggressor. The trigger for the shift in response patterns is the physical presence of the victim, which likely activates acute anxiety, fear, or anger reactions. It is further possible that the reverse sequence may take place, namely, that predatory aggression may follow affective aggression as a means of causing more punishment to the victim. This behavioral pattern may be particularly true with psychopaths who express sadistic impulses (Meloy, 1988, 1997).

A second characteristic of predatory aggression is that there appears to be little conscious awareness of emotion. If there is any emotion at all, it is associated with positive reinforcement, in which case the individual may have possessed feelings such as exhilaration. The aggressive act will also heighten self-esteem, resulting in a greater sense of self-confidence and sadistic pleasure. Such feelings contrast dramatically with affective defense, which is associated with aversive feelings.

A third property is that, similar to the cat, the behavior is purposeful and planned. The attack is purposeful in that the aggressor chooses the target, the manner of attack, and the magnitude of the response. In this way, it parallels predatory attack in nonhuman species. One major difference, however, is that in nonhuman species such as the cat, predatory attack is typically directed against an animal of another species. In humans, the attack is directed against other humans. The exception here would be the human "sport" of hunting, which of course is directed against other species. Concerning the motivation underlying such behaviors, in animals such as felines, the purpose is to seek food. But, what is the motivation in humans? Certainly, it would seem that food-seeking behavior should play little or no role since food is generally available in supermarkets or local grocery stores. Meloy (1997) suggests, instead, that predatory behavior "may be used to gratify certain vengeful or retributive fantasies. It may be subjectively experienced as a necessary behavior that would be clinically assessed as compulsive."

A fourth property described by Meloy is that there is no perceived threat. Instead, the aggressor rather than his responding to a threat by an opponent, which occurs during affective defense, actively seeks the target. Of interest here is that the aggressor's active approach to the target can be considered a form of "stalking," which may represent an homologous form of behavior to that elicited by the cat in its "stalking" of a prey object. A fifth property is that predatory aggression may be triggered by a variety of objectives such as gratification of sadistic desires and fantasies, and relief from compulsive drives. This contrasts with affective defense where there is a single objective of reducing the perceived threat.

Meloy describes other properties of predatory aggression. There is limited or no displacement of the target of aggression in contrast to affective defense, where considerable displacement may occur, as the aggressor will tenaciously pursue the victim. The predatory response may take place over a period of minutes or extend over years and may be preceded or followed by some form of private rituals. Such rituals may involve the selection of certain items of clothing, nationalistic or religious symbols, weapons, or masks. The aggressor may anthropomorphize these objects in order to fantasize control over them and thus provide a basis for exerting control over the actual victim. In comparison to affective defense, there appears to be an important cognitive component to the response in which fantasy may play a key role. The human predator also has the ability to focus on the target by filtering out other sensory information in much the same way that a cat will do in focusing on its prey object.

HOSTILE VERSUS INSTRUMENTAL AGGRESSION

There is still yet another way in which aggression can be conceptualized. In this approach, aggression can be viewed from the perspective of two models — instrumental and hostile aggression (Aronson, 1992; Kingsbury et al., 1997). *Instrumental aggression* can be understood in terms of operant conditioning. In this view, an act of aggression occurs because of the expectancy of the positive reinforcement or reward that is to follow. One can see from this model that the likelihood of committing an aggressive act may increase as a function of social reinforcement emanating from an environment where gangs or mobs are present (Kingsbury et al., 1997). Alternatively, in cases that possibly involve certain types of brain-injured or psychopathic personalities, such individuals may not experience negative or guilt feelings after committing an act of aggression. Accordingly, the absence of negative reinforcement could increase the probability of aggression.

While the goal objective of instrumental aggression is the acquisition of some form of reward, in which harm inflicted on an individual serves as a tool for this end, the goal objective of *hostile aggression* is to specifically harm another individual (Kingsbury et al., 1997). It is triggered by specific stimuli in the environment such as the presence of a gun or other threatening objects, including a person, and is facilitated by conditions that include negative reinforcement, fear, anxiety, and frustration. Associated with hostile aggression is a state of heightened behavioral and autonomic arousal. Hostile aggression may occur in an individual who has just been assaulted or even in a person after perceiving that she was threatened or insulted.

One can argue that there are striking parallels between instrumental and hostile aggression, and predatory attack and affective defense behaviors, respectively. There are distinct similarities between instrumental aggression and predatory attack in that the goal of predatory attack in both animals and humans is positively reinforcing. In predatory attack, the response is planned with its purpose to achieve a specific goal — a prey object for food (in animals) and a symbolic or practical objective (in humans). In addition, few autonomic signs are present with instrumental aggression, as is the case with predatory attack. Similarities between hostile aggression and affective defense include the following: both forms of aggression are activated by specific or perceived stimuli that are viewed as threatening; both occur in an impulsive manner; both are directed at producing harm to the object of the attack; and both are associated with marked autonomic signs.

SUMMARY

The purpose of this chapter was to consider the different ways in which aggression has been defined by investigators in the fields of both animal and human research. Moyer (1968) and others originally proposed seven categories of aggression. Other models of aggression, such as the competition model and infanticide, do not readily fit into the classification scheme proposed by Moyer and may be viewed independently of them.

This chapter also considered the conceptual schemes proposed by various investigators, which categorizes aggression in a bimodal manner. One commonly used scheme is to categorize aggression in terms of *offense* and *defense*. The advantages and disadvantages of the use of these terms were considered. The main advantage is that much work has been conducted in which behavioral properties of offense and defense have been described and categorized. The disadvantage is that different forms of aggressive behavior often include both components of offense and defense, thus complicating interpretation of the data, especially when attempts are made to relate such behaviors to specific neuronal regions and functions. A second categorization scheme is to classify aggressive behavior as either affective defense or predatory attack. One strength of this approach is that it is possible to fit all of the categories of aggression proposed by Moyer into either affective defense or predatory attack. We pointed out that most of Moyer's categories can fit within the rubric of affective defense. A second advantage of this categorization scheme is that human aggression can also be described in terms of affective or predatory modes. One possible problematic feature with the use of this bimodal classification scheme is that in humans, affective components of aggression may appear during behavioral sequences that would normally be classified as predatory. Because these forms of aggression typically remain segregated in lower species such as cats, there may be logical as well as practical complications in attempting to apply these models in a universal manner at the human level. Finally, the bimodal categorization of hostile and instrumental aggression was considered. Here, we also pointed out that hostile and instrumental aggression could be viewed as fitting into the categories of affective defense and predatory attack, respectively.

REFERENCES

Adams, D.B., Brain mechanisms for offense, defense, and submission, *Behav. Brain Sci.*, 2, 1979, 201–241.

Adams, D.B., Motivational systems of agonistic behavior in muroid rodents: a comparative review and neural model, *Aggress. Behav.*, 6, 1980, 295–346.

Aronson, E., *The Social Animal*, W.H. Freeman, New York, 1992.

Blanchard, D.C. and R.J. Blanchard, Affect and aggression: an animal model applied to human behavior, in R.J. Blanchard and D.C. Blanchard, Eds., *Advances in the Study of Aggression*, vol. 1, Academic Press, New York, 1984, pp. 2–63.

Blanchard, D.C., R.J. Blanchard, L.K. Takahashi, T. Takahashi, Septal lesions and aggressive behavior, *Behav. Biol.*, 21, 1977a, 157–161.

Blanchard, D.C., R.J. Blanchard, T. Takahashi, M.J. Kelley, Attack and defensive behaviour in the albino rat, *Anim. Behav.*, 25, 1977b, 622–634.

Blanchard, R.J. and D.C. Blanchard, Attack and defense in rodents as ethoexperimental models for the study of emotion, *Prog. Neuro-Psychopharmacol. Biol-Psychiatry*, 13, 1989, S3–S14.

Blanchard, R.J., D.C. Blanchard, L.K. Takahashi, Reflexive fighting in the albino rat: aggressive or defensive behaviour? *Aggress. Behav.*, 3, 1977c, 145–155.

Blanchard, R.J., L.K. Takahashi, D.C. Blanchard, The development of intruder attack in colonies of laboratory rats, *Anim. Learn. Behav.*, 5, 1977d, 365–369.

Blanchard, R.J., L.K. Takahashi, K.K. Fukunaga, D.C. Blanchard, Functions of the vibrissae in the defensive and aggressive behavior of the rat, *Aggress. Behav.*, 3, 1977e, 231–240.

Buss, A.H. and A. Durkee, An inventory for assessing different kinds of hostility, *J. Consult. Psychol.*, 21, 1957, 343–349.

Buss, A.H. and M. Perry, The aggression questionnaire, *J. Personal. Soc. Psychol.*, 63, 1992, 452–459.

Chi, C.C. and J.P. Flynn, Neural pathways associated with hypothalamically elicited attack behavior in cats, *Science*, 171, 1971, 703–706.

Eron, L.D., The development of aggressive behavior from the perspective of a developing behaviorism, *Am. Psychologist*, 42, 1987, 435–442.

Flynn, J.P., Neural basis of threat and attack, in R.G. Grenell and S. Gabay, Eds., *Biological Foundations of Psychiatry*, Raven Press, New York, 1976, pp. 111–133.

Flynn, J.P., H. Vanegas, W.E. Foote, S. Edwards, Neural mechanisms involved in a cat's attack on a rat, in R. Whalen, Ed., *The Neural Control of Behavior*, Academic Press, New York, 1970, pp. 135–173.

Fuchs, S.A.G., H.M. Edinger, A. Siegel, The organization of the hypothalamic pathways mediating affective defense behavior in the cat, *Brain Res.*, 330, 1985a, 77–92.

Fuchs, S.A.G., H.M. Edinger, A. Siegel, The role of the anterior hypothalamus in affective defense behavior elicited from the ventromedial hypothalamus of the cat, *Brain Res.*, 330, 1985b, 93–108.

Hennes, H., A review of violence statistics among children and adolescents in the United States, *Pediatr. Clin. North Am.*, 45, 1998, 269–280.

Hess, W.R. and M. Brugger, Das subkortikale Zentrum der affektiven Abwehrreaktion, *Helv. Physiol. Pharmacol. Acta*, 1, 1943, 33–52.

Kingsbury, S.J., M.T. Lambert, W. Hendrickse, A two-factor model of aggression, *Psychiatry*, 60, 1997, 224–232.

Kruk, M.R., C.E. Van der Laan, J. Mos, A.M. Van der Poel, W. Meelis, B. Olivier, Comparison of aggressive behaviour induced by electrical stimulation in the hypothalamus of male and female rats, *Prog. Brain Res.*, 61, 1984, 303–314.

Kruk, M.R., C.E. Van der Laan, J. Van der Spuy, A.M.M. Van Erp, W. Meelis, Strain differences in attack patterns elicited by electrical stimulation in the hypothalamus of male CPBWEzob and CPBWI rats, *Aggress. Behav.*, 16, 1990, 177–190.

Kruk, M.R. and A.M. Van der Poel, Is there evidence for a neural correlate of an aggressive behavioural system in the hypothalamus of the rat? *Prog. Brain Res.*, 53, 1980, 385–390.

Kruk, M.R., A.M. Van der Poel, T.P. De Vos-Frerichs, The induction of aggressive behaviour by electrical stimulation in the hypothalamus of male rats, *Behaviour*, 70, 1979, 292–322.

Kruk, M.R., A.M. Van der Poel, W. Meelis, J. Hermans, P.G. Mostert, J. Mos, A.H.M. Lohman, Discriminant analysis of the localization of aggression-inducing electrode placements in the hypothalamus of male rats, *Brain Res.*, 260, 1983, 61–79.

Leyhausen, P., *Cat Behavior: The Predatory and Social Behavior of Domestic and Wild Cats*, Garland STPM Press, New York, 1979.

Meloy, J.R., *The Psychopathic Mind: Origins, Dynamics, and Treatment*, Jason Aronson, Inc., Northvale, N.J., 1988.

Meloy, J.R., Predatory violence during mass murder, *J. Forensic Sci.*, 42, 1997, 326–329.

Monroe, R.R., *Brain Dysfunction in Aggressive Criminals*, Lexington Books, Lexington, Mass., 1978.

Mos, J., B. Olivier, J.H.C.M. Lammers, A.M. Van der Poel, M.R. Kruk, T. Zethof, Postpartum aggression in rats does not influence threshold currents for EBS-induced aggression, *Brain Res.*, 404, 1987, 263–266.

Mos, J., B. Olivier, R. Van Oorschot, H. Dijkstra, Different test situations for measuring offensive aggression in male rats do not result in the same wound patterns, *Physiol. Behav.*, 32, 1984, 453–456.

Moyer, K.E., Kinds of aggression and their physiological basis, *Communic. Behav. Biol.*, 2, 1968, 65–87.

Palanza, P., R.J. Rodgers, D. Della Seta, P.F. Ferrari, S. Parmigiani, Analysis of different forms of aggression in male and female mice: ethopharmacological studies, in S.J. Cooper and C.A. Hendrie, Eds., *Ethology and Psychopharmacology*, John Wiley & Sons, New York, 1994, pp. 179–189.

Panksepp, J., Aggression elicited by electrical stimulation of the hypothalamus in albino rats, *Physiol. Behav.*, 6, 1971, 321–329.

Panksepp, J. and J. Trowill, Electrically induced affective attack from the hypothalamus of the albino rat, *Psychon. Sci.*, 16, 1969, 118–119.

Reiss, A.J., K.A. Miczek, J.A. Roth, *Understanding and Preventing Violence*, vol. 2, National Academy Press, Washington, D.C., 1994, pp. vii–viii.

Sandnabba, N.K., Predatory aggression in male mice selectively bred for isolation-induced intermale aggression, *Behav. Genet.*, 25, 1995, 361–366.

Siegel, A. and M. Brutus, Neural substrates of aggression and rage in the cat, in A.N. Epstein and A.R. Morrison, Eds., *Progress in Psychobiology and Physiological Psychology*, Academic Press, San Diego, Calif., 1990, pp. 135–233.

Siegel, A. and C.B. Pott, Neural substrate of aggression and flight in the cat, *Prog. Neurobiol.*, 31, 1988, 261–283.

Vitiello, B., D. Behar, J. Hunt, D. Stoff, A. Ricciuti, Subtyping aggression in children and adolescents, *J. Neuropsychiatry*, 2, 1990, 189–192.

Whalen, R.E. and F. Johnson, Aggression in adult female mice: chronic testosterone treatment induces attack against olfactory bulbectomized male and lactating female mice, *Physiol. Behav.*, 43, 1988, 17–20.

2 History of the Neurology of Aggression and Rage

In contrast to other areas of neuroscience research that have evolved only over the past three decades, roots of the neurobiology of aggression extend back into the 19th century. The basic objective of this chapter is to provide a historical background of the neurobiology of aggression and rage. A major purpose in applying a historical approach is that it can provide understanding of the rationale and strategies employed by noted investigators during the early years of such research. In doing so, it may likely offer insight into the directions that research in this field generated more recently.

Early periods of aggression research were characterized by attempts to determine the role of broad regions of the brain in aggressive behavior. This was generally done by observing the behavioral and physiological effects of gross ablation of certain regions in the cat (and occasionally dog). This approach was gradually refined over later periods of time in which gross regional ablations were replaced by the use of more focal lesions of brain regions, as well as by the application of electrical brain stimulation. Electrical stimulation procedures have been applied principally to studies in cats, but over time demonstrations of aggressive forms of behavior elicited by electrical brain stimulation have been expanded to include chicken, ring dove, rat, and monkey. More recently, various investigators have administered drugs, either locally into specific regions of the brain, or peripherally, in order to induce aggression or to modulate its occurrence. In this manner, such methodologies have given rise to the field of neurobiology of aggressive behavior. Still other investigators have manipulated endocrine function either by removal of specific endocrine organs or by drug manipulation, and these procedures have provided the framework for the subdiscipline of the endocrinology of aggression. The following discussion will attempt to provide from a historical perspective the growth and development of the field of the neurobiology of aggression and rage.

EFFECTS OF ABLATIONS

Perhaps the first recorded study of the neurobiology of aggression was carried out in dogs by Goltz (1892). In these preparations, all parts of the cerebral cortex were either removed or disconnected from the rest of the brain. Some additional damage involved parts of the neostriatum and dorsal aspect of diencephalon. The animals displayed stereotyped barking and growling coupled with attempts to bite as well as struggle when efforts were made to lift them out of their cages. When the animals were moved back into their cages, they calmed down. Similar results were described by Rothmann (1923), who showed that snarling and growling could be elicited in decorticate dogs by gently scratching their backs. Parallel observations were also described by Dusser de Barenne (1920), who demonstrated that two decorticate cats would elicit rage responses such as hissing, growling, spitting, and piloerection in response to innocuous stimulation. In fact, 15 years earlier, Woodworth and Sherrington (1904) showed in their classical study that strong afferent nerve stimulation could even elicit elements of a rage response in a decerebrate cat preparation. *Thus, the results of these early studies indicated that the expression of rage behavior is not dependent on an intact cerebral cortex, but that these functions are likely mediated through subcortical structures.*

ATTEMPTS TO IDENTIFY ANATOMICAL LOCUS OF RAGE MECHANISM, AND NOTION OF "SHAM RAGE"

If the mechanism underlying rage behavior is of subcortical origin, then a natural question that one might be tempted to ask is: What is the level along the neuraxis of the brainstem or forebrain that is critical for the expression of this form of aggression? This issue was directly addressed by a number of early investigators, including Cannon (Cannon, 1929; Cannon and Britton, 1925), Bard et al. (Bard, 1939; Bard and McK. Rioch, 1937; Yasukochi, 1960), and McK. Rioch (1938).

Cannon (1929) believed that rage and related emotional responses were held in check by the inhibitory actions of certain forebrain regions such as the cerebral cortex and thalamus. This was based on observations from their studies as well as those described above. Cannon succeeded in removing the cortex of cats and separating it from the thalamus. Such preparations periodically displayed emotional excitement such as biting and clawing in spite of the removal of the cortex and underlying thalamus. In a related set of studies, Bard (1928) sought to determine the specific region mediating rage by employing a series of progressive ablations of levels of the cat forebrain. Rage responses were elicited by pinching the tip of the tail, foot, or the skin of the flank and back, or by restraining the cat. Bard discovered that rage responses could be elicited as long as the posterior hypothalamus remained intact. Failure to induce rage occurred only after ablations included the posterior hypothalamus, which in effect, produced complete destruction of the hypothalamus. The significance of this finding was that it was the first of its kind to demonstrate the importance of the hypothalamus in the expression of rage behavior.

As a result of their findings, Cannon and Britton (1925) introduced the term "sham rage," which was later employed by Bard (1934) and many other investigators in later years. The following reasoning was given by these investigators for applying this term to their experimental animal preparations. First, it is reasonable to assume that animals have a subjective experience associated with the expression of rage behavior. Second, the behavioral pattern of the rage response in the decorticate (or decerebrate) preparation is highly similar to that displayed by a normal, unoperated animal when exposed to a threatening stimulus. Such response patterns were described in detail by McK. Rioch (1938). Therefore, on removal of the cerebral cortex and related regions, one may conclude that there is significant modification of the conscious aspects of emotion. These authors thus concluded that while the decorticate cat is quite capable of expressing the behavioral components of the rage response, it is likely that consciousness and the subjective aspects of rage are no longer present. Thus, in order to characterize the behavior in this experimental preparation, the term "sham rage" was employed.

Over the years, a number of difficulties have arisen concerning the use and application of the term "sham rage." One difficulty is that it has been applied in a liberal manner to depict the rage response, even in experimental conditions where surgical ablation procedures were not employed. This problem suggests a theoretical bias in the attempt to understand the neural basis for this response, namely, that one may be easily led to conclude that the rage behavior (even in an intact animal) may not be a true form of emotion. In fact, rage behavior, such as that elicited by electrical stimulation of the medial hypothalamus, is a true expression of emotional behavior which closely mimics that which occurs in nature under conditions where a real or perceived threat is involved. (See Chapter 1 for further discussion of this model of aggression.) A second difficulty is that it gave rise to the notion that the rage response is merely reflexive in nature, which implies that higher parts of the brain, including the cerebral cortex, normally have little or no role in the patterning mechanism for this behavior. Unfortunately, to this date, our understanding of how the cerebral cortex and other higher regions of the brain interact with the hypothalamic and brainstem mechanisms for rage remains largely incomplete. Nevertheless, it would be very premature to rule out possible functions of higher regions of the brain in the expression and regulation of rage behavior. In fact, further examination of the possible role played by higher regions such as the limbic system are considered in Chapters 5 and 6.

MORE SELECTIVE FOREBRAIN AND BRAINSTEM LESIONS

From the ablation studies described above and, in particular, those of Bard et al., it became apparent that the next step required for identifying the loci of the mechanisms regulating aggression would necessitate the application of methods that could more selectively destroy or activate specific regions of the brain. Such methods included the use of methods that produced more restrictive lesions and the application of electrical stimulation of specific sites in the forebrain and brainstem.

Over different periods of time, investigators have sought to identify and place lesions in what they thought were regions of the brain that are critical for expressing rage. These regions can be divided into three categories: hypothalamus, periaqueductal gray (PAG), and limbic structures. One of the first studies employing deep brain lesions was carried out and summarized by Ranson (1934).

HYPOTHALAMUS

In one study, large lesions were placed into the posterior aspect of the hypothalamus and adjoining aspects of the midbrain in cats (Ranson and Ingram, 1932). In a second study, lesions were placed into similar regions of the hypothalamus in the monkey

(Ranson, 1939). In both studies, such lesions produced a marked reduction in activity with an absence of aggressive signs, and which culminated in somnolence. In later studies, Wang and Akert (1962) and Emmers et al. (1965) provided data consistent with those described above by showing that decorticate cats whose thalami had largely been destroyed but whose hypothalami remained intact were capable of displaying rage reactions in response to noxious stimulation. At about the same time, Carli et al. (1966) showed that lesions involving the posterior hypothalamus could block rage elicited by noxious stimulation. Studies such as these thus suggested the possible critical importance of the posterior hypothalamus for expressing rage behavior. This line of research involving the use of selective lesions of the brain to block aggression has continued into more recent times. An example of one of the many lesion studies that have been conducted over the past 40 years is that of Shivers and Edwards (1978) who provided evidence that lesions of the medial preoptic-anterior hypothalamic region can eliminate the display of aggressive behavior in male mice while having little or no effect on testicular hormone function. This study suggests that the importance of different parts of the hypothalamus may be species specific.

In our discussion thus far, we have considered how lesions of specific regions of the hypothalamus could reduce or eliminate the expression of rage behavior elicited by the onset of noxious stimulation. However, in a study conducted by Wheatley (1944) and later confirmed and extended by Glusman (1974), it was discovered that lesions placed bilaterally into the ventromedial hypothalamus of the cat could produce a permanently savage animal. Obviously, this finding was certainly not predicted from the results of earlier ablation studies. Neither could it have been predicted from studies (described below) which demonstrated that rage can be elicited by electrical or chemical stimulation of similar regions of the hypothalamus. In attempting to account for this paradoxical finding, Glusman (1974) conducted a series of experiments that tested several hypotheses. One hypothesis stated that a lesion of the ventromedial nucleus produces a denervation supersensitivity of the remaining neurons. In support of this hypothesis, Glusman showed that lesions of the PAG (see below) produced a temporary loss of rage behavior. Glusman argued that the eventual return of rage behavior could be understood in terms of the development of a supersensitivity in more caudal links of the neuronal chain mediating this form of aggression. Glusman also suggested a further possibility — namely, that there occurred axonal sprouting from neurons in the hypothalamus that made functional connections with the PAG, thus providing the anatomical basis for the activation of the final common path subserving rage behavior.

A third possibility should also be mentioned here. Since it has been shown that electrolytic lesions produce an irritative focus around the lesion (Rabin, 1968, 1972; Reynolds, 1963), it is possible that such lesions ultimately generate a stimulation-like effect within the ventromedial nucleus. It is of interest to note that rage responses in the human have been associated with the presence of a tumor in the ventromedial hypothalamus (Reeves and Plum, 1969). Here, too, one may speculate that the manifestation of rage may have been the result of a stimulation-like effect of the tumor within the ventromedial hypothalamus. Further evidence that lesions of the posterior hypothalamus can reduce aggression in humans was provided from several reports by Sano et al. (Sano and Mayanagi, 1988; Sano et al., 1970). These authors initially identified sites within the ergotropic triangle of the hypothalamus (i.e., the region of the hypothalamus between the mammillary bodies and cerebral aqueduct), which, upon stimulation, produced a marked sympathetic activation. Subsequent to stimulation, bilateral lesions placed in this region resulted in a calming effect in most patients over a 2-year period in which they were observed. This finding thus provided additional evidence at the human level that underscores the importance of the posterior hypothalamus in mediating rage behavior.

Midbrain Periaqueductal Gray

In addition to the hypothalamus, a number of investigators have pointed to the importance of the PAG in the expression of rage behavior. Most PAG studies have involved the use of electrical or chemical stimulation, and are described below. From a historical perspective, it is useful to note one of the first studies that focused on the importance of the PAG. Kelly et al. (1946) showed that central midbrain lesions that destroy the PAG and adjacent parts of the tegmentum could eliminate the facio-vocal component of the rage response in the cat. Moreover, this paper further showed that destruction of the hypothalamus at the level of the mammillary bodies failed to eliminate growling, crying, and spitting following the application of tail pressure. The significance of this study is that it brought attention to the importance of the PAG as an essential component in the neural circuit mediating rage behavior in the cat.

Limbic Structures

Over the past half-century, an extensive literature has emerged in which lesions were placed into different limbic structures in order to assess their functions in aggressive and related forms of behavior. Some of these studies will be examined in a later chapter concerning the role of the limbic system and aggression. In this section, the basic purpose is to indicate from a historical perspective several of the seminal papers that provided the impetus for significant developments in this aspect of aggression research.

Several laboratories have attempted to delineate the role of temporal lobe structures in the regulation of rage behavior. In what has become one of the most frequently cited studies of its kind, Kluver and Bucy (1939) demonstrated that lesions of the pyriform cortex and adjoining amygdaloid tissue produce a rather bizarre syndrome characterized by hypersexuality, abnormal docility, lack of recognition of objects, and marked oral tendencies. This constellation of symptoms has since come to be known as "Kluver-Bucy syndrome." For many years, this finding had a significant impact in thinking about how limbic structures relate to aggression. In particular, this report established the view that the amygdala normally serves as a potent facilitator of rage behavior. This view prevailed until Bard and Mountcastle (1948) conducted a study aimed at identifying the locus of the subcortical region mediating rage and aggression in the decorticate cat. In contrast to the Kluver and Bucy finding, Bard and Mountcastle observed a marked increase in rage behavior following removal of the amygdala in decorticate cats. From these observations,

Bard and Mountcastle concluded that the amygdala serves as an *inhibitory* funnel through which the mechanism controlling rage is mediated. Thus, these two studies suggested opposing views concerning how the amygdala related to the control of rage and aggression. However, the contradictory conclusions concerning functions of the amygdala generated from these studies provided an important stimulant that gave rise to an extensive body of literature which utilized electrical stimulation and lesions of the amygdala in order to resolve this paradox. A more detailed analysis of these studies is considered in Chapters 5 and 6. Among the many other important papers that stimulated considerable research was a study by Brady and Nauta (1953), which established the importance of the septal area as a significant limbic structure in the regulation of aggression and rage. The essential feature of this study was that rats sustaining lesions of the septal area became very irritable, were difficult to handle, and showed hyperemotionality and heightened aggressiveness. This study served as a catalyst for many other investigations that utilized lesions or stimulation methods to characterize the role of the septal area in the control of aggression and rage. Such studies have continued into the present.

STIMULATION OF REGIONS MEDIATING AGGRESSION AND RAGE

Two laboratories, one in Europe, and the other in United States, were instrumental in the development and use of electrical stimulation for the study of aggression and related forms of behavior. In Europe, Hess and Brugger (1943) were the first to examine in a systematic manner the physiological and behavioral effects of electrical stimulation of the diencephalon. In this manner, they were able to identify the sites in the medial hypothalamus of the cat from which rage behavior can be elicited. The response pattern that they observed from perifornical (medial) hypothalamic stimulation included hissing, spitting, growling, pupillary dilatation, retraction of the ears, the appearance of a bushy tail, and piloerection. Hess then coined the term "affective defense reaction" to describe this form of aggression. For his efforts in mapping the regions of the diencephalon from which rage and related responses can be elicited by electrical stimulation, Hess received the Nobel Prize in 1949. It should be added that the accuracy of his work (and those of other investigators at that time) was made all the more remarkable by the fact that the methods of electrical stimulation in the 1930s and 1940s were rather crude, as they lacked such features as constant current conditions and isolation of electrical pulses.

At about the same time as Hess was conducting his research, Magoun and colleagues (Magoun et al., 1937; Ranson and Magoun, 1939) adopted a similar approach. These investigators were able to show response patterns in the cat following hypothalamic stimulation that were very similar to those described by Hess (Hess and Akert, 1955; Ranson and Magoun, 1939). In addition, they further demonstrated that in decerebrate cats, signs of rage behavior could be elicited by stimulation of the ventral tegmental fields (Magoun et al., 1937).

A significant feature of the research conducted in both of these laboratories was that it brought attention to the value of electrical stimulation as a tool for studying the neurobiology of aggression. The following discussion briefly summarizes the direction of this line of research over the past half-century.

The use of electrical brain stimulation for the study of aggressive behavior followed various directions. In the period spanning the 1950s until approximately 1970, numerous investigators applied this methodology in order to identify specific limbic forebrain and brainstem structures that mediate rage and aggression. One of the first investigators to identify the role of neural structures in mediating rage behavior was Hunsperger (1956, 1959, 1969), who mapped out the sites in the hypothalamus, stria terminalis, and PAG in the cat by eliciting the behavior with electrical stimulation. In addition, Hunsperger also identified the functional relationship between the medial hypothalamus and PAG by demonstrating that rage elicited from the medial hypothalamus can be blocked following the placement of a lesion in the PAG. Other investigators used electrical stimulation methods to elicit components of rage from the pontine tegmentum of the cat, thus providing evidence suggesting that the pathway mediating rage descends through the pontine tegmentum (Bizzi et al., 1963; Kanai and Wang, 1962).

In what turned out be a landmark study, Wasman and Flynn (1962) published a paper which demonstrated that a directed attack against an anesthetized rat (i.e., predatory attack) by a cat can be elicited along a rostral-ventral continuum extending through much of the lateral or perifornical hypothalamus, while affective defense (rage) is elicited from sites along a similar rostral-ventral continuum within the medial hypothalamus. The discovery that two clearly different forms of aggression can be elicited from neighboring regions of hypothalamus demonstrated the strengths inherent in the utilization of these models of aggression. It clearly provided the basis for many other studies that were able to characterize the anatomical, physiological, and pharmacological properties (see Roberts and Kiess, 1964; Romaniuk, 1965; Sato and Wada, 1974; Siegel and Brutus, 1990; Siegel and Pott, 1987; Siegel et al., 1999; Yasukochi, 1960) of these forms of aggression, and gave considerable insight into the neural bases of aggression in humans. Thus, our current understanding of the neural basis of aggression and rage is due in part to the impact of this and subsequent findings by Flynn et al. (Chi and Flynn, 1971a, 1971b; Flynn and Smith, 1979, 1980a, 1980b). Several examples are briefly indicated here to illustrate how this model was used to gain valuable information about the neural bases of aggression and rage. With respect to the anatomy of aggression, the laboratory of John Flynn was the first to apply neuroanatomical methods for the tracing of the neural pathways mediating aggression. This laboratory initially used the Fink-Heimer silver impregnation method of degenerating axons to trace the ascending and descending fibers from predatory and affective attack sites in the hypothalamus (Chi and Flynn, 1971a, 1971b), and later applied retrograde tracing procedures to identify afferent sources of fibers that project to attack sites in the cat (Flynn and Smith, 1979, 1980a, 1980b). Other studies followed on this approach, using newer anterograde tracing methods to further elucidate the pathways that mediate aggression and rage, as well as those that modulate it (see Siegel and Brutus, 1990, for details). More recent studies have used immunocytochemistry coupled with retrograde tracing to characterize the neurochemical properties of neurons that mediate or modulate aggression and rage in the cat (see Siegel et al., 1997a; Siegel and Shaikh, 1997b). The physiological mechanisms mediating aggression and rage in the cat have been examined in detail initially by Flynn (1972; Flynn et al., 1970) in which the sensory and motor properties of the attack response were identified and characterized, and the modulating properties of the limbic system and related structures have been elucidated by Siegel and others (see Siegel and Brutus, 1990, for details). A number of early pharmacological studies were conducted using these cat models of aggression that attempted to test how various drugs such as benzodiazepines, antipsychotics, and specific neurotransmitter agonists and antagonists affect aggressive behavior. A summary of these findings is provided in a review by Siegel and Pott (1988) and Siegel et al. (1999). A detailed analysis of the anatomical, physiological, and pharmacological properties of aggression and rage are considered in Chapters 3 to 7.

This work also served as a stimulus for investigators to explore the nature of stimulation induced aggression in other species such as rats (Kruk et al., 1979; Panksepp and Trowill, 1969); ring doves (Harwood and Vowles, 1967); chickens (Putkonen, 1966); and monkeys (Ciofalo and Malick, 1969; Lipp, 1978; Perachio and Alexander, 1975). Over the past 25 years, the body of research concerning the biology of aggression has shifted significantly from cat models of aggression to those of rodents. However, the most common approach to the study of aggression in rodents has utilized the resident–intruder model rather than models employing brain stimulation. This has been most apparent for studies examining both the pharmacology of aggression as well as hormones and aggression where an extensive literature has emerged over this period of time. An extensive review of the pharmacology of aggression in rodents and other species has been provided by Miczek et al. (1994a, 1994b, 1994c), and the literature concerning hormones and aggression has been summarized by Siegel and Demetrikopoulos (1993). The pharmacology of aggression and hormonal systems are considered in greater detail in Chapters 7 and 8, respectively.

REFERENCES

Bard, P., A diencephalic mechanism for the expression of rage with special reference to the sympathetic nervous system, *Am. J. Physiol.*, 84, 1928, 490–515.

Bard, P., On emotional expression after decortication with some remarks on certain theoretical views: Part 1, *Psychol. Rev.*, 41, 1934, 309–329.

Bard, P., Central nervous mechanisms for emotional behavior patterns in animals, *Res. Publ. Assoc. Res. Nerv. Ment. Dis.*, 19, 1939, 190–218.

Bard, P. and D. McK. Rioch, A study of four cats deprived of neocortex and additional portions of the forebrain, *Bull. Johns Hopkins Hosp.*, 60, 1937, 73–124.

Bard, P. and V.B. Mountcastle, Some forebrain mechanisms involved in expression of rage with special reference to suppression of angry behavior, *Yale J. Biol. Med.*, 14, 1948, 362–399.

Bizzi, E., A. Malliani, J. Apelbaum, A. Zanchetti, Excitation and inhibition of sham rage behavior by lower brain stem stimulation, *Arch. Ital. Biol.*, 101, 1963, 614–631.

Brady, J.V. and W.J.H. Nauta, Subcortical mechanisms in emotional behavior: affective changes following septal forebrain lesions in the albino rat, *J. Comp. Physiol. Psychol.*, 46, 1953, 339–346.

Cannon, W.B., *Bodily Changes in Pain, Fear, Hunger and Rage*, Appleton-Century-Crofts, New York, 1929.

Cannon, W.B. and S.W. Britton, Studies on the conditions of activity in endocrine glands: XV. Pseudoaffective medulliadrenal secretion, *Am. J. Physiol.*, 72, 1925, 283–294.

Carli, G., A. Malliani, A. Zanchetti, Lesioni selettive di varie strutture ipotalamiche e comportamento spontaneo e provocato di falsa rabbia del gatto decorticato acuto, *Boll. Della Soc. Ital. Di Biol. Speriment.*, 42, 1966, 291–294.

Chi, C.C. and J.P. Flynn, Neural pathways associated with hypothalamically elicited attack behavior in cats, *Science*, 171, 1971a, 703–706.

Chi, C.C. and J.P. Flynn, Neuroanatomic projections related to biting attack elicited from hypothalamus in cats, *Brain Res.*, 35, 1971b, 49–66.

Ciofalo, V.B. and J.B. Malick, Evoked aggressive behavior in Cebus Apella: a New World primate, *Life Sci.*, 8, 1969, 1117–1122.

Dusser De Barenne, J.G., Recherches experimentales sur les fonctions du systeme nerveux central, faites en particulier sur deux chats dont le neopallium avait ete enleve, *Arch. Neer. Phys. del'Homme et des Animaux*, 4, 1920, 31–123.

Emmers, R., R.W.M. Chun, G.H. Wang, Behavior and reflexes of chronic thalamic cats, *Arch. Ital. Biol.*, 103, 1965, 178–193.

Flynn, J.P., Patterning mechanisms, patterned reflexes, and attack behavior in cats, *Nebraska Symposium on Motivation*, 20, 1972, 125–153.

Flynn, J.P. and D.A. Smith, Afferent projections related to attack sites in the pontine tegmentum, *Brain Res.*, 164, 1979, 103–119.

Flynn, J.P. and D.A. Smith, Afferent projections to affective attack sites in cat hypothalamus, *Brain Res.*, 194, 1980a, 41–51.

Flynn, J.P. and D.A. Smith, Afferent projections to quiet attack sites in cat hypothalamus, *Brain Res.*, 194, 1980b, 29–40.

Flynn, J.P., H. Vanegas, W.E. Foote, S. Edwards, Neural mechanisms involved in a cat's attack on a rat, in R. Whalen, Ed., *The Neural Control of Behavior*, Academic Press, New York, 1970, pp. 135–173.

Glusman, M., The hypothalamic "savage" syndrome, *Res. Publ. Assoc. Res. Nerv. Ment. Dis.*, 52, 1974, 52–92.

Fr. Goltz, P., The dog without cerebrum, *Pfluger's Archiv fur die gesamte, Physiol. Menschen Tiere*, 51, 1892, 1–2.

Harwood, D. and D.M. Vowles, Defensive behaviour and the after effects of brain stimulation in the ring dove, Streptopelia Risoria, *Neuropsychologia*, 5, 1967, 345–366.

Hess, W.R. and K. Akert, Experimental data on role of hypothalamus in mechanism of emotional behavior, *Arch. Neurol. Psychiatry*, 73, 1955, 127–129.

Hess, W. R. and M. Brugger, Das subkortikale Zentrum der affektiven Abwehrreaktion, *Helv. Physiol. Pharmacol. Acta*, 1, 1943, 33–52.

Hunsperger, R.W., Affektreactionen auf elektrische Reizung im Hirnstamm der Katze, *Helv. Physiol. Acta*, 14, 1956, 70–92.

Hunsperger, R.W., Les representation centrales des reactions affectives dans le cerveau anterieur et dans le tronc cerebral, *Neuro-Chirurgie*, 5, 1959, 207–233.

Hunsperger, R.W., Die Asynchrone und Lokal umschriebene Reizwirkung von Mittelfrequenz-Dauerstromen im Hypothalamus der Katze, *Exp. Brain Res.*, 9, 1969, 164–182.

Kanai, T. and S.C. Wang, Localization of the central vocalization mechanism in the brain stem of the cat, *Exp. Neurol.*, 6, 1962, 426–434.

Kelly, A.H., L.E. Beaton, H.W. Magoun, A midbrain mechanism for facio-vocal activity, *J. Neurophysiol.*, 9, 1946, 181–189.

Kluver, H. and P.C. Bucy, Preliminary analysis of functions of the temporal lobes in monkeys, *Arch. Neurol., Psychiatry*, 42, 1939, 979–1000.

Kruk, M.R., A.M. Van der Poel, T.P. De Vos-Frerichs, The induction of aggressive behaviour by electrical stimulation in the hypothalamus of male rats, *Behaviour*, 70, 1979, 292–322.

Lipp, H.P., Aggression and flight behaviour of the Marmoset monkey Callithrix jacchus: an ethogram for brain stimulation studies, *Brain Behav. Evol.*, 15, 1978, 241–259.

Magoun, H. W., D. Atlas, E.H. Ingersoll, S.W. Ranson, Associated facial, vocal and respiratory components of emotional expression: an experimental study, *J. Neurol. Psychopathol.*, 17, 1937, 241–255.

McK. Rioch, D., Certain aspects of the behavior of decorticate cats, *Psychiatry*, 1, 1938, 339–345.

Miczek, K.A., J.F. DeBold, M. Haney, J. Tidey, J. Vivian, E. Weerts, Alcohol, drugs of abuse, aggression, and violence, in A.J. Reiss and J.A. Roth, Eds., *Understanding and Preventing Violence*, vol. 3, National Academy Press, Washington, D.C., 1994a, pp. 377–570.

Miczek, K.A., M. Haney, J. Tidey, J. Vivian, E. Weerts, Neurochemistry and pharmacotherapeutic management of aggression and violence, in A. Reiss, K. Miczek, J. Roth, Eds., *Understanding and Preventing Violence*, vol. 2, National Academy Press, Washington, D.C., 1994b, pp. 245–514.

Miczek, K.A., E. Weerts, M. Haney, J. Tidey, Neurobiological mechanisms controlling aggression: Preclinical developments for pharmacotherapeutic interventions, *Neurosci. Biobehav. Rev.*, 18, 1994c, 97–110.

Panksepp, J. and J. Trowill, Electrically induced affective attack from the hypothalamus of the albino rat, *Psychon. Sci.*, 16, 1969, 118–119.

Perachio, A.A. and M. Alexander, The neural bases of aggression and sexual behavior in the Rhesus, in R.C. Bourne, Ed., *The Rhesus Monkey*, Academic Press, New York, 1975, pp. 381–409.

Putkonen, P.T.S., Attack elicited by forebrain and hypothalamic stimulation in the chicken, *Experientia*, 22, 1966, 405–407.

Rabin, B.M., Effects of lesions of the ventromedial hypothalamus on the electrical activity of the ventrolateral hypothalamus, *Electroencephalogr. Clin. Neurophysiol.*, 25, 1968, 344–350.

Rabin, B.M., Ventromedial hypothalamic control of food intake and satiety: a reappraisal, *Brain Res.*, 43, 1972, 317–342.

Ranson, S.W., The hypothalamus: its significance for visceral innervation and emotional expression, *Coll. Physicians Trans. (Phila.)*, 2, 1934, 222–242.

Ranson, S.W., Somnolence caused by hypothalamic lesions in the monkey, *Arch. Neurol., Psychiatry*, 41, 1939, 1–23.

Ranson, S.W. and W.R. Ingram, Catalepsy caused by lesions between the mammillary bodies and third nerve in the cat, *Am. J. Physiol.*, 101, 1932, 690.

Ranson, S.W. and H. W. Magoun, The hypothalamus, *Ergebn. Physiol.*, 41, 1939, 56–163.

Reeves, A.G. and F. Plum, Hyperphagia, rage, and dementia accompanying a ventromedial hypothalamic neoplasm, *Arch. Neurol.*, 20, 1969, 616–624.

Reynolds, R.W., Ventromedial hypothalamic lesions without hyperphagia, *Am. J. Physiol.*, 204, 1963, 60–62.

Roberts, W.W. and H.O. Kiess, Motivational properties of hypothalamic aggression in cats, *J. Comp. Physiol. Psychol.*, 58, 1964, 187–193.

Romaniuk, A., Representation of aggression and flight reactions in the hypothalamus of the cat, *Acta Biol. Exp.*, 25, 1965, 177–186.

Rothmann, H. Zusammenfassender Bericht uber den Rothmannschen groshirnlosen Hund nach klinischer und anatomischer Untersuchung, Zeit. gesamte, *Neurol. Psychiatry*, 87, 1923, 247–313.

Sano, K. and Y. Mayanagi, Posteromedial hypothalamotomy in the treatment of violent, aggressive behaviour, *Acta Neurochirurgica Suppl.*, 44, 1988, 145–151.

Sano, K., Y. Mayanagi, H. Sekino, M. Ogashiwa, B. Ishijima, Results of stimulation and destruction of the posterior hypothalamus in man, *J. Neurosurg.*, 33, 1970, 689–707.

Sato, M. Md. and J.A. Md. Wada, Hypothalamically induced defensive behavior and various neuroactive agents, *Folia Psychiatr. Neurol. Jpn.*, 28, 1974, 101–106.

Shivers, M. and D.A. Edwards, Hypothalamic destruction and mouse aggression, *Physiol. Psychol.*, 6, 1978, 485–487.

Siegel, A. and M. Brutus, Neural substrates of aggression and rage in the cat, in A.N. Epstein and A.R. Morrison, Eds., *Progress in Psychobiology and Physiological Psychology*, Academic Press, San Diego, Calif., 1990, pp. 135–233.

Siegel, A. and M. Demetrikopoulos, Hormones and aggression, in J. Shulkin, Ed., *Hormonally Induced Changes in Mind and Brain*, Academic Press, San Diego, Calif., 1993.

Siegel, A. and C. Pott, Neural regulation of hypothalamically-elicited aggression and flight behavior in the cat, *Prog. Neurobiol.*, 31, 1987, 261–283.

Siegel, A. and C.B. Pott, Neural substrate of aggression and flight in the cat, *Prog. Neurobiol.*, 31, 1988, 261–283.

Siegel, A., T.A.P. Roeling, T.R. Gregg, M.R. Kruk, Neuropharmacology of brain-stimulation-evoked aggression, *Neurosci. Biobehav. Rev.*, 23, 1999, 359–389.

Siegel, A., K.L. Schubert, M.B. Shaikh, Neurotransmitters regulating defensive rage behavior in the cat, *Neurosci. Biobehav. Rev.*, 21, 1997a, 733–742.

Siegel, A. and M.B. Shaikh, The neural bases of aggression and rage in the cat, *Aggress. Viol. Behav.*, 2, 1997b, 241–271.

Wang, G.H. and K. Akert, Behavior and reflexes of chronic striatal cats, *Arch. Ital. Biol.*, 100, 1962, 48–85.

Wasman, M. and J.P. Flynn, Directed attack elicited from hypothalamus, *Arch. Neurol.*, 6, 1962, 220–227.

Wheatley, M.D., The hypothalamus and affective behavior in cats: a study of the effects of experimental lesions, with anatomic correlations, *Arch. Neurol. Psychiatry*, 52, 1944, 296–316.

Woodworth, R. . and C.S. Sherrington, A pseudaffective reflex and its spinal path, *J. Physiol.*, 31, 1904, 234–243.

Yasukochi, G., Emotional responses elicited by electrical stimulation of the hypothalamus in cat, *Folia Psychiatr. Neurol. Jpn.*, 14, 1960, 260–267.

3 The Neuroanatomy of Aggression and Rage

A basic question is whether it is necessary or even useful to attempt to understand the neurobiology of aggression and rage by first determining their neuroanatomical substrates. This is a philosophical question because a person who accepts this approach basically identifies with a strategy that is deemed necessary for understanding the nature of the behavioral processes linked to aggression and rage. This approach is adopted in this book and, hopefully, its advantages will be self-evident from a perusal of this chapter.

In Chapter 2, I reviewed from a historical perspective the knowledge base of the brainstem and forebrain structures that play important if not critical roles in the organization and expression of rage behavior. In this chapter, the primary focus is on the circuits in the brain that mediate the expression of defensive rage and predatory attack behavior. The studies described in Chapter 2 provided the background for the present studies. Accordingly, the conclusions of these studies again are summarized here as a means of providing the rational basis for the anatomical studies described below.

Results of ablation, lesion, and stimulation studies all point to the likelihood that the medial hypothalamus and periaqueductal gray (PAG) contain the key neuronal pools that are necessary for the expression of defensive rage behavior, and that the lateral hypothalamus is the central structure for the expression of predatory attack behavior.

NEUROANATOMICAL METHODS

The efferent pathways from the hypothalamus and regions of the PAG that mediate aggressive behavior were examined for the first time by Chi and Flynn (1971a, 1971b) and at a later times by Fuchs et al. (1981, 1985a, 1985b) and Shaikh et al. (1987). The study by Chi and Flynn was unique because it was the first of its kind to use a functional approach to trace fiber pathways in an anterograde direction in the central nervous system (CNS) with respect to behavioral systems. In their study, a stainless steel stimulating electrode was initially placed into a defensive rage or predatory attack site in the medial or lateral hypothalamus, respectively. After responses were firmly established, radio frequency lesions were placed at the tip of the electrode — the site at which the attack response was elicited. Following a suitable time period in which the damaged axons were allowed to degenerate, the animal was sacrificed and brain tissue was processed histochemically by the Fink–Heimer modification of the Nauta technique, which detects degenerating axons and their preterminals. The strength of this approach is that it allows tracing of the pathways associated with structures from which attack behavior had been elicited. This approach is deficient in that the lesion could very well have damaged fibers "of passage" that may have been incidental to the behavioral process in question, and thus generate a "false positive" finding.

In later years, several other anterograde methods were also employed to trace the functional pathways associated with aggression and rage. One method is tritiated amino acid radioautography. In this approach, an attack site is identified by applying current through the cannula electrode. Then, a fine microsyringe is inserted through the barrel of the cannula electrode and a tritiated amino acid such as ^{3}H-leucine or ^{3}H-proline is microinjected into the site at which the attack response had been elicited. The labeled amino acid is converted into protein by the neuronal cell body, and is then transported down the axon to its terminals. Traditional autoradiographic procedures, employed to visualize the label over axons, makes it possible to trace the pathway from the cell body to its terminal field. The advantage of this method is that it eliminates the problem of labeling of fibers of passage. The possible deficiency is that, if the behavioral response is elicited as a result of activation of some fibers of passage, then this method would not be able to identify such axons.

A third method is ^{14}C-2-deoxyglucose (2DG) autoradiography. In this method, 2DG is systemically administered through a vessel such as the femoral vein and the attack response is elicited at different times over the following 45 minutes. In brief, the rational for the use of 2DG is as follows: 2DG is transported through the blood–brain barrier, and is metabolized through part of the pathway of glucose metabolism at a known rate relative to that of glucose. When the glucose analog enters the cell, it competes with native glucose for phosphorylation by hexokinase. The resulting product is reduced to 2DG-6-phosphate where it cannot be further reduced. Thus, the labeled metabolite is essentially trapped in the cell, thus allowing the pattern of distribution to be visualized with autoradiographic techniques (Sokoloff, 1977). The basic assumption here is that the increased optical densities on the autoradiographs are associated with higher concentrations of 2DG-6-phosphate, and therefore with increased rates of glucose utilization.

Retrograde tracing techniques have also been used to identify the circuitry involved in the expression and modulation of aggression and rage. This method is based on the knowledge that when a tracer such as horseradish peroxidase (HRP) or Fluoro-Gold is microinjected into the terminal arborizations of a given pathway, it is taken up into the terminals of the neurons and retrogradely transported back to the cell body where it can be clearly identified under the microscope. In this manner, and depending on the strategy adopted, it is possible to identify the precise locations of the cell bodies of origin of a given pathway linked to aggression or rage, if one knows where the pathway terminates. For example, as indicated below, it is known that defensive rage is mediated over a pathway extending from parts of the medial hypothalamus to the PAG. Thus, if the retrograde label is microinjected into the PAG, one should be able to identify the loci of the cells in the medial hypothalamus, which project to the PAG. Another strategy that could be adopted with this method is to determine the origins of the fiber pathways that project to a site from which aggression or rage is elicited. This can easily be accomplished by microinjecting the retrograde label into an attack site, and then observing the locations of the retrogradely labeled neurons. Both of these approaches have been used in the analysis of the circuitry associated with aggression and rage.

PATHWAYS MEDIATING DEFENSIVE RAGE IN CATS

HYPOTHALAMUS

The first study was carried out by Chi and Flynn (1971a), who used the Fink–Heimer modification of the Nauta technique for impregnating degenerating axons to trace fibers passing through the hypothalamus, which presumably participate in the expression of defensive rage behavior. As noted above, this was achieved by initially implanting a stimulating electrode into the medial hypothalamus just above the ventromedial nucleus from which defensive rage could be elicited. Then, a lesion was placed at the tip of the electrode and the degenerating axons were traced from the site of the lesion. The most significant finding from this study was that the investigators were able to trace degenerating fibers directly into the PAG. This finding was to be confirmed in later studies (described below) where newer methodologies were employed.

Investigators using a variety of techniques examined the study of the projections from the medial hypothalamus to the PAG. Several investigators used retrograde labeling with HRP histochemistry. In one study, Bandler and McCulloch (1984) placed HRP injections into PAG sites from which defensive rage had been elicited. These authors observed one group of labeled neurons in the tuberal region of the hypothalamus. Other groups of labeled neurons were located in the ventromedial, dorsomedial, and anterior medial hypothalamus. There is some difficulty in assessing which groups of neurons are actually critical for the expression of defensive rage. For example, our experience has been that stimulation of the tuberal region does not elicit defensive rage. Therefore, it is likely that the tuberal region projection to the PAG may underlie modulation by this region of hypothalamus on defensive rage. On the basis of further evidence provided below, the labeled cells situated in the anterior medial hypothalamus may constitute the most likely site for projection to the PAG mediating defensive rage behavior.

In a study that was designed to directly determine more precisely the origin and distribution of the pathway mediating the expression of defensive rage from the hypothalamus, Fuchs et al. (1985a, 1985b) conducted the following studies. The strategy here was to combine the methods of tritiated amino acid radioautography and 2DG autoradiography. In the first phase of the study, cannula-electrodes were implanted into sites throughout the anterior–posterior extent of the medial hypothalamus from which defensive rage could be elicited. Then, (^{3}H) leucine was microinjected into each of these sites in separate cats and the distribution of fibers in these regions was determined. In a parallel manner, stimulation was applied to different sites throughout the anterior–posterior extent of the medial hypothalamus in separate cats and the pathways activated by stimulation were identified following the systemic administration of 2DG. Most interestingly, the distribution of label following microinjections of tritiated amino acids was highly similar to the patterns of metabolic activation observed after electrical stimulation of the medial hypothalamus when coupled with administration of 2DG. The correspondence in the distribution of label for each of these methods gives one confidence that the structures labeled reflect the functional pathways activated by medial hypothalamic stimulation of defensive rage sites. The first key observation was that the label related to an attack site in or around the ventromedial nucleus was primarily directed in a rostral direction to the anterior medial hypothalamus and medial preoptic region, while there was little evidence of label distributed in a caudal direction (Figure 3.1 and Figure 3.2). At first, this observation would appear to be at odds with the reports of other laboratories, which demonstrated the presence of ventromedial hypothalamic projections to the PAG (Krieger et al., 1979; Sakuma and Pfaff, 1982; Saper et al., 1976; Bandler and McCulloch, 1984). One likely explanation for the differences between our observations and those of others is that the fibers arising from the ventromedial nucleus that project to the PAG do not serve as a substrate for defensive rage behavior, although it is quite possible that such fibers may enable the ventromedial nucleus to modulate the activity of PAG neurons.

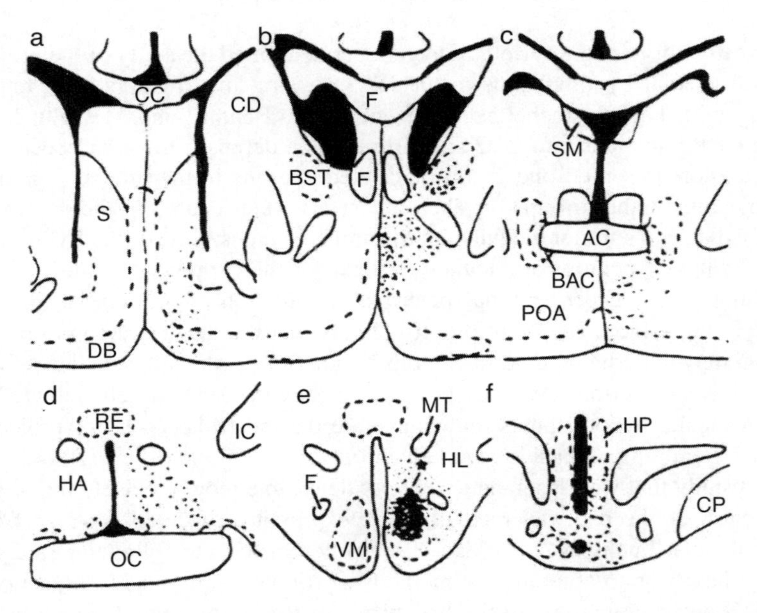

FIGURE 3.1 Outline of the distribution of activated regions following stimulation of the ventromedial hypothalamus and systemic injection of 2-DG. Note that the areas exhibiting an increase in optical density are located primarily rostral to the level of the ventromedial nucleus with a relative absence of label caudal to the level of stimulation. The site of stimulation is indicated by a star and (a through f) displays the sections in a rostral-caudal sequence. (From S.A.G. Fuchs, H.M. Edinger, A. Siegel, *Brain Res.*, 330, 1985a, 77–92. With permission.)

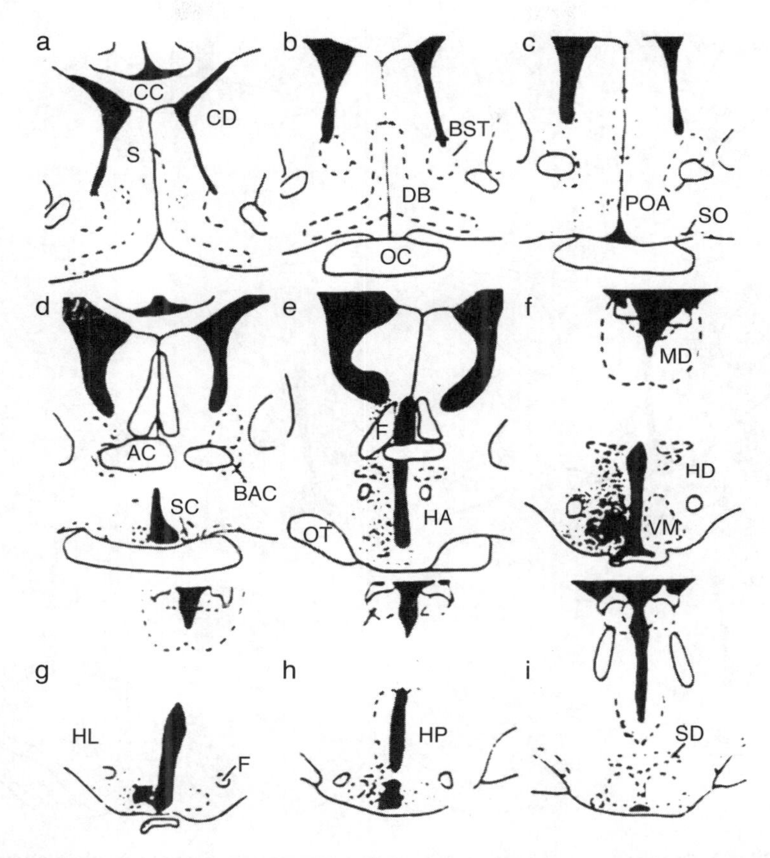

FIGURE 3.2 Outline of projections associated with a (^{3}H) leucine injection placed into the ventromedial hypothalamus. Note the similarity of the pattern of labeling with that shown in Figure 3.1 where labeled axons travel mainly in a rostral direction to terminate in large quantities in the anteromedial hypothalamus. (From S.A.G. Fuchs, H.M. Edinger, A. Siegel, *Brain Res.*, 330, 1985a, 77–92. With permission.)

These observations indicated that fibers originating from the ventromedial nucleus and associated with defensive rage behavior are primarily directed rostrally to the anterior medial hypothalamus and adjoining preoptic region. This finding suggests that the major descending pathway from the medial hypothalamus to the PAG likely arises instead from the anterior medial hypothalamus. In fact, when the descending pathways from the anterior medial hypothalamus were examined, it was noted that significant projections could be identified to reach the dorsolateral PAG (Figure 3.3 and Figure 3.4). Thus, this study established that the principal sites of origin of the pathway to the PAG mediating defensive rage behavior arise from anterior medial hypothalamus and not the ventromedial hypothalamus. The question may then be asked: What is the role of the ventromedial nucleus? My interpretation of these data is that the ventromedial nucleus may serve as a trigger zone for the

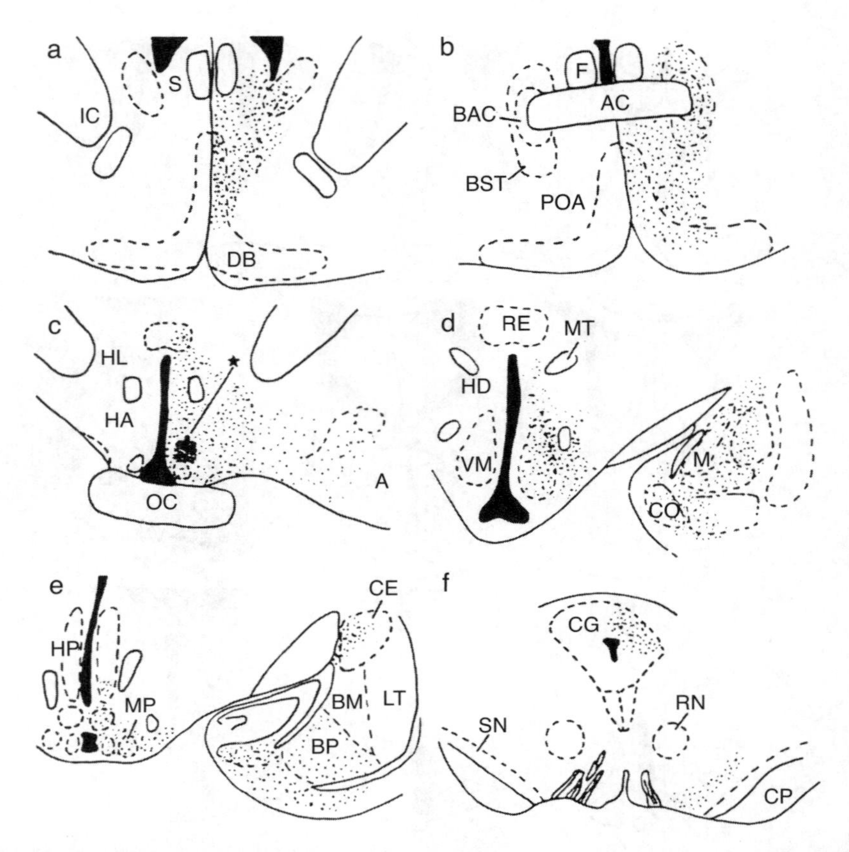

FIGURE 3.3 Outline of a distribution of activated regions associated with stimulation of the ventral aspect of the anterior hypothalamus following a systemic injection of 2-DG. Note the distribution of label within the midbrain periaqueductal gray. (From S.A.G. Fuchs, H.M. Edinger, A. Siegel, *Brain Res.*, 330, 1985a, 77–92. With permission.)

initiation of defensive rage by having the capacity to powerfully excite neurons in the anterior medial hypothalamus, which project to defensive rage neurons in the PAG.

As I have indicated, the results of these studies clearly point to the view that the anterior hypothalamus is the principal output source to the PAG for the expression of defensive rage. This hypothesis was put to a further test in the following experiment conducted by Fuchs et al. (1985b). In this study, defensive rage sites were initially identified in both anterior medial hypothalamus and ventromedial nucleus. Then, dual stimulation procedures were employed in which stimulation of either of these sites at a current value below threshold for elicitation of the response was found to facilitate the occurrence of the attack response elicited from the other site. The purpose of this aspect of the study was to show that responses elicited from both sites were functionally similar and thus comprise separate components of the same system for elicitation of defensive rage behavior.

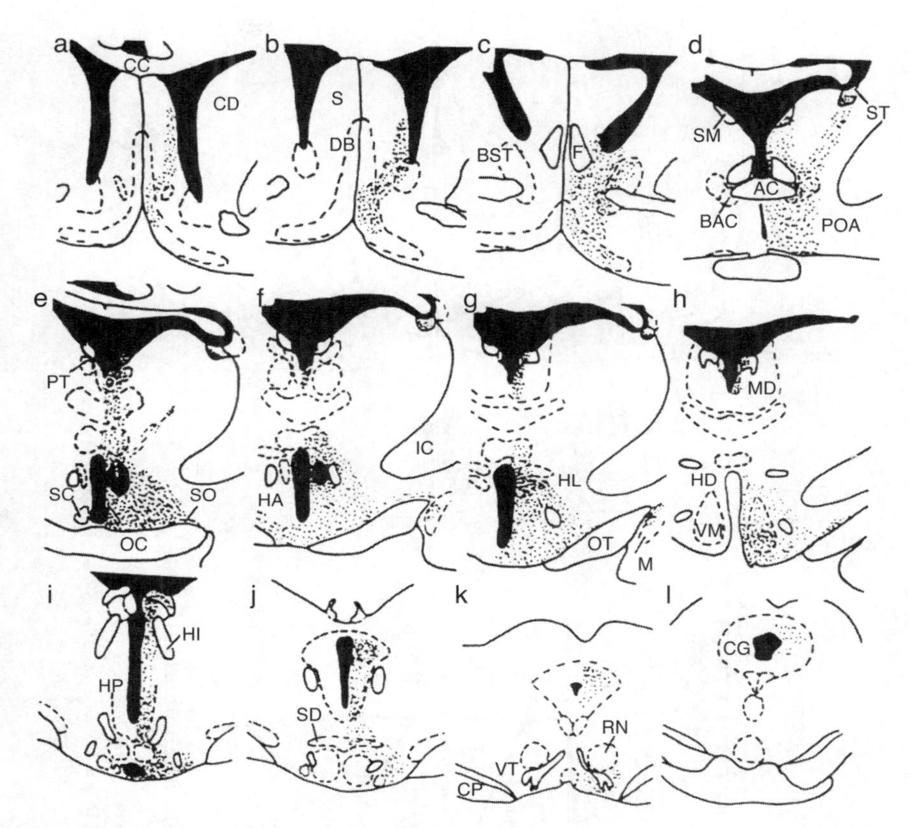

FIGURE 3.4 Outline of the distribution of label following a (^{3}H) leucine injection placed within the anterior dorsal hypothalamus. Note the similarity in labeling pattern to that shown in Figure 3.3 with label present in the midbrain periaqueductal gray. (From S.A.G. Fuchs, H.M. Edinger, A. Siegel, *Brain Res.*, 330, 1985b, 93–108. With permission.)

In principle, the hypothesis states that the expression of defensive rage behavior originating from the ventromedial nucleus requires the presence and activation of neurons in the anterior medial hypothalamus, which then project to the PAG. This hypothesis was then directly tested by the placement of lesions in sites within the anterior medial hypothalamus from which defensive rage behavior could be elicited, and then observing how such lesions could modify the threshold for elicitation of the defensive rage response from the ventromedial nucleus. Here we also injected 2DG systemically following stimulation of the ventromedial nucleus in order to correlate the effects of the anterior medial hypothalamic lesions on defensive rage with the distribution of activated fibers from the ventromedial nucleus to the anterior hypothalamus. Results of this study are shown in Figure 3.5. In brief, lesions of the anterior medial hypothalamus were effective in blocking the defensive rage response elicited from the ventromedial nucleus. In addition, the presence of ascending fibers passing from the ventromedial nucleus to the region of the lesion in the anterior hypothalamus was further noted in the 2DG analysis. This experiment

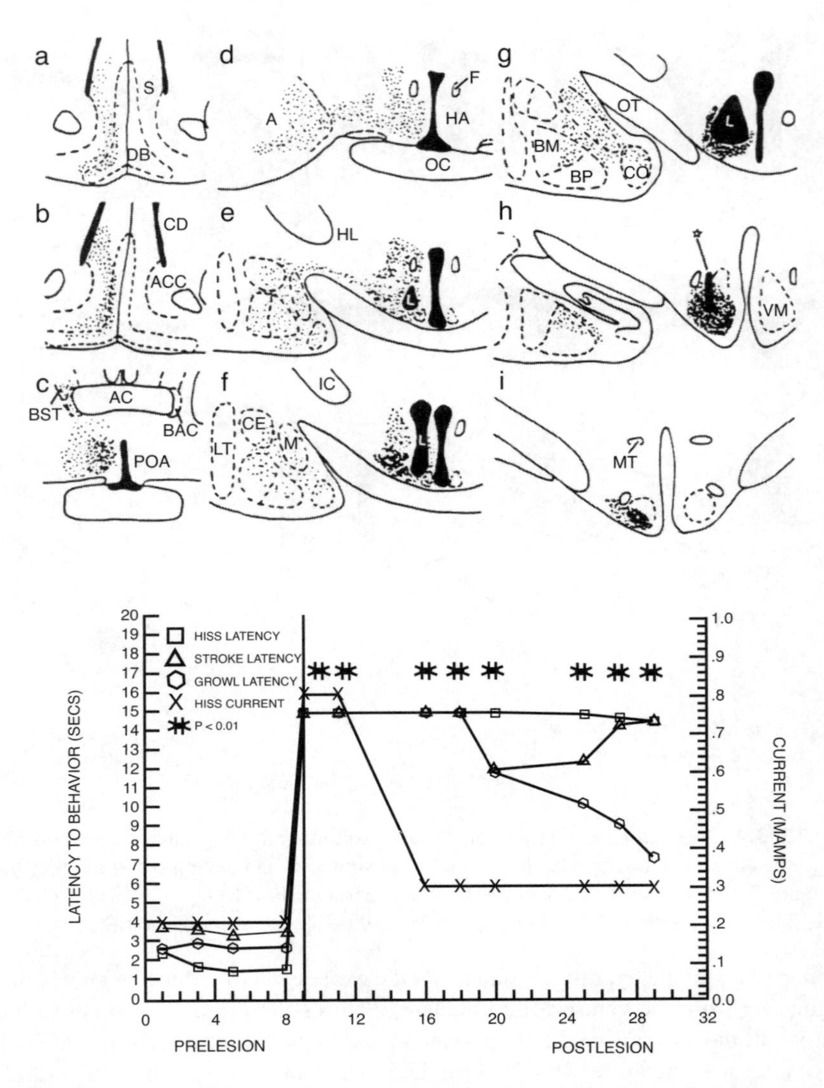

FIGURE 3.5 A lesion placed within the anteromedial hypothalamus resulted in an increase in latency and threshold for defensive rage behavior elicited from the ventromedial hypothalamus. Note the increase in optical density within the areas situated rostral to the site of stimulation in the region of the lesion in the anterior hypothalamus. (From S.A.G. Fuchs, H.M. Edinger, A. Siegel, *Brain Res.*, 330, 1985b, 93–108. With permission.)

also demonstrated that when lesions spared significant parts of the anterior hypothalamus, indicating that the pathway from the ventromedial nucleus to the anterior hypothalamus remained largely intact, ventromedial hypothalamic stimulation remained effective in eliciting the defensive rage response.

Results of these studies strongly imply that the pathways mediating defensive rage behavior from the ventromedial nucleus involve at least one synapse in the

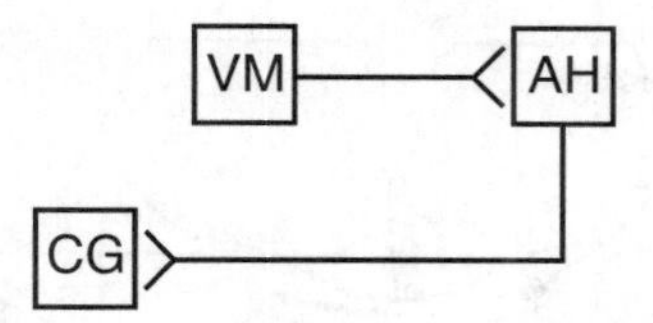

FIGURE 3.6 Summary of the principal hypothalamic pathways mediating defensive rage behavior. The organization of the descending pathway from the ventromedial hypothalamus (VM) to the brainstem involves neurons whose axons travel almost exclusively in a rostral direction to terminate largely in the anterior medial hypothalamus (AH). The second-order neurons of this pathway arise in the anterior medial hypothalamus and have more extensive descending projections, the most significant of which supply the PAG (CG). (From S.A.G. Fuchs, H.M. Edinger, A. Siegel, *Brain Res.*, 330, 1985b, 93–108. With permission.)

anterior hypothalamus and a second synapse in the PAG. This circuit is shown in Figure 3.6. Fuchs et al. (1985b) also point to the importance of the anterior hypothalamus in the process of defensive rage, as it occupies a key position within the basal forebrain. It is strategically interposed between the limbic system, brainstem, tuberohypophyseal system and ventromedial nucleus. In this scheme, the anterior medial hypothalamus could serve as a central integrating mechanism for defensive rage because of its direct inputs to the PAG, and because it receives important "modulating" inputs from such limbic regions as the septal area and amygdala. Moreover, the defensive rage response, which includes marked autonomic activity such as respiratory (Kastella et al., 1974) and cardiovascular changes (Stoddard-Apter et al., 1983), have also been linked to anterior hypothalamus. Thus, it is reasonable to conclude that further studies detailing the nature of the input–output relationships of the anterior medial hypothalamus might provide significant new information about the neural circuitry underlying defensive rage behavior.

Midbrain Periaqueductal Gray

The studies described above demonstrate that the primary target of medial hypothalamic neurons mediating defensive rage behavior is the dorsolateral aspect of the PAG. Thus, the next step in the analysis of the pathways mediating defensive rage was to identify the projection pathways arising from sites in the PAG in cats from which this form of aggressive behavior is elicited. The methods used were identical to those described above for the study of hypothalamic pathways mediating defensive rage behavior (Shaikh et al., 1987).

In this study, defensive rage behavior was elicited by both electrical and chemical (i.e., D,L-homocysteic acid) stimulation, and these sites were located mainly in the rostral half of the dorsolateral PAG — the principal receiving area of medial hypothalamic fibers mediating defensive rage. Then, either 2DG was injected systemically following electrical stimulation of a defensive rage site in the PAG of an awake or lightly anesthetized cat preparation, or (^{3}H) leucine was injected through the cannula-electrode into a defensive rage site within the PAG. Similar to findings observed in the previous experiment, similar patterns of labeling were observed after each of the tracing procedures. From the PAG, labeled descending axons were followed into

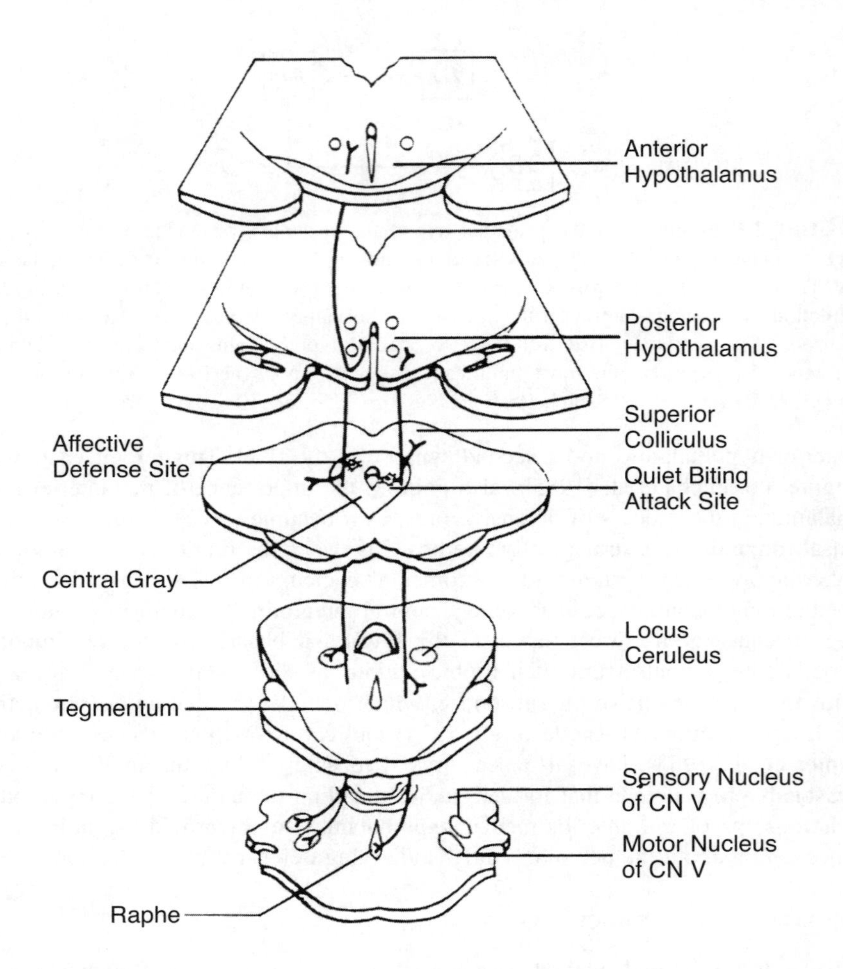

FIGURE 3.7 Diagrams indicating the principal efferent projections from the PAG associated with defensive rage (affective defense) (left) and quiet biting (predatory) attack behavior (right). Ascending fibers associated with defensive rage are distributed to the medial preopticohypothalamus, and caudally directed projections supply the locus ceruleus, tegmental fields and trigeminal complex. In contrast, fibers associated with quiet biting attack have a more limited distribution in which the rostrally directed fibers supply the posterior lateral hypothalamus, while the caudally directed fibers supply the central tegmental fields and median raphe. (From A. Siegel and C.B. Pott, *Prog. Neurobiol.*, 31, 1988, 261–283. With permission.)

the central tegmental fields of the midbrain, and pons, locus ceruleus, and motor and sensory nuclei of the trigeminal complex (Figure 3.7). It has been suggested that the projections to the reticular formation of the midbrain and pons serve as substrates for regulating cardiovascular responses associated with defensive rage as well as the paw strike, which presumably involves descending reticulospinal fibers to spinal cord motor neurons. In particular, activation of the sympathetic nervous system may come about through several routes. One route might involve projections to the locus cereleus, whose descending fibers have been shown to pass to sympathetic neurons in the intermediolateral cell column of the thoracic and lumbar spinal cord (Nygren and Olson, 1977). Alternatively, other investigators have described projections from the PAG to the solitary nucleus (Bandler and Tork, 1987), whose axons pass via the ventrolateral medulla to the intermediolateral cell column of the thoracic and lumbar spinal cord. Concerning the projections to the trigeminal complex, this pathway would account for the jaw opening (digastric) reflex that characterizes the hissing component of the defensive rage response.

In addition to the descending projections from the PAG, there is also an ascending projection whose primary target is the rostro-caudal extent of the medial hypothalamus. The unique feature of this projection is that the fibers pass to all regions of the medial hypothalamus associated with defensive rage behavior (Figure 3.7). In this manner, this ascending projection may serve as a substrate for a positive feedback mechanism. Such a mechanism could increase the likelihood that defensive rage would be expressed over an extended period of time following the presence of a threatening stimulus. It is reasonable to assume that by increasing duration of the response under dangerous conditions, survival value to the animal is likewise increased.

In summary, the anterior medial hypothalamus constitutes the primary output to the PAG for the expression of defensive rage behavior. Inasmuch as only a fragment of defensive rage components can be elicited below the level of the PAG, such as in the pons and medulla, it is probable that the PAG represents the lowest level of the CNS, where integration of somatomotor and autonomic components of this form of aggression takes place. These conclusions are summarized in Figure 3.7.

PATHWAYS MEDIATING PREDATORY ATTACK IN CATS

In Chapter 2, the sites in the hypothalamus from which predatory attack behavior could be elicited were identified (Wasman and Flynn, 1962). These and subsequent findings (e.g., Siegel and Pott, 1988; Siegel and Brutus, 1990; and Siegel et al., 1999) demonstrated that predatory attacks in cats could be elicited from the rostro-caudal extent of the preopticohypothalamus.

After having mapped the regions of the hypothalamus associated with predatory attack, Chi and Flynn (1971b) used the design described earlier in this chapter to trace the pathways mediating predatory attack. Again, predatory attack sites were identified and then a lesion was placed at the tip of the electrode. Following suitable survival times, animals were sacrificed and the brain processed by the Fink Heimer modification of the Nauta technique for identifying degenerating axons and their terminals. The major finding of this study was the presence of a large bundle of descending fibers that could be traced into the ventral tegmental area, and a second bundle that terminated in the midbrain PAG. In a second study based on these findings, Chi et al. (1976) used similar methods to trace descending fibers from the ventral tegmental area in association with predatory attack behavior. The major finding of this study suggested a projection to motor and sensory nuclei of the trigeminal nerve and to the motor nucleus of the facial nerve, which would play an important part in the biting component of the attack. While some of the data should be interpreted with caution because of the possibility of damage to fibers of passage, it is still likely that the projections from the lateral hypothalamus to the ventral tegmental area and PAG are important for the expression of predatory attack.

Several studies were conducted using retrograde tracing procedures, where HRP was injected into the pontine tegmentum and ventral tegmental area of the midbrain from which components of attack could be elicited (Smith and Flynn, 1979; Schoel et al., 1981). The key point to note from both studies was that labeled cells were identified in the lateral perifornical hypothalamus — the region from which predatory attack is typically elicited. Thus, this finding suggests the presence of a direct projection from the perifornical hypothalamus to the tegmentum of the midbrain and pons in association with predatory attack.

These findings were soon followed by an anterograde tracing study by Fuchs et al. (1981) in which the combined methods of 2DG and (^{3}H)-leucine radioautography were employed to re-examine the descending pathways from the perifornical and lateral hypothalamus to the brainstem. Again, the distribution of label from the perifornical hpothalamus was similar with both methodologies. In brief, fibers arising from perifornical hypothalamus were traced caudally through the medial forebrain bundle into the ventral tegmental area. Additional groups of fibers passed dorsally from the ventral tegmentum into the midbrain PAG and laterally into the central tegmental fields. Other fibers could be followed further caudally into the lateral tegmental fields of the pons, locus ceruleus, and motor nucleus of the trigeminal nerve. When compared with the projections from the perifornical region, it was observed that fibers arising from more laterally situated aspects of the hypothalamus, from which predatory attack was also elicited, displayed a much more limited projection pattern. The key point here is that the principal projection was directed to the perifornical region. *This finding, taken together with the findings obtained from the anterograde tracing methods and the retrograde labeling described by Smith and Flynn (1979, 1980a) and Schoel et al. (1981), provides support for the view that the perifornical region represents the central focus of the descending projection system of fibers for the expression of predatory attack behavior.* It should be pointed out that the functional relationship between hypothalamic attack sites and the midbrain tegmentum has also been documented by Huang and Flynn (1975). These authors showed that electrical stimulation of predatory attack sites in the awake cat generated excitation of neurons throughout this aspect of the midbrain.

A key challenge here is to determine the significance of these descending pathways with respect to predatory attack. For example, it is possible that the projection to the ventral tegmental area may play an important role by serving as the first neuron in a descending pathway to the lower brainstem tegmentum associated with predatory attack (Bernston, 1973), and nuclei of cranial nerves V and VII associated with the jaw-closing reflex used in biting (Chi et al., 1976). However, the more significant pathway is likely the one that arises directly from the perifornical hypothalamus, and which projects directly to the PAG, pontine tegmentum and motor nucleus of the trigeminal nerve. Such a projection enables the perifornical hypothalamus to directly control the masseteric reflex that is intrinsic to predatory attack. It is of interest to note that neurophysiological support for this connection was established by MacDonnell and Flynn (1966) and by Landgren and Olsson (1980), who showed that hypothalamic stimulation can facilitate jaw reflexes in cats.

An ascending component of the perifornical region was also noted. Ascending fibers from the perifornical hypothalamus could be followed rostrally into the preoptic region, bed nucleus of the stria terminalis, and lateral septal area (Figure 3.8). The possible significance of this projection to limbic regions is that it may provide the basis for a feedback mechanism to the limbic structures, which play an important role in modulating the attack mechanism.

RELATIONSHIP BETWEEN PREDATORY ATTACK AND DEFENSIVE RAGE: LINKAGE BETWEEN MEDIAL AND LATERAL HYPOTHALAMUS

It has been my experience over the course of conducting research on the neurobiology of aggression in cats that when experimental procedures are introduced which result in facilitation of one form of aggression, the same manipulations will cause suppression of the other form of aggression. A case in point, which is described in greater detail in Chapter 4, concerns amygdaloid modulation of aggression and rage. It has been shown that stimulation of the medial amygdala facilitates defensive rage (Shaikh et al., 1993) and suppresses predatory attack (Han et al., 1996a, 1996b). Similar observations have been found following administration of drugs such as naloxone (Brutus and Siegel, 1989) and cholecystokinin (CCK) compounds (Luo et al., 1998).

Evidence for the relationship between the medial and lateral hypothalamus was originally provided by Millhouse (1973) in a Golgi study, which demonstrated lateral hypothalamic processes passing toward the medial hypothalamus in rodents, and from a retrograde tracing study by Smith and Flynn (1980b), who identified retrograde-labeled neurons in the lateral hypothalamus following HRP injections that were placed at defensive rage sites in the medial hypothalamus of cats.

Several recent experiments have begun to systematically address this important issue (Han et al., 1996b; Cheu and Siegel, 1998). One study considered the projections from the medial to lateral hypothalamus as part of an overall analysis of amygdaloid modulation of aggression and rage (Han et al., 1996a, 1996b), which is considered in more detail in Chapter 4. For purposes of the present chapter, the analysis is limited to the relationship between the medial and lateral hypothalamus. In one experiment, Han et al. (1996b) showed that microinjection of the GABA agonist, muscimol, into a predatory attack site suppressed this response in a dose- and time-dependent manner (Figure 3.9). This finding, coupled with pretreatment with the $GABA_A$ receptor antagonist, bicuculline, which blocked the suppressive effects of muscimol, demonstrated the presence of $GABA_A$ receptors in the region of the lateral hypothalamus from which predatory attack is elicited. In a second experiment, it was initially shown that microinjection of the substance P agonist $(Sar^9,Met(O_2)^{11})$-substance P, into a site in the medial hypothalamus from which

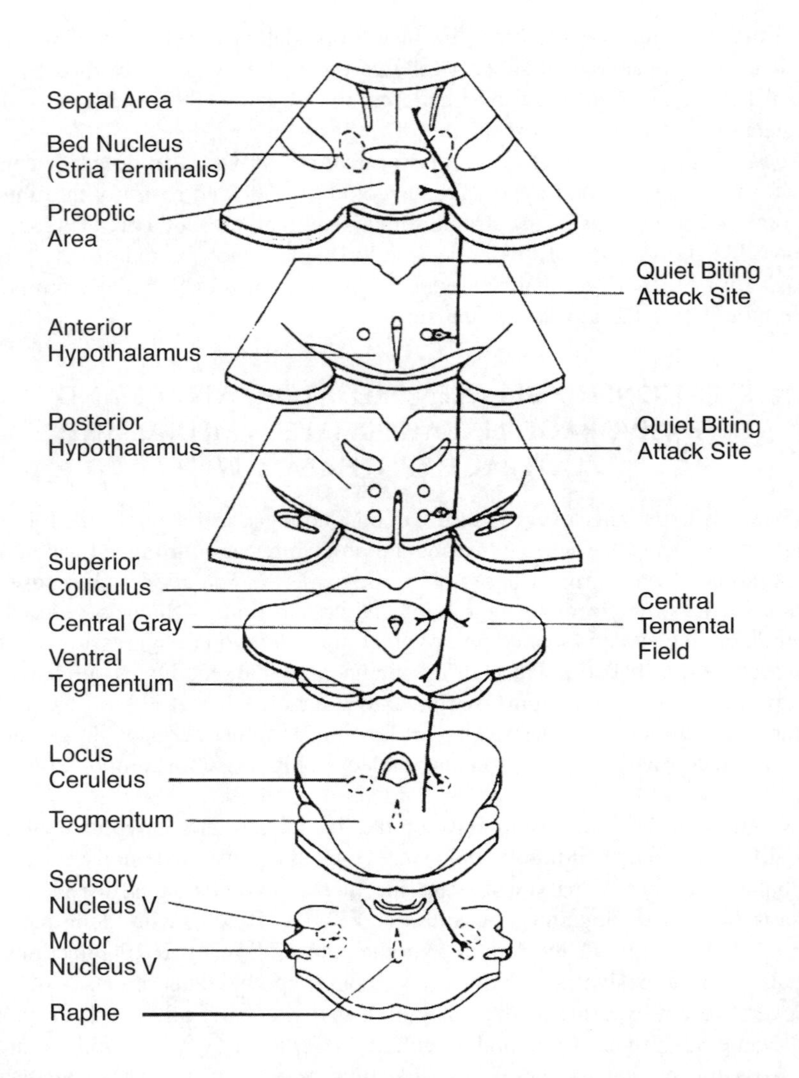

FIGURE 3.8 Diagram indicating the principal ascending and descending projections of the perifornical hypothalamus associated with quiet biting predatory attack. Note the projections to the PAG, tegmental fields, locus ceruleus, and motor nucleus of the trigeminal nerve. (From A. Siegel and C.B. Pott, *Prog. Neurobiol.*, 31, 1988, 261–283. With permission.)

defensive rage had been elicited, suppressed the occurrence of predatory attack elicited from the lateral hypothalamus. The most parsimonious way to explain this finding is that the medial hypothalamus contains extensive quantities of substance P (SP) neurokinin-1 (NK_1) receptors, which when activated, excite neurons in that region. Moreover, this finding would further suggest that one class of neurons containing NK_1 receptors is a GABA neuron that projects from the medial to lateral hypothalamus. This hypothesis was directly supported by the subsequent observation

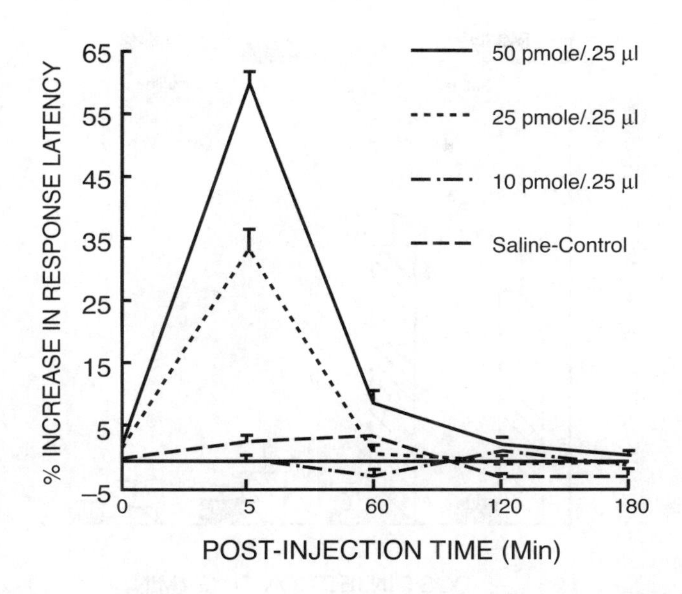

FIGURE 3.9 Suppressive effects of microinjections of muscimol into the lateral hypothalamus on predatory attack behavior elicited by stimulation of the lateral hypothalamus. (From Y. Han, M.B. Shaikh, A. Siegel, *Brain Res.*, 716, 1996b, 72–83. With permission.)

that microinjection of bicuculline into a lateral hypothalamic predatory attack site blocked the suppressive effects of SP (Figure 3.10). Control microinjections of bicuculline into the lateral hypothalamus had no effect on the attack response, ruling out any nonspecific effects that could have been attributed to bicuculline administration. In a final aspect of this study, the retrograde tracer Fluoro-Gold was microinjected into the lateral hypothalamic predatory attack site. The brain tissue was then processed for Fluoro-Gold and GABA immunohistochemical labeling. The results indicated the presence of GABA-positive neurons situated in the medial hypothalamus, which project to the regions of the lateral hypothalamus from which predatory attack can be elicited (Figure 3.11).

This finding, then, provides additional support for the view that communication from the medial to lateral hypothalamus involves a GABAergic neuron. The likely functional significance of this relationship is that activation of medial hypothalamic neurons, such as in the case when defensive rage is expressed, results in a concomitant suppression of predatory attack. The ethological value of this relationship is apparent, namely, that since these two responses are basically mutually exclusive, the effective expression of one of them requires the absence or suppression of the other. In the following experiment, I further demonstrate that the inhibitory GABAergic connection between the medial and lateral hypothalamus is a reciprocal one.

Utilizing a similar approach, Cheu and Siegel (1998) first demonstrated that lateral hypothalamic stimulation suppressed defensive rage behavior elicited from the medial hypothalamus. It was further demonstrated that this effect was mediated

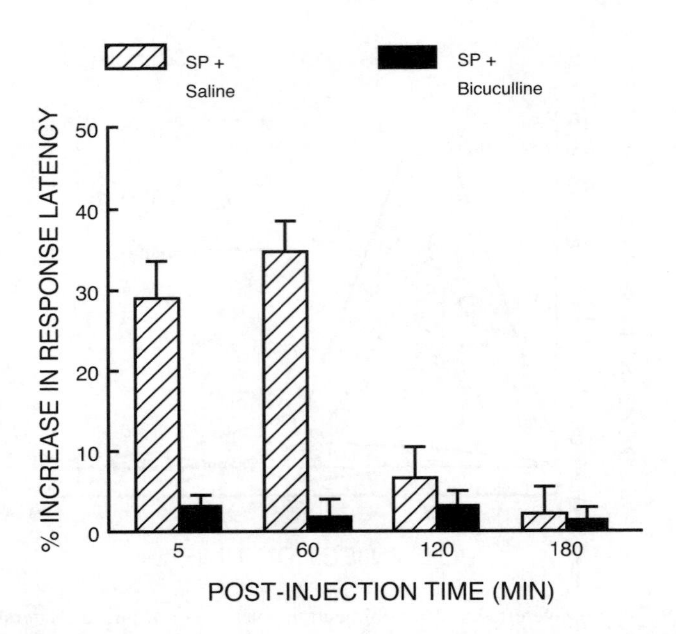

FIGURE 3.10 Blocking effects of bicuculline upon substance P induced suppression of predatory attack. Hatched bars represent effect of substance P (2.0 nM/0.25 l) infusion into the medial hypothalamus following pretreatment with saline into the lateral hypothalamus. Closed bars reflect the effects of bicuculline (0.15 nM) pretreatment into lateral hypothalamic sites from which predatory attack was elicited upon suppression of this response following infusion of SP into the medial hypothalamus. (From Y. Han, M.B. Shaikh, A. Siegel, *Brain Res.*, 716, 1996b, 72–83. With permission.)

by $GABA_A$ receptors in the medial hypothalamus since bicuculline administration into the medial hypothalamic defensive rage site blocked the suppressive effects of lateral hypothalamic stimulation. This study also demonstrated through double-labeling procedures involving retrograde Fluoro-Gold labeling and immunocytochemical staining for GABA neurons that GABA positive neurons are present in the lateral hypothalamus, which project to the medial hypothalamus. These reciprocal relationships between the medial and lateral hypothalamus are illustrated in Figure 3.12, and further reinforce the dynamic interaction between defensive rage and predatory attack.

HYPOTHALAMIC AGGRESSION IN RATS

In attempting to develop an understanding of the nature of neural bases of aggression and rage, it becomes useful to compare and contrast the forms of hypothalamically elicited aggression elicited from rats relative to those of cats. Likewise, it is also instructive to compare the neuroanatomical circuitry for the respective forms of aggression in rats and cats.

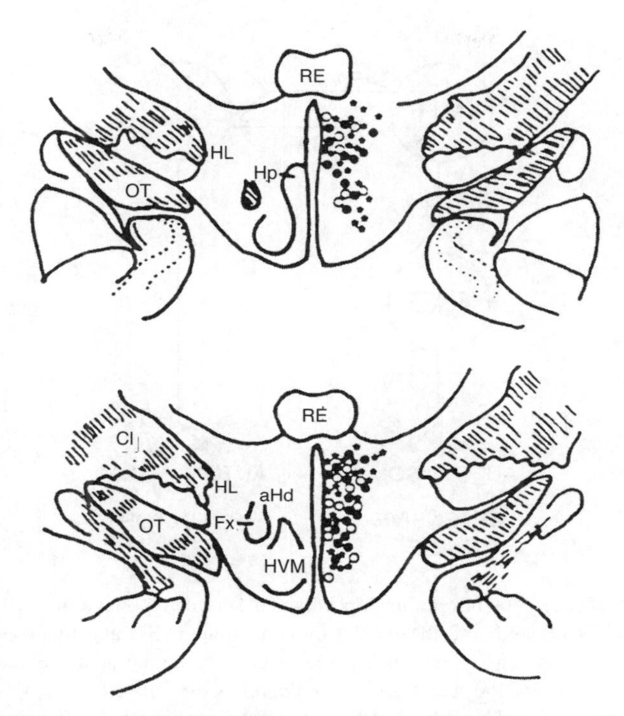

FIGURE 3.11 Maps indicate the distribution of cells retrogradely labeled with Fluoro-Gold (open circles) microinjected into the lateral hypothalamus; GABA-immunopositive neurons (closed circles) and double-labeled neurons for both GABA and Fluoro-Gold (stars). Note the presence of double-labeled neurons in the medial hypothalamus. (From Y. Han, M.B. Shaikh, A. Siegel, *Brain Res.*, 716, 1996b, 72–83. With permission.)

Nature of Attack Response

Many investigators have shown that electrical stimulation of the rat hypothalamus can produce diverse aggressive responses that may include affective attack (Koolhaas, 1978), quiet attacks (Panksepp, 1971), bite attacks (Woodworth, 1971), attack jumps, and clinch fights (Kruk et al., 1979; Lammers et al., 1988; Van der Poel et al., 1982). In recent years, a greater emphasis has been placed on the study of electrical stimulation–induced attack involving forms of biting attack of the neck, head, or back, coupled with paw kicks to the flank, clinch fights, and attack jumps (Kruk et al., 1979, 1983, 1984, 1990). In a jump attack, the attacker jumps toward the opponent with the aim of biting its head. In a clinch fight, the attacker typically becomes locked with the top or back of the opponent, which often occurs when the attacker loses its footing after attempting to attack the target animal (Kruk et al., 1990). Stimulation-induced attack from the hypothalamus can be directed against a dominant, subordinate, anesthetized, or dead rat. Sex does not seem to be a major factor, as a male can attack a female, while a female may attack another female or male. Variety in the forms of attack is a function of factors such as the intensity of

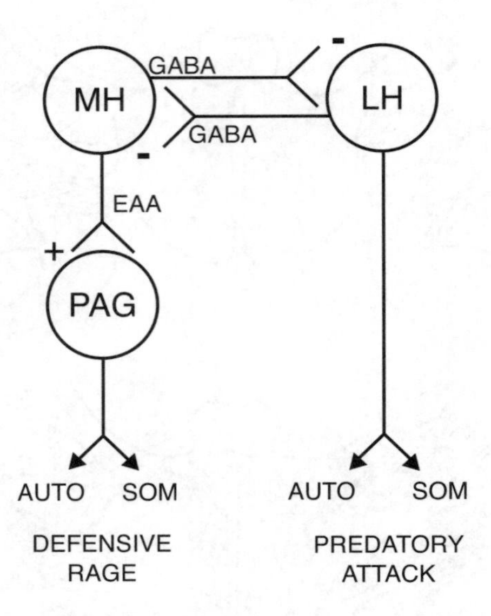

FIGURE 3.12 Model depicts relationship between lateral hypothalamus (LH) from which predatory attack is elicited, and the medial hypothalamus (MH) and its projection target in the PAG over which defensive rage is elicited; excitatory amino acids (EAA) comprise the neurotransmitter. The medial and lateral hypothalamus are related by reciprocal inhibitory pathways that are mediated by GABA. Auto, autonomic component; som, somatic component of the attack response.

stimulation (i.e., the higher the intensity of current, the shorter the latency for attack) and reaction of the opponent.

When comparing responses in rats with those of cats, it should be pointed out that with respect to defensive rage, a male may attack a female or another male, and vice versa. In contrast, predatory attack is never directed at another cat, but is instead limited to rodents or other smaller animals.

Attack Sites in Rats

In contrast to cats, it would appear that a single area, called the hypothalamic aggressive area (HAA), constitutes the neural substrate for the release of attacks that can take various forms, which include both offensive and defensive components. As noted above, the variety of response forms appears to be a function of variables such as current strength and environmental conditions. The HAA extends from the ventral aspect of the lateral hypothalamus to the medial aspect of the ventromedial nucleus and arcuate nucleus, and anteriorly to the anterior hypothalamic nucleus (Figure 3.13). Based on the results of a recent morphological analysis of this region, the region eliciting attack has been classified as an intermediate hypothalamic area (IHA) (Geeraedts et al., 1990a, 1990b). It should be noted that this region in rats does not specifically correspond to the regions eliciting predatory attack or defensive rage in cats, although there is some overlap. For example, the medial aspects of IHA from

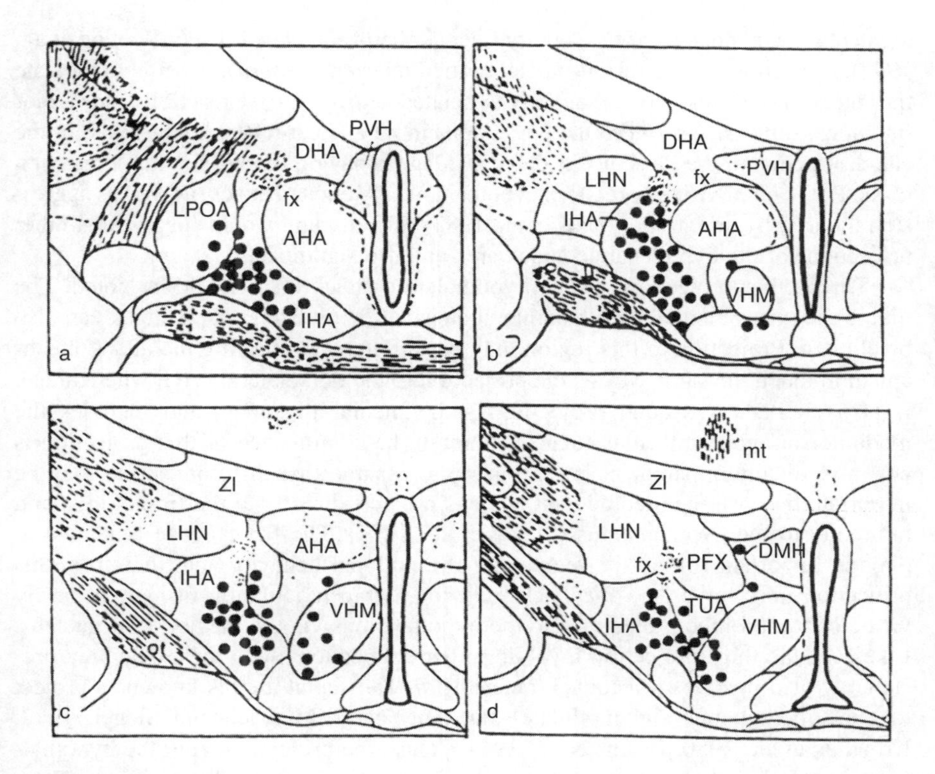

FIGURE 3.13 The attack area coincides with the intermediate hypothalamic area and the ventrolateral aspect of the ventromedial nucleus. AHA, anterior hypothalamic area; ci, internal capsule; DHA, dorsal hypothalamic area; DMH, dorsomedial hypothalamic nucleus; fx, fornix; LHN, lateral hypothalamus; LPOA, lateral preoptic area; mt, mammillothalamic tract; ot, optic tract; PFX, perifornical nucleus; PVH, paraventricular nucleus; VMN, ventromedial nucleus, ZI, zona incerta. (From A. Siegel, T.A.P. Roeling, T.R. Gregg, M.R. Kruk, *Neurosci. Biobehav. Rev.*, 23, 1999, 359–389. With permission.)

which aggressive responses are elicited correspond to sites in cats where defensive rage can be elicited. Likewise, some of the laterally displaced sites in the IHA, which are located adjacent to the fornix, correspond to sites in cats where predatory attack can be elicited. The interpretation of this finding is that in cats, there is a greater development of precise "labeled" pathways mediating different forms of aggression compared to that in rats, where the pathway mediating aggression may be less precise. The next section describes recent studies that have attempted to identify these pathways.

Efferent Pathways from Hypothalamus Mediating Attack in Rats

There are many similarities between projection patterns associated with aggression in rats and cats. For example, the hypothalamic attack area in rats projects heavily to the PAG, ventral tegmental area, locus ceruleus, raphe magnus, A5 region of the brainstem tegmentum, and nucleus of the solitary tract (Roberts and Nagel, 1996; Roeling et al., 1994). Projections to the various nuclei of the brainstem presumably serve to mediate the autonomic and endocrine functions associated with aggressive reactions. One major functional difference between the projections in rat and cat relates to the PAG. In the cat, destruction of the PAG blocks defensive rage behavior. In rats, similar destruction of the PAG only transiently reduces hypothalamically elicited attack, as well as aggression induced by an intruder (Mos et al., 1983). This finding would suggest that other projections of the hypothalamic attack area are also significant.

Three other projections of the hypothalamic attack area should be noted. The first is the ventromedial hypothalamic nucleus, a region from which attack can also be elicited. Projection to this region may be important for driving the attack mechanism in much the same way as the projection to the dorsolateral PAG, where attack is also elicited. A second projection is the medial thalamus, and includes the mediodorsal and parataenial nuclei. Although the significance of this projection is not immediately apparent, it is reasonable to assume that this connection may be important because the mediodorsal nucleus projects directly to the frontal lobe and indirectly to the hypothalamus (Siegel et al., 1972). The frontal lobe is known to play an important role in the regulation of emotional behavior and, therefore, this projection may serve as a feedback loop for limbic modulation of aggression by virtue of frontal lobe projections to the hypothalamus via the mediodorsal nucleus. Likewise, the third projection involving fiber connections with the septal area presumably also involves a feedback mechanism. The septal area is known to project to the hypothalamus and modulate attack behavior (Meibach and Siegel, 1977; Krayniak et al., 1980; Brutus et al., 1984). Thus, the projection from the hypothalamus to the dorsal septal area would provide the necessary feedback circuit. Here, activation of the hypothalamic attack area can excite the dorsal septal area, which, in turn, would modulate activity of the hypothalamic neurons governing attack in rats. Feedback circuits constitute an important feature of CNS function, and are certainly present in sensory and motor systems. Accordingly, these feedback circuits may also constitute an important overall mechanism in regulating aggression. The role of limbic system structures in the regulation of aggression and rage is considered in detail in Chapter 5.

It should be pointed out that forms of aggression in rats other than biting attack have not been systematically examined, and thus, there is no basis for a comparison with the circuits governing biting attack and affective forms of aggression in cats. Hopefully, future studies will be able to delineate the nature of specific pathways mediating different forms of aggression in rats. Such studies would allow for a more effective comparison of the similarities and differences between species concerning these forms of aggression, and perhaps would provide new insights into the neural bases of these behaviors.

SUMMARY

In this chapter, I have considered the anatomical circuitry governing defensive rage and predatory attack in cats and biting attack in rats. Studies in cats have revealed that the key neuronal groups mediating defensive rage include the anterior medial hypothalamus and dorsolateral PAG, which receives its input from the anterior medial hypothalamus. Projections from the PAG descend to a variety of lower brainstem nuclei that control the somatomotor and autonomic components of the defensive rage response. With respect to predatory attack, the most important cell groups essential for the expression of this response are situated in the perifornical hypothalamus. Fibers from this region descend to nuclei such as the PAG, ventral tegmental area, tegmental fields of the midbrain and pons, locus ceruleus, and motor nucleus of the trigeminal nerve. These pathways also provide the basis for the somatomotor and autonomic components of this form of aggression. This chapter also considered the relationship between the neural circuitry governing predatory attack and defensive rage. It was discovered that reciprocal inhibitory connections are present between the lateral hypothalamic predatory attack zone and the medial hypothalamic defensive rage zone. The functional significance of these reciprocal connections was considered in this chapter. Finally, the nature of the pathways governing biting attack in the rat was also considered, and was shown to parallel those described in cats. The significance of identifying the anatomical circuitry underlying various forms of aggression and rage is that by providing the basic substrates for their expression, a rational basis is established for advancing new hypotheses concerning the functional and neurochemical mechanisms that produce and control these responses.

REFERENCES

Bandler, R. and T. McCulloch, Afferents to a midbrain periaqueductal grey region involved in the 'defense reaction' in the cat as revealed by horseradish peroxidase: II. The diencephalon, *Behav. Brain Res.*, 13, 1984, 279–285.

Bandler, R. and I. Tork, Midbrain periaqueductal grey region in the cat has afferent and efferent connections with solitary tract nuclei, *Neurosci. Lett.*, 74, 1987, 1–6.

Berntson, G.G., Attack, grooming, and threat elicited by stimulation of the pontine tegmentum in cats, *Physiol. Behav.*, 11, 1973, 81–87.

Brutus, M., M.B. Shaikh, A. Siegel, H.E. Siegel, An analysis of the mechanisms underlying septal area control of hypothalamically elicited aggression in the cat, *Brain Res.*, 310, 1984, 235–248.

Brutus, M. and A. Siegel, Effects of the opiate antagonist naloxone upon hypothalamically elicited affective defense behavior in the cat, *Behav. Brain Res.*, 33, 1989, 23–32.

Cheu, J.W. and A. Siegel, GABA receptor mediated suppression of defensive rage behavior elicited from the medial hypothalamus of the cat: role of the lateral hypothalamus, *Brain Res.*, 783, 1998, 293–304.

Chi, C.C., R. Bandler, J.P. Flynn, Neuroanatomic projections related to biting attack elicited from ventral midbrain in cats, *Brain Behav. Evol.*, 13, 1976, 91–110.

Chi, C.C. and J.P. Flynn, Neural pathways associated with hypothalamically elicited attack behavior in cats, *Science*, 171, 1971a, 703–706.

Chi, C.C. and J.P. Flynn, Neuroanatomic projections related to biting attack elicited from hypothalamus in cats, *Brain Res.*, 35, 1971b, 49–66.

Fuchs, S.A.G., M. Dalsass, H.E. Siegel, A. Siegel, The neural pathways mediating quiet-biting attack behavior from the hypothalamus in the cat: a functional autoradiographic study, *Aggress. Behav.*, 7, 1981, 51–67.

Fuchs, S.A.G., H.M. Edinger, A. Siegel, The organization of the hypothalamic pathways mediating affective defense behavior in the cat, *Brain Res.*, 330, 1985a, 77–92.

Fuchs, S.A.G., H.M. Edinger, A. Siegel, The role of the anterior hypothalamus in affective defense behavior elicited from the ventromedial hypothalamus of the cat, *Brain Res.*, 330, 1985b, 93–108.

Geeraedts, L.M.G., R. Nieuwenhuys, J.G. Veening, Medial forebrain bundle of the rat: III. Cytoarchitecture of the rostral, telencephalic, part of the medial forebrain bundle bed nucleus, *J. Comp. Neurol.*, 294, 1990a, 507–536.

Geeraedts, L.M.G., R. Nieuwenhuys, J.G. Veening, Medial forebrain bundle of the rat: IV. Cytoarchitecture of the caudal, lateral hypothalamic, part of the medial forebrain bundle bed nucleus, *J. Comp. Neurol.*, 294, 1990b, 537–568.

Han, Y., M.B. Shaikh, A. Siegel, Medial amygdaloid suppression of predatory attack behavior in the cat: I. Role of a substance P pathway from the medial amygdala to the medial hypothalamus, *Brain Res.*, 716, 1996a, 59–71.

Han, Y., M.B. Shaikh, A. Siegel, Medial amygdaloid suppression of predatory attack behavior in the cat: II. Role of a GABAergic pathway from the medial to the lateral hypothalamus, *Brain Res.*, 716, 1996b, 72–83.

Huang, Y.H. and J.P. Flynn, Unit activities in the hypothalamus and midbrain during stimulation of hypothalamic attack sites, *Brain Res.*, 93, 1975, 419–440.

Kastella, K.G., H.A. Spurgion, G.K. Weiss, Respiratory related neurons in anterior hypothalamus of the cat, *Am. J. Physiol.*, 227, 1974, 710–713.

Koolhaas, J.M. Hypothalamically induced intraspecific aggressive behaviour in the rat, *Exp. Brain Res.*, 32, 1978, 365–375.

Krayniak, P., S. Weiner, A. Siegel, An analysis of the efferent connections of the septal area in the cat, *Brain Res.*, 189, 1980, 15–29.

Krieger, M.S., L.C.A. Conrad, D.W. Pfaff, An autoradiographic study of the efferent connections of the ventromedial nucleus of the hypothalamus, *J. Comp. Neurol.*, 183, 1979, 785–816.

Kruk, M.R., C.E. Van der Laan, J. Mos, A.M. Van der Poel, W. Meelis, B. Olivier, Comparison of aggressive behaviour induced by electrical stimulation in the hypothalamus of male and female rats, *Prog. Brain Res.*, 61, 1984, 303–314.

Kruk, M.R., C.E. Van der Laan, J. Van der Spuy, A.M.M. Van Erp, W. Meelis, Strain differences in attack patterns elicited by electrical stimulation in the hypothalamus of male CPBWEzob and CPBWI rats, *Aggress. Behav.*, 16, 1990, 177–190.

Kruk, M.R., A.M. Van der Poel, T.P. De Vos-Frerichs, The induction of aggressive behaviour by electrical stimulation in the hypothalamus of male rats, *Behaviour*, 70, 1979, 292–322.

Kruk, M.R., A.M. Van der Poel, W. Meelis, J. Hermans, P.G. Mostert, J. Mos, A.H.M. Lohman, Discriminant analysis of the localization of aggression-inducing electrode placements in the hypothalamus of male rats, *Brain Res.*, 260, 1983, 61–79.

Lammers, J.H.C.M., M.R. Kruk, W. Meelis, A.M. Van der Poel, Hypothalamic substrates for brain stimulation-induced patterns of locomotion and escape jumps in the rat, *Brain Res.*, 449, 1988, 294–310.

Landgren, S. and K.A. Olsson, The effect of electrical stimulation in the hypothalamus on the monosynaptic jaw closing and the disynaptic jaw opening reflexes in the cat, *Exp. Brain Res.*, 39, 1980, 389–400.

Luo, B., J.W. Cheu, A. Siegel, Cholecystokinin B receptors in the periaqueductal gray potentiate defensive rage behavior elicited from the medial hypothalamus of the cat, *Brain Res.*, 796, 1998, 27–37.

MacDonnell, M. and J.P. Flynn, Sensory control of hypothalamic attack, *Anim. Behav.*, 14, 1966, 399–405.

Meibach, R. and A. Siegel, Efferent connections of the septal area in the rat: an analysis utilizing retrograde and anterograde transport methods, *Brain Res.*, 119, 1977, 1–20.

Millhouse, O.E., Certain ventromedial hypothalamic afferents, *Brain Res.*, 55, 1973, 89–105.

Mos, J., J.H.C.M. Lammers, A.M. Van der Poel, B. Bermond, W. Meelis, M.R. Kruk, Effects of midbrain central gray lesions on spontaneous and electrically induced aggression in the rat, *Aggress. Behav.*, 9, 1983, 133–155.

Nygren, L.-G. and L. Olson, A new major projection from locus coeruleus: the main source of noradrenergic nerve terminals in the ventral and dorsal columns of the spinal cord, *Brain Res.*, 132, 1977, 85–93.

Panksepp, J. Aggression elicited by electrical stimulation of the hypothalamus in albino rats, *Physiol. Behav.*, 6, 1971, 321–329.

Roberts, W.W. and J. Nagel, First-order projections activated by stimulation of hypothalamic sites eliciting attack and flight in rats, *Behav. Neurosci.*, 110, 1996, 509–527.

Roeling, T.A.P., J.G. Veening, M.R. Kruk, J.P.W. Peters, M.E.J. Vermelis, R. Nieuwenhuys, Efferent connections of the hypothalamic "aggression area" in the rat, *Neuroscience*, 59, 1994, 1001–1024.

Sakuma, Y. and D.W. Pfaff, Properties of ventromedial hypothalamic neurons with axons to midbrain central gray, *Exp. Brain Res.*, 46, 1982, 292–300.

Saper, C.L., Swanson, W. Cowan, The efferent connections of the ventromedial nucleus of the hypothalamus of the rat, *J. Comp. Neurol.*, 169, 1976, 409–442.

Schoel, W.M., C.A. Opsahl, J.P. Flynn, Afferent projections related to attack sites in the ventral midbrain tegmentum of the cat, *Brain Res.*, 204, 1981, 269–282.

Shaikh, M.B., J.A. Barrett, A. Siegel, The pathways mediating affective defense and quiet biting attack behavior from the midbrain central gray of the cat: an autoradiographic study, *Brain Res.*, 437, 1987, 9–25.

Shaikh, M.B., A. Steinberg, A. Siegel, Evidence that substance P is utilized in medial amygdaloid facilitation of defensive rage behavior in the cat, *Brain Res.*, 625, 1993, 283–294.

Siegel, A. and M. Brutus, Neural substrates of aggression and rage in the cat, in A.N. Epstein and A.R. Morrison, Eds., *Progress in Psychobiology and Physiological Psychology*, Academic Press, San Diego, Calif., 1990, pp. 135–233.

Siegel, A., J.P. Flynn, R. Bandler, Thalamic sites and pathways related to elicited attack, *Brain Behav. Evol.*, 6, 1972, 542–555.

Siegel, A. and C.B. Pott, Neural substrate of aggression and flight in the cat, *Prog. Neurobiol.*, 31, 1988, 261–283.

Siegel, A., T.A.P. Roeling, T.R. Gregg, M.R. Kruk, Neuropharmacology of brain-stimulation-evoked aggression, *Neurosci. Biobehav. Rev.*, 23, 1999, 359–389.

Smith, D.A. and J.P. Flynn, Afferent projections to attack sites in the pontine tegmentum, *Brain Res.*, 164, 1979, 103–119.

Smith, D.A. and J.P. Flynn, Afferent projections to affective attack sites in cat hypothalamus, *Brain Res.*, 194, 1980a, 41–51.

Smith, D.A. and J.P. Flynn, Afferent projections to quiet attack sites in cat hypothalamus, *Brain Res.*, 194, 1980b, 29–40.

Sokoloff, L. Relation between physiological function and energy metabolism in the central nervous system, *J. Neurochem.*, 29, 1977, 13–26.

Stoddard-Apter, S. B., Levin, A. Siegel, A sympathoadrenal and cardiovascular correlates of aggressive behavior in the awake cat, *J. Auton. Nerv. Syst.*, 8, 1983, 343–360.

Van der Poel, A.M., B. Olivier, J. Mos, M.R. Kruk, W. Meelis, J.H.M. Van Aken, Anti-aggressive effect of a new phenylpiperazine compound, DU27716, on hypothalamically induced behavioural activities, *Pharmacol. Biochem. Behav.*, 17, 1982, 147–153.

Wasman, M. and J.P. Flynn, Directed attack elicited from hypothalamus, *Arch. Neurol.*, 6, 1962, 220–227.

Woodworth, C.H., Attack elicited in rats by electrical stimulation of the lateral hypothalamus, *Physiol. Behav.*, 6, 1971, 345–353.

4 Physiological Processes and Mechanisms

DIENCEPHALON AND BRAINSTEM

Various approaches to the study of the neurobiology of aggression and rage have evolved. In this chapter, consideration is given to those studies that have attempted to address the following questions:

1. What are the relative roles of the hypothalamus and brainstem in the expression of aggression and rage?
2. Can aggression be elicited from regions other than the hypothalamus and brainstem?
3. What are the physiological characteristics associated with electrical stimulation of attack sites in the hypothalamus and brainstem?
4. How does stimulation at attack sites affect the sensory and motor processes associated with attack?
5. Which sensory processes are critical for the expression of attack and which are not?
6. What is the key behavioral properties associated with attack?

Some of the key issues that have been studied include the specificity of the attack response, whether the attack response is positively or negatively reinforcing, and whether there can be conditioning of the attack response. Several of these issues have been addressed briefly in Chapters 2 and 3. However, because of their importance, further consideration is given to them in this chapter.

RELATIVE ROLES OF HYPOTHALAMUS, BRAINSTEM, AND OTHER REGIONS IN EXPRESSION OF AGGRESSION AND RAGE

As indicated in previous chapters, there have been two basic approaches to the importance of specific regions in the brain in the expression of aggression and rage. One has been to place a selective lesion in a given region and then observe the effects of that lesion on the propensity to elicit rage or aggression by stimulation or occurring spontaneously in response to environmental conditions. The second has been to use electrical or chemical stimulation of specific sites in the brain as a means of eliciting an attack response. This approach would thus permit identification of a given site as a component of a region associated with attack.

HYPOTHALAMUS

Studies conducted mainly in cats and rodents include those that have been directed at different hypothalamic and thalamic nuclei, related regions of the forebrain, and aspects of the brainstem, which include the periaqueductal gray (PAG) and adjoining parts of the tegmentum. In studies using cats, several different approaches have been used.

Ellison and Flynn (1968) asked a basic question: Is the hypothalamus necessary for the expression of rage and aggression? Their strategy was to create a surgical island of the hypothalamus by using a specialized knife. This knife was stereotaxically placed into the hypothalamus and, by rotating it 360 degrees, much of hypothalamus was isolated from the rest of the brain. One set of experiments used cats that spontaneously attacked rodents prior to the surgical isolation of the hypothalamus, and demonstrated that the surgical procedure did not prevent spontaneous predatory attack, in particular when the rodent approached the cat. Other cats displayed no spontaneous attack after the surgical procedure. However, in these animals, a biting attack could be elicited by stimulation of the midbrain tegmentum lateral to the PAG. Likewise, cats did not show spontaneous defensive rage after isolation of the hypothalamus. However, defensive rage such as an attack on the experimenter could be elicited after pinching the tail. These authors thus concluded that aggression and rage can be elicited in the absence of the hypothalamus. But they further stated that normally the hypothalamus plays an important role in the expression of both defensive rage and predatory attack.

In addition to knife cuts that separated the hypothalamus from brainstem, Paxinos and Bindra (1972) examined the effects of knife cuts separating the medial from lateral hypothalamus in rats, as determined by the number of bites delivered to a mouse (as an index of aggression) or the degree to which a rat would bite a glove pushed against its snout (as an index of irritability). These authors reported that knife cuts, in particular at the level of the anterior hypothalamus, increased levels of irritability, and food and water intake, but failed to affect biting attack. If, as indicated in Chapter 1, it is assumed that irritability and defensive behavior reflect the same or related processes, then these results are consistent with those provided by Cheu et al. (1998), which were described in Chapter 3. Cheu et al. provided evidence that the lateral hypothalamus can inhibit defensive rage in the cat by virtue of GABAergic neurons that project from the lateral to medial hypothalamus. Therefore, it is possible and perhaps likely that the increased irritability observed by Paxinos and Bindra may have resulted from the disruption of inhibitory GABAergic neurons that connect the lateral with the medial hypothalamus.

Another basic question concerns the relative importance of specific nuclei or regions of the hypothalamus in the expression of rage and aggression. Several investigators have addressed this issue. In one group of studies, the effects of lesions placed in specific regions of the hypothalamus on predatory attack behavior were determined. Fonberg and Serduchenko (1980) placed bilateral lesions in the ventrolateral posterior hypothalamic lesions in cats. These investigators observed that such lesions did not diminish naturally occurring mouse-killing behavior, and thus are not critical for this form of aggression. However, these lesions did diminish mouse consumption, thus reflecting a neurally based distinction between predatory attack and eating, a subject that will be considered later in this chapter.

In another study, Halliday and Bandler (1981) sought to determine the role of the anterior hypothalamus in predatory attack in cats. Here, knife cuts separating the anterior aspect from more posterior aspects of the hypothalamus blocked this form of attack, which had been elicited from either the posterior hypothalamus or ventral midbrain. The results of this study and the one described above indicate the importance of the anterior hypothalamus in mediating predatory attack relative to other parts of the hypothalamus. These data suggest that the larger numbers of neurons which integrate the attack response are situated in the anterior hypothalamus, and that the expression of this behavior is mediated over axons arising from these cells that pass both caudally to the brainstem tegmentum and rostrally to the basal forebrain and limbic structures in the medial forebrain bundle.

It is apparent that the anterior hypothalamus-preoptic region plays an important role not only in the expression of predatory attack but in mediating affective forms of attack in several species as well. As indicated in Chapter 3, Fuchs et al. (1985a, 1985b) showed that fibers from the ventromedial hypothalamus in association with defensive rage project to the anterior medial hypothalamus-preoptic region. These authors further demonstrated that lesions of the anterior medial hypothalamus-preoptic region block or significantly suppress this form of aggression. Likewise, bilateral lesions of the anterior hypothalamus-preoptic region decrease intermale aggression in mice (Edwards et al., 1993). Thus, these studies demonstrated the importance of the region of the anterior hypothalamus in different forms of aggression across several species, including defensive rage and fighting behavior.

While the anterior hypothalamus is of obvious importance in the expression of rage and aggression, Canteras et al. (1997) provided evidence that the dorsal premammillary nucleus of the rat may also be of importance in defensive behavior for this species. In this study, these authors first showed that c-Fos levels increased significantly in the dorsal premammillary nucleus when the rat displayed defensive behavior characterized by freezing, escape, and jump attacks in response to the presence of a cat. Then, ibotenic acid lesions of the dorsal premammillary nucleus were made. The lesions eliminated defensive behavior while exploratory behavior was increased, indicating that the suppressive effects were not due to a nonspecific blockade of motor responses. These authors suggested that the dorsal premammillary nucleus is involved in defensive behavior as a result of the inputs that it receives from the anterior and ventromedial hypothalamus and because it projects to the PAG.

Midbrain PAG and Related Structures

A number of studies, the majority of which were with cats, have attempted to assay the importance of the PAG and adjoining regions of the brainstem tegmentum in aggression and rage. In an early study, Carli et al. (1963) elicited rage by stimulation of the hypothalamus in acutely prepared cats in which much of the forebrain had previously been removed. Then, initially lesions were placed in the anterior aspect of the PAG but, surprisingly, they had little effect on the rage response. Other lesions separately involving either the medial or lateral aspects of the midbrain tegmentum also did not significantly impair the response. It was only when these investigators placed extensive lesions involving much of midbrain tegmentum did they notice impairment of the rage response. They concluded that the pathways mediating rage are quite diffuse and likely involve a variety of descending pathways from the hypothalamus to the brainstem. However, as described below, there is a much larger body of evidence pointing to the likelihood that defensive rage is mediated over descending pathways in the cat that largely are directed to the PAG. In another attempt to use a lesion approach for identification of the neural pathways mediating defensive rage, Kanai and Wang (1962) elicited feline hissing associated with this form of aggression by stimulation of sites either in the medial hypothalamus, PAG, and parts of the tegmentum of the midbrain and pons. They then placed discrete lesions just ventral to the medial lemniscus at the level of the caudal medulla, and observed that such lesions blocked vocalization elicited from the pontine tegmentum.

At approximately the same time, Skultety (1963) conducted an experiment in which he elicited defensive rage by stimulation of the anterior hypothalamus and then placed extensive lesions in the PAG. In contrast to the above findings observed by Carli et al. (1963), Skultety found that such lesions produced extensive and prolonged blocking of the rage response. It was also shown that when stimulation was applied to aspects of the PAG from which rage could be elicited and which were located caudal to the lesion, few or no effects of the lesions were noted, thus demonstrating the anatomical and functional specificity of the lesions on hypothalamically elicited rage.

As described in Chapter 3 and in chapters to follow, there are other studies that used lesion, stimulation, neuroanatomical, and neuropharmacological approaches to demonstrate the importance of the PAG in the expression of defensive rage. In particular, the research of Hunsperger is a case in point. Hunsperger (1956) demonstrated that defensive rage elicited by medial hypothalamic stimulation in the cat was initially blocked following a lesion of the PAG and, after several weeks, the current threshold for elicitation of the response was still significantly elevated. Similarly, Berntson (1972) also observed the blockade of the vocalization component of the defensive rage response following a lesion placed in the PAG of the cat. In contrast, Mos et al. (1983) observed that PAG lesions did not block biting attacks in a rat (on another rat), although such lesions did raise the current threshold for elicitation of the attack response. This finding would suggest that in the rat, the pathways mediating biting attack do not pass exclusively through the PAG and do involve other regions of the midbrain tegmentum as well.

The use of lesions to study both predatory attack and rage was applied by Berntson (1972) who examined the effects of medial, lateral, and ventral tegmental lesions on hypothalamically elicited biting attack and components of rage. In this study, it was shown that medial tegmental lesions blocked hypothalamically elicited biting attacks, while spontaneous attacks and reflexive biting appeared after lesions were placed in the lateral tegmentum. Other lesions involving the dorsal tegmental region blocked the ear-flattening component of the rage response. These findings led Berntson to conclude that different pathways passing through the brainstem mediate different sensorimotor components of the attack response.

Additional evidence of involvement of the hypothalamus and midbrain PAG has been determined in studies using electrical and chemical stimulation. The preponderance of studies described above involved the use of lesions to demonstrate the roles of the hypothalamus and PAG in rage and aggression. The studies described below, considered together with experiments described in the previous chapters, further document the importance of the hypothalamus, PAG, and adjoining regions in the expression of rage and aggression.

As previously noted, chemical stimulation of the hypothalamus such as the use of carbachol (see Golebiewski and Romaniuk, 1977) can induce rage behavior in the cat. Likewise, Eckersdorf et al. (1987) initially showed that carbachol injections into the anteromedial hypothalamus of the cat could induce rage behavior; and then when injected into the same sites, the neurotoxin, kainic acid, from which rage was elicited and which acts on glutamate receptors. These authors concluded that muscarinic receptors are either not located on the neurons responsible for the expression of rage or that such neurons are insensitive to glutamate. Since most if not all neurons contain glutamate receptors, it is somewhat difficult to understand these findings in terms of the explanations proposed by these authors. But the results do indicate the likely importance of cholinergic receptors in the excitation of hypothalamic neurons associated with defensive rage behavior in the cat.

As noted in Chapter 3, a number of studies employed chemical stimulation of the PAG as a means of demonstrating the importance of neurons in this neuropil in mediating defensive rage. In order to summarize this material, I will briefly review several of the papers that exhibited this phenomenon. In one earlier study, Golebiewski and Romaniuk (1977) placed injections of carbachol bilaterally into either the medial hypothalamus or PAG, and obtained defensive rage behavior from chemical stimulation at either site. It was further noted that the response latency was more rapid following microinjections into the PAG than hypothalamus, suggesting that microinjections of carbachol activated either more neurons or cholinergic receptors in the PAG relative to the medial hypothalamus. When carbachol stimulation in the PAG was preceded by microinjections of hexamethonium, carbachol-induced rage from that region was blocked. In addition, a similar blockade was also observed when rage was elicited from the hypothalamus, and hexamethonium was injected into the PAG. These findings are consistent with others described above or in previous chapters that demonstrated the functional linkage between the medial hypothalamus and PAG in the expression of defensive rage.

An extensive series of studies conducted by Bandler et al. (1982, 1984, 1985a, 1985b, 1988, 1991, 1992) attempted to characterize the functional anatomical organization of the PAG in both cats and rats. These studies made a number of contributions to our understanding of this region in the expression of defensive rage behavior. The first was a demonstration that excitation of PAG neurons by excitatory amino acids can induce defensive rage behavior. The second was that studies essentially provided a map of the sites and regions where defensive rage behavior is elicited. Here it was shown that the most effective sites where defensive rage is elicited are the dorsolateral aspect of the PAG, although adjoining regions can also elicit this response. That such sites extend in a caudal direction through the PAG led Bandler to propose that the PAG contains a columnar organization. The concept presented here, which was supported by experimental data, indicated that distinct segments (i.e., dorsal, dorsolateral, lateral, and ventral divisions) of the PAG mediate different functions as evidenced by their input–output relationships. As an example to illustrate this point, I indicated in Chapter 3 that the dorsolateral segment receives significant inputs from the anteromedial hypothalamus in association with defensive rage behavior (Fuchs et al., 1985a), and that this region provides a principal site of origin of descending pathways to the lower brainstem mediating this form of aggression (Shaikh et al., 1987).

The roles of the PAG and neighboring regions have also been studied with respect to predatory attack. Inselman and Flynn (1972) used the paradigm of dual stimulation of the preoptic region and more caudal aspects of the lateral hypothalamus to demonstrate that preoptic stimulation can facilitate the occurrence of predatory attack elicited from the lateral hypothalamus.

In several related studies, Proshansky and Bandler (1975) and Bandler (1979) used the findings of Chi and Flynn (1971) who had reported that descending fibers from the lateral hypothalamus in association with predatory attack behavior terminate or pass through the ventral tegmental area in extensive quantities. On the basis of this anatomical study, these investigators identified sites in the lateral hypothalamus and ventral tegmental area from which predatory attack behavior could be elicited. In one experiment, lesions of attack sites in the ventral tegmental area were shown to block predatory attack elicited from the lateral hypothalamus. This study also indicated that attack elicited from sites located in the anterior hypothalamus were more effectively blocked than those situated in the posterior hypothalamus. In a second study, Bandler (1979) demonstrated that qualitatively there exists a similarity in character of the predatory attack response elicited from the preoptic area, lateral hypothalamus, or ventral tegmental area. However, the current threshold required for elicitation of the attack response was highest in the preoptic area, somewhat lower in the lateral hypothalamus, and lowest in the ventral tegmental area. In addition to the ventral tegmental area, Bandler (1977) further showed that predatory attack behavior could also be elicited by electrical stimulation of the ventral aspect of the PAG. Moreover, the current threshold for elicitation of this response from the ventral PAG was found to be three to four times lower than that required to elicit biting attack from the lateral hypothalamus. These findings may be interpreted to mean that each of these sites represents part of a functional circuit linking the preoptic area of the lateral hypothalamus with parts of the midbrain that include the ventral tegmental area and PAG. The data also suggest that the pathways mediating predatory attack converge in the vicinity of the ventral tegmental area, thus enabling stimulation to activate more neurons associated with this response than at the level of the hypothalamus.

Evidence from Studies Conducted in Rats

As described in earlier chapters, a variety of studies have been conducted in the rat in which brain stimulation was used to demonstrate the functional role of the hypothalamus in predatory and affective forms of attack in this species. In my attempt to document the importance of the hypothalamus in the expression of aggressive reactions across species, a number of these studies are briefly reviewed at this time.

Approximately 30 years ago, Panksepp (1971) was the first to report that predatory and affective forms of attack could be elicited from electrical stimulation of the hypothalamus. Panksepp differentiated predatory from affective attack on the basis of whether a rat would attack a mouse (defined as predatory) or another rat (defined as affective). Panksepp observed that the sites yielding affective attack were situated, in general, medial to those producing predatory attack, thus demonstrating a functional anatomical relationship parallel to that described for the cat. More recently, Roberts and Nagel (1996) reported that jump and prey kicking against mice could be elicited mainly from stimulation of the ventromedial hypothalamus. On the other hand, intraspecific aggression was reported following stimulation of the lateral hypothalamus (Koolhaas, 1978). Still a third set of studies (Kruk et al., 1981; Mos et al., 1982; Kruk et al., 1984) provided maps of the sites in the hypothalamus at which electrical stimulation could elicit biting attack of one rat on another rat in which it was indicated that such attack sites were located principally in the intermediolateral aspect of the hypothalamus. Thus, there appears to be some variation concerning the loci of the sites in the hypothalamus from which intraspecific attack occurs. However, the overall data presented from these studies suggest that it is likely that sites linked to this form of aggressive behavior are concentrated mainly in the medial and intermediolateral aspect of the hypothalamus. The third set of studies also suggested that brain stimulation experiments can be used to determine (1) the neurophysiological properties of the response such as chronaxie and rheobase, (2) the modulatory effects of one region on another, and (3) that there seems to be little or no sex differences with respect to intraspecific attack.

Aggression in Other Species Induced by Brain Stimulation

As indicated previously, aggressive reactions have also been elicited from brain stimulation in other species. These findings are briefly reviewed here. In the lizard, stimulation of the reticular formation could elicit gular extension, which reflects a component of aggressive and defensive behavior (Sugarman and Demski, 1978). Since both defensive behavior and aggression could be elicited at the same site, it is suggested that at lower levels of the phylogenetic tree, the pathways for defense and aggression are not as clearly separated as it has been with respect to felines. Stimulation of the hypothalamus in the avian brain could likewise elicit defensive and threat postures (Putkonen, 1966, 1967) as well as a combination of threat and attack as indicated by pecking, wing raising, slapping, fending, and leaning (Harwood and Vowles, 1967). A number of investigators have examined the effects of electrical stimulation on aggression and defense. In citing a few of these studies, it has been shown that hypothalamic stimulation can elicit aggressive calls in both the squirrel (Jürgens et al., 1967) and rhesus monkeys (Perachio and Alexander, 1975). Threat displays and short attacks can also be elicited from stimulation of the PAG in marmoset monkeys (Lipp and Hunsperger, 1978). Thus, in spite of wide species differences in brain structures, the hypothalamus and PAG play important roles in aggression and rage at different phylogenetic levels, leading Putkonen to conclude that the general plan for the regulation of these responses is expressed early in vertebrate phylogenesis.

PHYSIOLOGICAL PROPERTIES OF STIMULATION-INDUCED AGGRESSION AND RAGE

MODULATION FROM HYPOTHALAMUS AND PERIAQUEDUCTAL GRAY

Several studies provide evidence that different parts of the hypothalamus can also modulate aggression and rage, in addition to constituting a region from which these responses can be directly elicited. An early study by Brown et al. (1969) showed that stimulation of a site eliciting flight behavior facilitated defensive rage elicited from the hypothalamus. The first study to demonstrate modulation of predatory attack from the preoptico-hypothalamus was carried out by Inselman and Flynn (1972) who showed that dual stimulation involving the preoptic region facilitated the occurrence of this form of aggression. However, it was noted that facilitation was not uniformly observed from the basal forebrain since stimulation of the diagonal band of Broca suppressed predatory attack. Regions that suppress attack extend to other parts of the preoptico-hypothalamus. In a study conducted by Adamec (1976), electrical stimulation of either the medial hypothalamus including the ventromedial nucleus, or mammillary bodies suppressed spontaneously occurring predatory attacks in the cat. As noted in Chapter 3, a series of studies by Han et al. (1996a, 1996b) provided data consistent with the findings of Adamec by demonstrating that sites in the medial hypothalamus from which defensive rage had been elicited could (at subthreshold stimulation for elicitation of defensive rage) powerfully suppress predatory attack elicited from the lateral hypothalamus. Likewise, stimulation of sites in the lateral hypothalamus from which predatory attack had been elicited could (subthreshold stimulation) block defensive rage elicited from the medial hypothalamus (Cheu and Siegel, 1998). Again, as I previously noted, these mutually inhibitory effects were found to be mediated by reciprocal inhibitory GABAergic neurons connecting the medial with the lateral hypothalamus.

A number of studies have identified the modulating properties of various aspects of the brainstem. The first study to demonstrate modulation from the brainstem was reported by Bizzi et al. (1963). In their study, rage was elicited by midbrain stimulation in thalamic cats and when stimulation was applied to the lower brainstem reticular formation, separate sites produced either facilitation or suppression of this response.

Concerning predatory attack, an initial study of the midbrain was carried out by Sheard and Flynn (1967). In this study, the authors used a dual stimulation paradigm in which stimulation was applied to a given site in the midbrain reticular formation concurrently with stimulation of the lateral hypothalamus from which predatory attack was elicited in the cat. Results indicated that stimulation of wide aspects of the reticular formation facilitated the occurrence of predatory attack. A study by Goldstein and Siegel (1980) examined the effects of ventral tegmental stimulation on both predatory attack and defensive rage in the cat and observed that the modulatory effects were frequency sensitive. Low-frequency stimulation (6 Hz) suppressed these responses while high-frequency stimulation (60 Hz) was largely ineffective. It should be noted that this finding is somewhat inconsistent with those reported by Bandler (1977) and Proshansky and Bandler (1975) who demonstrated the importance of the ventral tegmental area in the expression of predatory attack. It is difficult to account for these inconsistent findings at this time. In a further study, Shaikh et al. (1985) provided a map of the sites in the midbrain from which predatory attack and defensive rage could be modulated. It was discovered that stimulation of the medial and lateral aspects of the midbrain tegmentum differentially modulated predatory attack and defensive rage elicited from the lateral and medial hypothalamus of the cat, respectively. Facilitation of predatory attack and suppression of defensive rage followed stimulation of the lateral tegmentum, while suppression of predatory attack and facilitation of defensive rage followed stimulation of its medial aspect. The effects of tegmental stimulation on predatory attack are consistent with those described by Sheard and Flynn (1967), and further point to the likelihood that the sites of interaction for midbrain tegmental modulation occurs in the hypothalamus, where fibers from this region are known to project.

Other Effects of Electrical Stimulation of Attack Sites in Cats

Sledjeski and Flynn (1972) attempted to identify key characteristics of the behavioral process that is initiated by stimulation and which results in predatory attack behavior. Because they had noted that the autonomic components associated with stimulation of the attack site outlast the behavioral component, they specifically focused on the effect of prior stimulation on excitability at attack sites. To address this question, they used 5 seconds of priming stimulation that were followed by a train of stimulation present until the attack occurred at 2, 4, 8, 16, and 32 seconds after the priming stimulation. These investigators observed that stimulation initiated at all but the longest intervals after the priming stimulation led to a reduction in response latency. It was of interest to note that priming at other sites (from which attack could also be elicited in two of five cats after longer stimulation) were not effective in shortening response latencies. Because of the localized nature of the excitatory effects by stimulation, it was concluded that electrical stimulation of attack sites in the hypothalamus excites specific parts of the central nervous system. These authors pointed out that this conclusion contrasts with a prior view — namely, that excitatory hypothalamic stimulation results in an "affective" state such as anger, mood change, or drive, and that such a change in the psychological state of the animal determines the responsiveness of the central nervous system and organism.

In a related study, Bandler and Fatouris (1978) extended this approach to determine whether the principal of "priming" could be applied to the study of attack situated in different regions of the brain as a means of establishing the functional neural relationship between two given regions. These authors observed that when priming was generated from the ventral tegmental area, response latencies for predatory attack following stimulation of the lateral hypothalamus were shortened. However, when the procedures were reversed (i.e., priming stimulation was administered to the lateral hypothalamus and attack elicited from the ventral tegmental area), little change in response latencies was observed. These authors suggested that their findings could be understood by the likelihood that the descending components of the medial forebrain bundle (which mediate predatory attack) are more tightly arranged at the level of the ventral tegmental area than within the lateral hypothalamus. In this manner, stimulation of the ventral tegmental area would excite greater quantities of medial forebrain fibers than stimulation of the lateral hypothalamus. This study thus provides further evidence of the functional linkage between the lateral hypothalamus and ventral tegmental area with respect to predatory attack behavior in the cat.

Neurophysiological Relationships

Several studies have been conducted using single-cell recordings as a means of identifying neurons that are functionally related to aggression or rage. I have previously described the findings of Adams (1968) who showed that single cells in the PAG of the cat respond selectively to the presence of a second cat at the time at which it is expressing defensive rage behavior, and the cat in question is responding in kind. The results of two other experiments are described in this chapter. One experiment carried out by Vanegas et al. (1970) is described later in this chapter in the section concerning sensory processes. The second study, conducted by Huang and Flynn (1975), concerned analysis of single-cell responses identified in the midbrain in response to elicitation of predatory attack by stimulation of the lateral hypothalamus in the cat. The major findings reported were that stimulation at sites that elicited predatory attack affected more units, yielded a greater change in unit activity, and significantly increased the firing rates in the dorsal tegmentum just lateral to the PAG as well as the PAG itself. Accordingly, these findings provide additional evidence linking the PAG and adjoining regions of the tegmentum in the functional regulation of aggression and rage.

Other Regions Associated with Rage and Aggression

There have been scattered reports in the literature that have indicated that rage and/or aggression may also be associated with regions other than the hypothalamus, PAG, and ventral tegmental area. With respect to defensive rage, several forebrain regions have been implicated. In an early study, Emmers et al. (1965) was able to induce cats to elicit hissing and growling when provoked after forebrain ablation that spared the hypothalamus and amygdala. This finding would suggest that the thalamus is not essential for the expression of defensive rage behavior. Fernandez de Molina and Hunsperger (1962) reported that defensive rage in the cat could be elicited by electrical stimulation of the medial amygdala, stria terminalis, medial hypothalamus, or PAG. An extensive body of literature documenting the roles of the hypothalamus and PAG in the expression of defensive rage behavior was described earlier. What is of interest in this study is the observation that amygdaloid or stria terminalis stimulation can induce defensive rage. One likely interpretation of these findings is that medial amygdaloid stimulation directly excites the stria terminalis and that stimulation of the stria terminalis drives the neurons in the medial hypothalamus (by virtue of its direct projections to this region) much the same as if stimulation were directly applied to the medial hypothalamus.

Concerning predatory attack, MacDonnell and Flynn (1967) provided evidence that electrical stimulation of sites within the medial thalamus, mainly including the midline of the mediodorsal thalamic nucleus, could produce predatory attack. However, in contrast to attack elicited from the lateral hypothalamus (see discussion below), loss of vision results in a loss of locomotion necessary for the occurrence of the attack response. It is reasonable to assume that midline thalamic induced–attack is mediated through projections, either direct or indirect, from this region to the lateral hypothalamus. In addition to the medial thalamus, as noted in Chapter 3, stimulation of the central tegmentum in the caudal half of the midbrain adjacent to the decussation of the superior cerebellar peduncle was also reported to elicit predatory attack (Berntson, 1973). It is difficult to determine from this kind of study whether the brainstem sites from which the attack was elicited constituted cell bodies or fibers of passage arising from more rostral levels of the brainstem or hypothalamus and which pass caudally in the reticular formation.

Role of Sensory Processes in Aggressive Behavior

One of the most well-studied and insightful lines of research conducted with respect to the neural basis of aggressive behavior has been the analysis of sensory mechanisms in regulating this behavior. The senses studied include olfactory, tactile, and visual processes.

Olfaction

A number of investigators have attempted to identify the role of olfaction in the expression of aggression and rage in several species. However, the results of these studies, which have involved olfactory bulb ablations, have been less than uniform. Various studies conducted in rodents have provided evidence that such ablations enhance both muricidal behavior in rats (Thorne et al., 1974; Vergnes et al., 1974; Oishi and Ueki, 1978), as well as intraspecific aggression in rats (Malick, 1970). However, several studies have provided opposite findings. Adams (1976) reported that the combination of smell from a strange male (rat) in the presence of a familiar environment could trigger offensive mechanisms in the resident rat. More recently, Edwards et al. (1993) obtained results consistent with those of Adams when they reported that olfactory bulb lesions can significantly decrease intermale aggression in mice. That the mechanisms governing olfactory mechanisms are complex was suggested from an earlier study by Bandler and Chi (1972). In their study, olfactory bulb removal suppressed natural frog-killing behavior in non–mouse-killing rats, but such lesions did induce killing in some rats that did not naturally kill mice. These authors noted that those rats whose aggression was facilitated sustained only partial olfactory bulb lesions, whereas those rats whose aggression was inhibited sustained more complete olfactory bulb lesions. These findings led Bandler and Chi to suggest that olfactory cues may act in quite different ways in natural mouse-killing and non-mouse-killing rats.

Although the studies described above are certainly not entirely consistent, they do point to the complexity of the olfactory system in significantly modulating aggression in the rodent. However, in the cat, the olfactory system appears to play little or no role in modulating predatory attack in the cat. The evidence for this conclusion is based on a study by MacDonnell and Flynn (1966) who demonstrated that olfactory bulb removal in the cat does little to alter its ability to locate and attack a target rat during stimulation of an attack site in the lateral hypothalamus.

TACTILE STIMULATION

Perhaps one of the most significant series of studies addressing the role of sensory processes in the regulation of predatory attack was undertaken by Flynn et al. In one experiment, MacDonnell and Flynn (1966) sought to identify the respective importance of different sensory processes. In this study, the authors examined how olfaction, and tactile and visual stimulation contributed to the expression of the attack response. As noted above, following the placement of lesions in the olfactory bulbs, the authors noted that olfactory bulb removal had no significant effect on the capacity of the cat to locate and attack an anesthetized rat when stimulation was applied to an attack site in the lateral hypothalamus. After loss of vision, there was clear-cut impairment of the cat to locate the target when it was placed in a large enclosure. However, if a smaller experimental chamber was employed, impairment was only minor because the probability that the cat would come into physical contact with the rat during arousal by stimulation was high. When tactile sensory nerves associated with the trigeminal nerve, which innervate the snout, were experimentally damaged, impairment of the attack response was paramount. These authors concluded that the trigeminal innervation to the area around the lipline plays a critical role in the afferent limb of a reflex leading to the jaw-closing response, which comprises an essential element of the attack response. It was further concluded that collectively, trigeminal and visual inputs are critical for the expression of predatory attack because, if both sensory inputs are eliminated, then the attack is essentially eliminated.

Findings from this study provided the basis for the following studies involving the analysis of sensory processes in predatory attack in the cat. In particular, it pointed to the fact that tactile inputs through the trigeminal nerve are inextricably linked to this form of aggression. In a seminal paper, MacDonnell and Flynn (1966) attempted to elucidate this relationship. The experiment involved the following procedures. First, a cat was placed in a restraining device in which its head was kept stationary and stimulation was applied to a site in the lateral hypothalamus from which predatory attack was elicited by lateral hypothalamic stimulation. At the time at which brain stimulation was applied, tactile stimulation concurrently applied to different sites along the midline of the lip. It was observed that touching the perioral region results in movement of the head, ultimately bringing the tactile stimulus in proximity to the cat's mouth, at which time, jaw opening and closing (i.e., attempts at biting of the probe) occurs. These investigators discovered that the lateral extent along the lipline where probing could elicit jaw opening and attempts at biting was a function of the current applied to the lateral hypothalamic attack site (i.e., as the current was increased, there was also a corresponding increase in the lateral extent

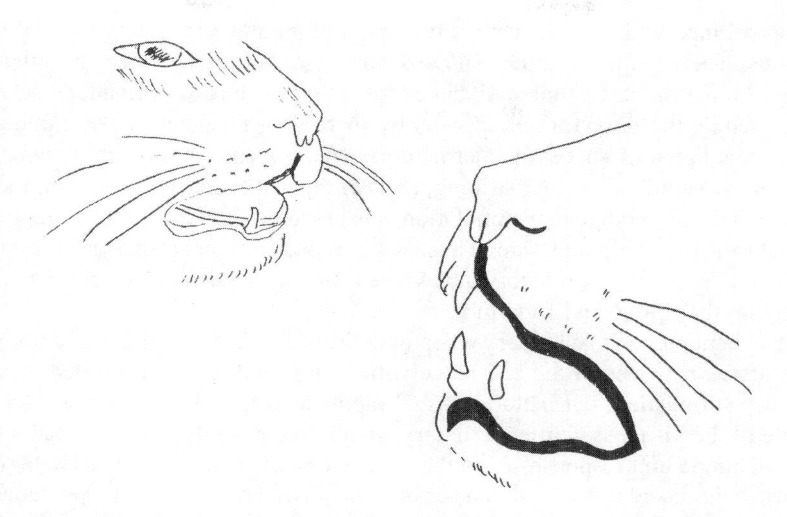

FIGURE 4.1 Extent of the "effective sensory fields" that were probed to elicit jaw opening and closing on a given trial. This field was generally probed from the lateral extent toward the midline. (From C. Block, Ph.D. thesis, College of Medicine and Dentistry of New Jersey, Newark, N.J., 1974, in M. MacDonnell and J.P. Flynn, *Anim. Behav.*, 14, 1966, 399–405. With permission.)

along the lipline from which this reflex could be elicited) (Figure 4.1). They further noted that upon using the distance from the midline of the lip where tactile stimulation elicited jaw opening as a measure, tactile stimulation of the side of the lipline contralateral to that of hypothalamic stimulation was more effective than ipsilateral stimulation. These findings were interpreted to mean that hypothalamic stimulation at an attack site alters the trigeminal sensory fields along the lipline. The change in the sensory fields in association with hypothalamic stimulation thus facilitates triggering of the reflex biting component of predatory attack by tactile stimulation. Therefore, when predatory attack occurs, either under natural conditions or by electrical stimulation of the lateral hypothalamus, it is the tactile stimulation experienced by the cat when its lipline touches the body of the prey object that activates the biting component and, when these trigeminal inputs are disrupted, as described by MacDonnell and Flynn (1966), the biting component is also disrupted. It should be noted that parallel findings were reported in a study in which electrical stimulation of the lateral hypothalamus of the rat was coupled with the application of tactile stimulation to the perioral region (Smith, 1972). Here, it was observed that only when electrical stimulation of various sites in the lateral hypothalamus elicited a specific behavior such as attack, eating, gnawing, or drinking tactile stimulation did tactile stimulation of the perioral region elicit turning toward the stimulus and biting. Thus, the role of sensory receptive field activation in elicitation of biting attack extends across species.

In a related study, Bandler and Flynn (1972) extended the findings of MacDonnell and Flynn by demonstrating that light tactile stimulation of the forelimb involving the fifth and sixth cervical dermatomes of the cat elicited striking by that limb when combined with low-intensity stimulation of an attack site in the lateral hypothalamus. These authors further showed other similarities between the effects of tactile stimulation of the limb and that of the lipline with respect to attack behavior. These include the observations that (1) by increasing the intensity of stimulation, tactile stimulation of successive dermatomes (i.e., seventh and eighth cervical and first thoracic) could also elicit striking; and (2) the tactile receptive field on the side contralateral to hypothalamic stimulation was greater than the corresponding receptive field on the ipsilateral side. Thus, with respect to somatosensory stimulation and its role in initiating predatory attack, the same mechanism appears to be present both along the lipline and forelimb.

The phenomenon that sensory receptor activation can trigger the final component of the attack response in the cat, namely, the biting reflex, is not limited to somatosensory stimulation, but includes visual inputs as well. Several studies addressed the role of the visual system in predatory attack. In one study, Vanegas et al. (1970) recorded single unit responses from the visual cortex to oriented slits of light, edges of a bar of light, and movement of a bar of light, as well as single cells spontaneously elicited or activated by diffuse light. These investigators reported that stimulation applied to lateral hypothalamic predatory attack sites caused modification of these unit responses. The implication from this study is that activation of neurons in the lateral hypothalamus in association with predatory attack can modify the firing patterns of neurons in the visual cortex, which likely play an important role in the cat's attempt to locate the target.

A second study that addressed the question of the role of vision in predation was conducted by Bandler and Flynn (1971). In their study, predatory attack was initially elicited by electrical stimulation of the lateral hypothalamus of the cat. Then, an anesthetized mouse was placed in the visual field of either the eye ipsilateral or contralateral to that of stimulation of the lateral hypothalamus in which the appearance of the mouse was available to one of the two eyes at a given time. The results revealed that a cat lunged toward a mouse more frequently following stimulation of a lateral hypothalamic attack site contralateral to the eye viewing the mouse than toward a mouse presented to the ipsilateral eye. The authors concluded that this finding is consistent with previous studies described above. They further suggested that biting involves both a head turning component and another of mouth opening, which are manifest first along the lipline and limb on the side contralateral to that of stimulation. As indicated above, Flynn et al. (1970) concluded from these studies that motivated (attack) behavior induced by brain stimulation engages the sensory and motor systems to act in highly specific ways as "patterned" reflexes whose ultimate expression is the attack response.

I would like to point out that parallel tactile sensory modulation of other behavioral processes has also been documented. Wolgin and Teitelbaum (1978) placed lesions in the lateral hypothalamus of the cat and noted the presence of aphagia and adipsia, deficiencies in arousal, and *sensory neglect*. During recovery, these functions returned, including responsiveness to sensory stimuli. These observations led these investigators to conclude that activation of sensory processes is essential for the occurrence of feeding, much in the same way that they are important for the occurrence of aggression.

With respect to sensory processes influencing aggression and rage, one final study should be briefly noted. Bizzi et al. (1961) prepared cats with ablations of the thalamus and subjected them to stimulation of the carotid body, thus activating chemoreceptors located in this region. The authors indicated that chemoreceptor activation evoked a pattern of rage behavior in these cats similar to that induced by tactile or noxious stimulation or that which occurred spontaneously. This early finding would thus suggest that chemoreceptor activation could facilitate the hypothalamic process regulating rage. Unfortunately, I am not aware of any follow-up studies to determine the possible mechanism for this effect.

Motor Aspects of Predatory Attack

As indicated above, Flynn presented the suggestion that predatory attack elicited either by electrical stimulation of the lateral hypothalamus or occurring spontaneously involves the activation of a patterning mechanism. I have indicated how this patterning mechanism may apply to sensory processing. However, the patterning mechanism required for the expression of predatory attack also requires the activation of motor processes that culminate in biting and/or striking at a prey object. Several studies summarized below illustrate how this mechanism may function.

The predatory attack response requires a biting component, and therefore necessitates activation of the motor division of the trigeminal nerve (for biting) and probably components of the motor nucleus of the facial nerve (for jaw opening). This was addressed in a study by MacDonnell and Flynn (cited by Flynn, 1967), in which predatory attack was elicited by electrical stimulation of the lateral hypothalamus of the cat. If stimulation of the attack site was coupled with stimulation of the motor nucleus of the trigeminal nerve, then there was a highly significant increase in the closing movement of the jaw. The significance of this finding is that it demonstrated the functional relationship between lateral hypothalamically elicited attack and trigeminal motor activity; namely, that this region of hypothalamus can excite the motor limb of the jaw-closing reflex that is essential for the biting component of the attack response. Anatomical observations have revealed that hypothalamic sites from which predatory attack can be elicited and the trigeminal motor nucleus are linked by a direct, monosynaptic projection (Fuchs et al., 1981).

While biting is a critical and final component of the predatory attack response, it is not the only motor component of this response. Quite frequently, the cat will strike a freely moving prey object when it is in the process of eliciting attack, either by electrical brain stimulation or spontaneously. Edwards and Flynn (1972) sought to identify the possible neural mechanisms by which stimulation-induced predatory attack can affect forelimb movements necessary for the occurrence of striking. In this experiment, striking was elicited by stimulation of an attack site in the PAG. They then made pyramidotomies, which abolished initial and terminal flexor movements essential for striking but which left extensor movements intact. It was shown that a single conditioning stimulus to the PAG facilitated pyramidal tract neurons in the ipsilateral motor cortex. These authors further demonstrated that a single conditioning stimulus to the PAG facilitated and then inhibited evoked potentials recorded from the biceps brachii muscle. Stimulation applied to the PAG at a rate that elicited striking facilitated cortically elicited biceps responses and shortened the latency for these responses. The results of this study suggest that PAG stimulation produces forelimb striking by modulating corticospinal pathway function at the level of both the cerebral cortex and spinal cord.

Results of recent data collected in our laboratory should also be mentioned here (Weiner et al., 2000; unpublished observations). The predatory attack response also appears to characterize the jaw muscle recruitment pattern. This was identified in a study of EMG responses in which the amplitude and mean power frequency of the masseter muscle, which controls jaw closure, was examined during predatory attack elicited from electrical stimulation of the lateral hypothalamus versus normal mastication in the absence of stimulation. During predatory attack, increases in the EMG amplitude and shifts to higher frequencies of muscle fiber contraction relative to normal mastication were observed (Figure 4.2). These findings suggest that during predatory attack the muscle contraction patterns may reflect the increased force necessary during predation relative to quiet masticatory activity. That hypothalamic stimulation associated with aggression (or other emotional processes) can influence motor activity such as jaw movements clearly has implications for a role of this region of the brain in certain motor disorders that include bruxism.

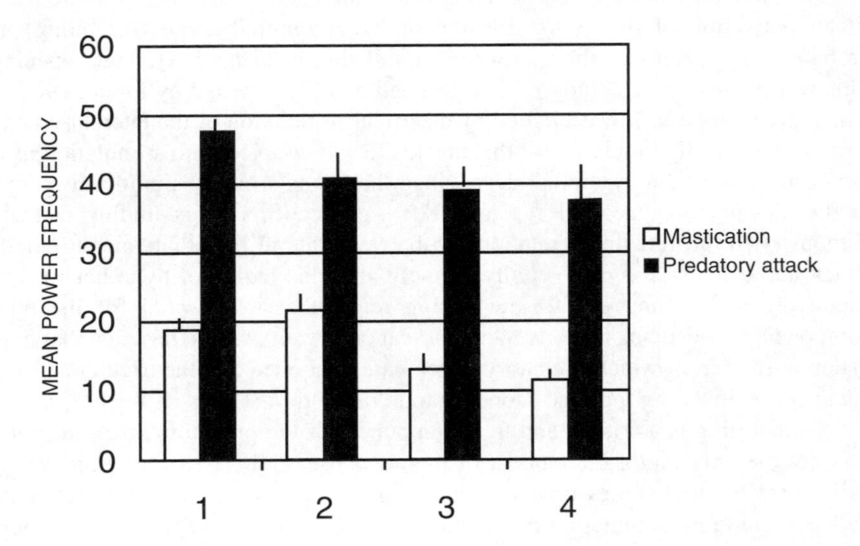

FIGURE 4.2 A series of trials in four cats in which the mean power frequency was calculated from EMG recordings of the masseter muscle during mastication and predatory attack elicited by electrical stimulation of hypothalamic sites. Paired t-tests showed significant differences at $p \leq 0.001$ between mastication during predatory attack and in the absence of predatory attack. (Courtesy of Dr. Saul Weiner.)

Behavioral Properties of Attack Behavior

A variety of studies have been conducted over the past 3–4 decades that have attempted to identify and characterize principal behavioral properties associated with aggression and rage most frequently elicited by electrical stimulation of the hypothalamus or PAG. These properties include: the capacity of the animal to identify a prey object and discriminate it from a non-prey object that has a similar appearance, the positive or negative reinforcing properties of different forms of attack, the possible relationship to feeding behavior, and the relationship of predation to avoidance and play.

Discriminative Properties of Aggression

A basic question that naturally arises with respect to predatory attack behavior is whether the attack is indiscriminate or directed to a specific prey object. This concern was addressed directly by Levison and Flynn (1965). In their study, a number of prey objects (or potential prey objects) were placed in a cage with a cat from which predatory attack behavior could be initiated by electrical stimulation of the lateral hypothalamus. The objects presented to the cat at any given time included an anesthetized rat, stuffed rat, and another object such as a hairy toy dog, Styrofoam block, or foam rubber block. The frequency of attack on these objects following brain stimulation was then determined. These investigators demonstrated that the highest frequency of attack was on the anesthetized rat with progressively fewer responses on the stuffed rat and other objects (i.e., hairy toy dog, Styrofoam block, and foam rubber block). Surprisingly, a relatively higher response frequency of attack was initiated in the presence of the hairy toy dog relative to the other inanimate objects. These findings indicate the specificity of predatory attack, which is directed primarily on a natural prey object (i.e., anesthetized rat) as opposed to a nonliving object. As these authors concluded, brain stimulation of an attack site is capable of producing behavior that limits the sensory objects triggering the completion of the attack response. Given the previous discussion, there are likely to be at least two sensory components active here. The first is the visual system, which allows the cat to discern the differences between a live and stuffed rat. The second is somatosensory stimulation associated with the trigeminal nerve. As indicated above, tactile stimulation along the lipline is a sufficient stimulus to trigger the biting response, thus accounting for the relative high frequencies of attack on the anesthetized rat and hairy toy dog.

Reinforcing Properties of Rage and Aggression

Various studies conducted in several species have provided evidence that predatory attack has positively reinforcing properties while defensive rage contains negatively reinforcing properties. One of the first studies to address this question was carried out by Roberts and Carey (1965). In their study, it was shown that when hypothalamic stimulation elicited gnawing behavior in rats, these objects could also act as rewards to induce learning of a position and black–white discrimination habits in a Y-maze. In rat studies, several other investigators described findings that were consistent with those of Roberts and Carey. DeSisto and Huston (1971) reported that stimulation of the same electrodes located in the posterior lateral hypothalamus from which frog killing had been elicited in rats also supported self-stimulation. Likewise, Van Hemel (1972) performed an experiment in which a rat was offered a choice to bar press for the presentation of a mouse or bar press where no mouse was presented. The results indicated that the rats consistently selected a bar-pressing response that produced a mouse, indicating that mouse killing had positive reinforcing properties.

In a study by Panksepp (1971), stimulation was applied to medial and lateral hypothalamic sites in rats, and affective- and biting attack-like responses were elicited. Panksepp differentiated the two forms of aggression on the basis that in biting attacks, the response was directed against live or dead mice but not against rats, while in defensive aggression, the rat would kill a mouse and bite another rat, but not a dead mouse. Panksepp then demonstrated that stimulation eliciting a biting attack also produced self-stimulation, while stimulation eliciting defensive aggression was also aversive by producing escape behavior.

Several studies were also conducted in cats. In one experiment, Nakao (1958) reported that hypothalamically elicited defensive rage could be used to produce conditioning of a (neutral) buzzer, which, when presented, induced crouching, ear retraction, and jumping. These responses were interpreted to suggest the presence of aversive properties of defensive rage. Further evidence in support of this notion was generated by Adams and Flynn (1966). In their study, predatory attack or defensive rage was elicited by brain stimulation. Then the animals were trained to escape from tail shock by jumping onto a stool. While sites from which predatory attack had been elicited could not induce such an escape response, stimulation of defensive rage sites did elicit the learned escape response, again indicating the presence of aversive properties in association with rage.

Using squirrel monkeys, Renfrew (1969) employed brain stimulation of the perifornical hypothalamus, which induced biting on a pneumatic tube. It further induced escape and avoidance but not self-stimulation, again indicating the aversive qualities of brain stimulation of this region of hypothalamus in the primate. In terms of the relationship of this study to the ones described previously, one problem exists — namely, it is difficult to determine whether the model employed (i.e., biting on a pneumatic tube) is more characteristic of predatory attack or defensive behavior.

With respect to predatory attack, Roberts and Kiess (1964) produced predatory attack by electrical stimulation of the lateral hypothalamus. Prior to brain stimulation, the cat was given cat food, which it ate. However, when brain stimulation was initiated, the cat stopped eating and attacked a rat. These authors interpreted their finding in terms of the positively reinforcing properties of brain stimulation in association with predatory attack. In addition, the results of this study also have implications for conclusions suggested from one study (considered below) regarding the possible linkage between predatory attack and feeding. Overall, the studies described in this section provide evidence in support of the view that predatory attack has positive reinforcing properties, while defensive rage and related responses are aversive in nature.

Aggression or Feeding?

Since predatory attack is directed at the killing of a prey object that is later used as a source of food, a question that naturally arises with respect to the study of predatory attack behavior is whether predatory attack and feeding are "part and parcel" of the same process. Several investigators examined this question. In one study, Hutchinson and Renfrew (1966) implanted electrodes into the lateral hypothalamus of the cat from which predatory attack could be elicited. These investigators reported that feeding behavior was induced when stimulation was applied to such sites at current levels below the threshold for elicitation of predatory attack. This observation led Hutchinson and Renfrew to conclude that feeding and aggression reflect the same process in which one group of neurons elicits attack when stimulated at a higher current, and feeding when stimulated at a lower current.

Support for the view that one group of neurons or neural pathways may mediate various functions was also suggested from a study by Cox et al. (1968). These authors reported that lateral hypothalamic stimulation in rats could elicit drinking, eating, or gnawing by changing the stimulus conditions of the environment. This observation suggests the presence of considerable plasticity in brain function and thus directly challenges the traditional view, and the one espoused in this book, that specific behavioral functions are mediated through fixed anatomical circuits.

However, other studies have not supported the findings of Hutchinson and Renfrew. Flynn (1970) observed that the overwhelming numbers of sites in the hypothalamus of the cat from which predatory attack had been elicited did not induce feeding, even when the cats had been subjected to fasting for 24 hours. In fact, when no stimulation was applied, the cats would eat, but when stimulation was applied to the lateral hypothalamus, they would stop eating, stalk, and then bite the rat until the stimulation was terminated. Since only approximately 10% of the sites showed overlap of aggression and feeding, Flynn concluded that there was no evidence to support the view of a neural linkage between feeding and aggression.

Support for the latter view also received support from several other investigators. As previously noted, Roberts and Kiess (1964) described the results of a similar experiment to that of Flynn. Here, when electrical stimulation was applied to the lateral hypothalamus from which predatory attack could be elicited at a time when the cat was feeding, the cat would stop eating and attack the rat. These authors concluded from their observations that attack behavior was not due to hunger and is, indeed, distinct from it. Findings parallel to those of Flynn were reported by De Sisto and Zweig (1974) who used rats instead of cat. In this study, bipolar stimulating electrodes were implanted in the posterior lateral hypothalamus from which either stimulus bound feeding or stimulus bound killing of a frog. Again, most sites produced one but not both types of responses and only several sites displayed mixed responses. Moreover, when the animals were subjected to self-stimulation tests using bar pressing as the operant. It was noted that stimulus-bound killers always held the bar down longer when a frog was present than if food or no goal object was present. In contrast, stimulus-bound feeders held down the bar longer in the presence of food than if a frog or no object was available. *Thus, when considered collectively, there is little evidence to support the view that eating and predatory attack are one and the same process. In fact, the existing data support the opposite view, further suggesting a distinctiveness of the neural circuitry mediating each of these behavioral processes.*

AVOIDANCE, PLAY, AND PREDATORY ATTACK

In the course of predatory attack, cats may engage in related behavior sometimes referred to as "predatory play." It typically involves repeated contact with the prey object without injuring it. Although the response pattern resembles that of attack, it can be viewed as constituting an independent behavioral pattern. Alternatively, it may also be considered as falling along a continuum with predatory attack, where defense represents the end of the continuum in opposition to predatory attack. In a study by Pellis et al. (1988), three lines of evidence were presented in support of the notion of a continuum (Figure 4.3). The first is that benzodiazepines administered to cats shifted the direction of behavior along the continuum by increased killing of mice which previously played with them and by increasing play in cats that previously avoided mice. Second, lateral hypothalamic lesions initially disrupted attack, but during recovery, an escalation towards killing was observed in the absence of play and avoidance. When tail pinches were applied to cats, a similar pattern of results occurred in both killer and nonkiller cats. These investigators thus argued that predatory play in reality represents predatory behavior that has not yet progressed to biting attack. They further suggested that this response may have survival value by reducing the likelihood of injury to the animal, as it may be construed as constituting a more cautious approach toward generating the ultimate attack on the prey object.

SUMMARY

In this chapter, I attempted to provide a review of the basic physiological and behavioral properties associated with aggression and rage. Although this analysis is not exhaustive, it does examine what I believe are the most important ones for the expression of aggression and rage.

The first point made considered the importance of the hypothalamus, PAG, and related regions of the brainstem in mediating these forms of aggression as well as the key interrelationships established both within and among these structures. I have tried to reinforce (based on previous chapters) the facts that the medial and lateral hypothalamus mediate defensive rage and predatory attack, and that the dorsolateral PAG constitutes a key brainstem component of the defensive rage response. It was shown that while the hypothalamus is normally quite important for the expression of aggression and rage, these responses might nevertheless take place after the hypothalamus has been disconnected from the rest of the brain. In contrast, destruction of the PAG generally eliminates defensive rage behavior and tegmental lesions significantly suppress or block predatory attack in cats. Studies conducted in rodents and other species have also pointed to the importance of the hypothalamus in the expression of aggressive reactions characteristic of such species. However, with

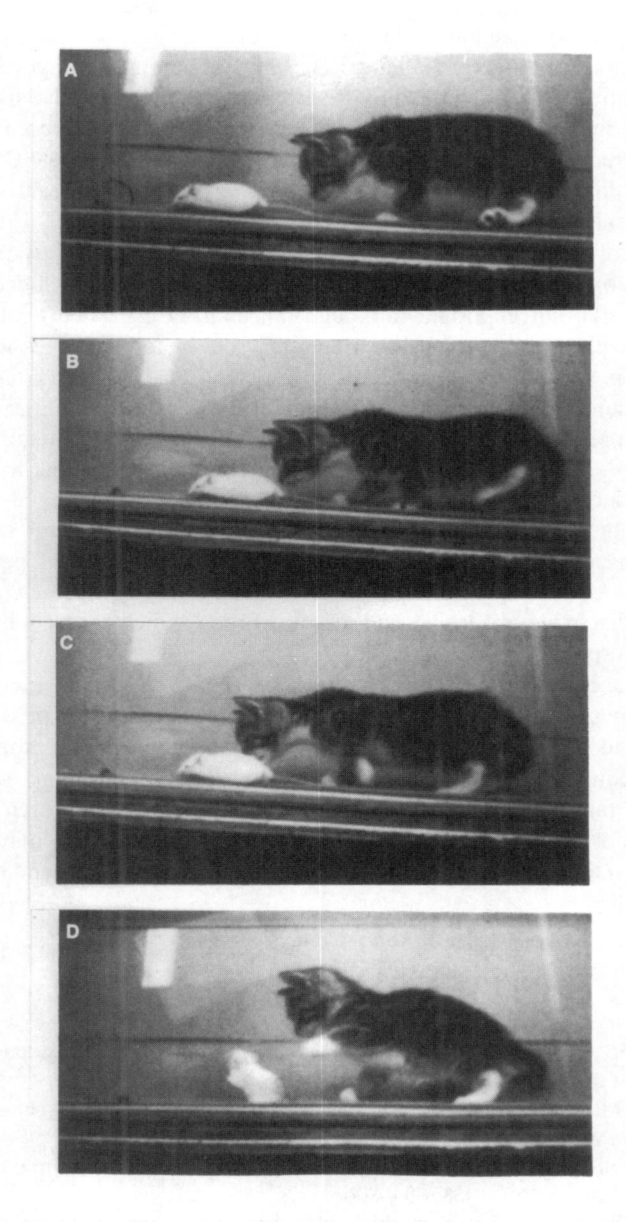

FIGURE 4.3 Illustration of the approach of a cat (A through C) toward an unanesthetized rat culminating in play attack (D), and ultimately biting attack (not shown in this figure). (Courtesy of Dr. David Wolgin.)

respect to the rat, the PAG may be important but perhaps not critical for the expression of stimulation-induced aggression against an opponent of the same species. Other studies have demonstrated the importance of different regions of the hypothalamus and brainstem tegmentum in modulating aggression and rage elicited from other regions of the hypothalamus. These findings indicate important anatomical and functional relationships among various regions of hypothalamus such as the excitatory interactions between the anterior and posterior hypothalamus and inhibitory interactions between the medial and lateral hypothalamus. Other studies identified the physiological properties that are characteristic of brain stimulation-induced aggression and rage.

A variety of studies have examined the role of different sensory processes in the expression of aggression and rage in several species. It was concluded that while olfaction may play an important (although not entirely consistent) role in rodent aggression, it appears to serve a less important role in feline aggression. Instead, the evidence is quite clear that vision and somatosensory inputs involving the trigeminal system around the lipline constitute the necessary stimuli critical for the expression of predatory aggression. Likewise, the motor components of aggression have also been examined. These studies have provided evidence that hypothalamic stimulation associated with predatory attack can significantly influence motor components of both the trigeminal nerve as well as spinal cord motor neurons that mediate biting and striking of the prey object, respectively. It was proposed that collectively the effects of brain stimulation–induced attack produce a change in the patterning of discharge of sensory and motor neurons, which ensure the occurrence of the attack response.

The final section of this chapter considered several key behavioral properties associated with aggression and rage: (1) discriminative properties of predatory attack that allow cat to attack the appropriate prey object; (2) reinforcing properties from which it was concluded that predatory attack is positively reinforcing, while defensive rage is negatively reinforcing; and (3) that predatory play represents a component that has not yet evolved into a full attack response and which may be viewed as lying along a continuum extending from predatory attack on one end to defensive rage on the other.

REFERENCES

Adamec, R.E., Hypothalamic and extrahypothalamic substrates of predatory attack. Suppression and the influence of hunger, *Brain Res.*, 106, 1976, 57–69.

Adams, D. and J.P. Flynn, Transfer of an escape response from tail shock to brain-stimulated attack behavior, *J. Exp. Anal. Behav.*, 9, 1966, 401–408.

Adams, D.B., Cells related to fighting behavior recorded from midbrain central gray neuropil of cat, *Science*, 159, 1968, 894–896.

Adams, D.B., The relation of scent marking, olfactory investigation, and specific postures in the isolation-induced fighting of rats, *Behaviour*, 56, 1976, 286–297.

Bandler, R., Predatory attack behavior in the cat elicited by preoptic region stimulation: a comparison with behavior elicited by hypothalamic and midbrain stimulation, *Aggress. Behav.*, 5, 1979, 269–282.

Bandler, R., Induction of 'rage' following microinjections of glutamate into midbrain but not hypothalamus of cats, *Neurosci. Lett.*, 30, 1982, 183–188.

Bandler, R., Identification of hypothalamic and midbrain periaqueductal grey neurones mediating aggressive and defensive behaviour by intracerebral microinjections of excitatory amino acids, in R. Bandler, Ed., *Modulation of Sensorimotor Activity During Alterations in Behavioral States*, Alan R. Liss, Inc., New York, 1984, pp. 369–391.

Bandler, R. and C.C. Chi, Effects of olfactory bulb removal on aggression: a reevaluation, *Physiol. Behav.*, 8, 1972, 207–211.

Bandler, R. and A. Depaulis, Elicitation of intraspecific defence reactions in the rat from midbrain periaqueductal grey by microinjection of kainic acid, without neurotoxic, *Neurosci. Lett.*, 88, 1988, 291–296.

Bandler, R.A., Depaulis, M. Vergnes, Identification of midbrain neurones mediating defensive behaviour in the rat by microinjections of excitatory amino acids, *Behav. Brain Res.*, 15, 1985, 107–119.

Bandler, R. and D. Fatouris, Centrally elicited attack behavior in cats: post-stimulus excitability and midbrain-hypothalamic inter-relationships, *Brain Res.*, 153, 1978, 427–433.

Bandler, R. and J.P. Flynn, Control of somatosensory fields for striking during hypothalamically elicited attack, *Brain Res.*, 38, 1972, 197–201.

Bandler, R., S. Prineas, T. McCulloch, Further localization of midbrain neurones mediating the defence reaction in the cat by microinjections of excitatory amino acids, *Neurosci. Lett.*, 56, 1985, 311–316.

Bandler, R. and M.T. Shipley, Columnar organization in the midbrain periaqueductal gray: modules for emotional expression, *Trends Neurosci.*, 17, 1994, 379–389.

Bandler, R.J., Predatory behavior in the cat elicited by lower brain stem and hypothalamic stimulation: a comparison, *Brain Behav. Evol.*, 14, 1977, 440–460.

Bandler, R.J. and A. Depaulis, Midbrain periaqueductal gray control of defensive behavior in the cat and the rat, in A. Depaulis and R.J. Bandler, Eds., *The Midbrain Periaqueductal Gray Matter*, Plenum Press, New York, 1991, pp. 175–198.

Berntson, G.G. Blockade and release of hypothalamically and naturally elicited aggressive behaviors in cats following midbrain lesions, *J. Comp. Physiol. Psychol.*, 81, 1972, 541–554.

Berntson, G.G. Attack, grooming, and threat elicited by stimulation of the pontine tegmentum in cats, *Physiol. Behav.*, 11, 1973, 81–87.

Bizzi, E., A. Libretti, A. Malliani, A. Zanchetti, Reflex chemoceptive excitation of diencephalic sham rage behavior, *Amer. J. Physiol.*, 200, 1961, 923–926.

Bizzi, E., A. Malliani, J. Apelbaum, A. Zanchetti, Excitation and inhibition of sham rage behavior by lower brain stem stimulation, *Arch. Ital. Biol.*, 101, 1963, 614–631.

Boehrer, A., M. Vergnes, P. Karli, Interspecific aggressiveness and reactivity in mouse-killing and nonkilling rats: compared effects of olfactory bulb removal and raphe lesions, *Aggress. Behav.*, 1, 1974, 1–16.

Brown, J.L., R.W. Hunsperger, H.E. Rosvold, Interaction of defence and flight reactions produced by simultaneous stimulation at two points in the hypothalamus of the cat, *Exp. Brain Res.*, 8, 1969, 130–149.

Canteras, N.S., S. Chiavegatto, V. Do, L.W. Swanson, Severe reduction of rat defensive behavior to a predator by discrete hypothalamic chemical lesions, *Brain Res. Bull.*, 44, 1997, 297–305.

Carli, G., A. Malliani, A. Zanchetti, Midbrain course of descending pathways mediating sham rage behavior, *Exp. Neurol.*, 7, 1963, 210–223.

Cheu, J.W. and A. Siegel, GABA receptor mediated suppression of defensive rage behavior elicited from the medial hypothalamus of the cat: role of the lateral hypothalamus, *Brain Res.*, 783, 1998, 293–304.

Chi, C.C. and J.P. Flynn, Neuroanatomic projections related to biting attack elicited from hypothalamus in cats, *Brain Res.*, 35, 1971, 49–66.

Cox, V.C., E.S. Valenstein, J.W. Kakolewski, Modification of motivated behavior elicited by electrical stimulation of the hypothalamus, *Science*, 159, 1968, 1119–1121.

Depaulis, A., K.A. Keay, R. Bandler, Longitudinal neuronal organization of defensive reactions in the midbrain periaqueductal gray region of the rat, *Exp. Brain Res.*, 90, 1992, 307–318.

DeSisto, M.J. and J.P. Huston, Aggression and reward from stimulating common sites in the posterior lateral hypothalamus of rats, *Commun. Behav. Biol.*, 6, 1971, 295–306.

DeSisto, M.J. and M. Zweig, Differentiation of hypothalamic feeding and killing, *Physiol. Psychol.*, 2, 1974, 67–70.

Eckersdorf, B., H. Golebiewski, M.K. Lewinska, Kainic acid lesions of the cat's antero-medial hypothalamus and emotional-defensive response evoked by carbachol injection to the same loci, *Behav. Brain Res.*, 24, 1987, 161–166.

Edwards, D.A., F.R. Nahai, P. Wright, Pathways linking the olfactory bulbs with the medial preoptic anterior hypothalamus are important for intermale aggression in mice, *Physiol. Behav.*, 53, 1993, 611–615.

Edwards, S.B. and J.P. Flynn, Corticospinal control of striking in centrally elicited attack behavior, *Brain Res.*, 41, 1972, 51–65.

Ellison, G.D. and J.P. Flynn, Organized aggressive behavior in cats after surgical isolation of the hypothalamus, *Arch. Ital. Biol.*, 106, 1968, 1–20.

Emmers, R., R.W.M. Chun, G.H. Wang, Behavior and reflexes of chronic thalamic cats, *Arch. Ital. Biol.*, 103, 1965, 178–193.

Fernandez De Molina, A. and R.W. Hunsperger, Organization of the subcortical system governing defence and flight reactions in the cat, *J. Physiol.*, 160, 1962, 200–213.

Flynn, J.P., The neural basis of aggression in cats, in D.C. Glass, Ed., *Neurophysiology and Emotion*, Rockefeller University Press and Russell Sage Foundation, New York, 1967, pp. 40–60.

Flynn, J.P. and R. Bandler, Visual patterened reflex present during hypothalamically elicited attack, *Science*, 171, 1971, 817–818.

Flynn, J.P. and M.F. MacDonnell, Control of sensory fields by stimulation of hypothalamus, *Science*, 152, 1966, 1406–1408.

Flynn, J.P. and M.F. MacDonnell, Sensory control of hypothalamic attack, *Anim. Behav.*, 14, 1966, 399–405.

Flynn, J.P., H. Vanegas, W.E. Foote, S. Edwards, Neural mechanisms involved in a cat's attack on a rat, in R. Whalen, Ed., *The Neural Control of Behavior*, Academic Press, New York, 1970, pp. 135–173.

Fonberg, E. and V.M. Serduchenko, Predatory behavior after hypothalamic lesions in cats, *Physiol. Behav.*, 24, 1980, 225–230.

Fuchs, S.A.G., M. Dalsass, H.E. Siegel, A. Siegel, The neural pathways mediating quiet-biting attack behavior from the hypothalamus in the cat: a functional autoradiographic study, *Aggress. Behav.*, 7, 1981, 51–67.

Fuchs, S.A.G., H.M. Edinger, A. Siegel, The organization of the hypothalamic pathways mediating affective defense behavior in the cat, *Brain Res.*, 330, 1985a, 77–92.

Fuchs, S.A.G., H.M. Edinger, A. Siegel, The role of the anterior hypothalamus in affective defense behavior elicited from the ventromedial hypothalamus of the cat, *Brain Res.*, 330, 1985b, 93–108.

Goldstein, J.M. and J. Siegel, Suppression of attack behavior in cats by stimulation of ventral tegmental area and nucleus accumbens, *Brain Res.*, 183, 1980, 181–192.

Golebiewski, H. and A. Romaniuk, Midbrain interaction with the hypothalamus in expression of aggressive behavior in cats, *Acta Neurobiol. Exp.*, 37, 1977, 83–97.

Halliday, R. and R. Bandler, Anterior hypothalamic knife cut eliminates a specific component of the predatory behaviour elicited by electrical stimulation of the posterior hypothalamus or ventral midbrain in the cat, *Neurosci. Lett.*, 21, 1981, 231–236.

Han, Y., M.B. Shaikh, A. Siegel, Medial amygdaloid suppression of predatory attack behavior in the cat: I. Role of a substance P pathway from the medial amygdala to the medial hypothalamus, *Brain Res.*, 716, 1996a, 59–71.

Han, Y., M.B. Shaikh, A. Siegel, Medial amygdaloid suppression of predatory attack behavior in the cat: II. Role of a GABAergic pathway from the medial to the lateral hypothalamus, *Brain Res.*, 716, 1996b, 72–83.

Harwood, D. and D.M. Vowles, Defensive behaviour and the after effects of brain stimulation in the ring dove *(Streptopelia risoria), Neuropsychologia*, 5, 1967, 345–366.

Huang, Y.H. and J.P. Flynn, Unit activities in the hypothalamus and midbrain during stimulation of hypothalamic attack sites, *Brain Res.*, 93, 1975, 419–440.

Hunsperger, R.W., Role of substantia grisea centralis mesencephali in electrically-induced rage reactions, *Prog. Neurobiol.*, 1956, 289–294.

Hutchinson, R.R. and J.W. Renfrew, Stalking attack and eating behaviors elicited from the same sites in the hypothalamus, *J. Comp. Physiol. Psychol.*, 61, 1966, 360–367.

Inselman, B.R. and J.P. Flynn, Modulatory effects of preoptic stimulation on hypothalamically-elicited attack in cats, *Brain Res.*, 42, 1972, 73–87.

Jurgens, U., M. Maurus, D. Ploog, P. Winter, Vocalization in the squirrel monkey, Saimiri sciureus, elicited by brain stimulation, *Exp. Brain Res.*, 4, 1967, 114–117.

Kanai, T. and S.C. Wang, Localization of the central vocalization mechanism in the brain stem of the cat, *Exp. Neurol.*, 6, 1962, 426–434.

Koolhaas, J.M. Hypothalamically induced intraspecific aggressive behaviour in the rat, *Exp. Brain Res.*, 32, 1978, 365–375.

Kruk, M.R., W. Meelis, A.M. Van der Poel, J. Mos, Electrical stimulation as a tool to trace physiological properties of the hypothalamic network in aggression, in P.F. Brain and D. Benton, Eds., *The Biology of Aggression*, Sythoff and Noordhoff International Publishers, The Netherlands, 1981, pp. 383–395.

Kruk, M.R., C.E. Van der Laan, J. Mos, A.M. Van der Poel, W. Meelis, B. Olivier, Comparison of aggressive behaviour induced by electrical stimulation in the hypothalamus of male and female rats, *Prog. Brain Res.*, 61, 1984, 303–314.

Levison, P.K. and J.P. Flynn, The objects attacked by cats during stimulation of the hypothalamus, *Anim. Behav.*, 13, 1965, 217–220.

Lipp, H.P. and R.W. Hunsperger, Threat, attack and flight elicited by electrical stimulation of the ventromedial hypothalamus of the Marmoset monkey Callithrix Jacchus, *Brain Behav. Evol.*, 15, 1978, 260–293.

MacDonnell, M. and J.P. Flynn, Sensory control of hypothalamic attack, *Anim. Behav.*, 14, 1966, 399–405.

MacDonnell, M.F. and J.P. Flynn, Attack elicited by stimulation of the thalamus and adjacent structures of cats. *Behaviour*, 31, 1967, 185–202.

Malick, J.B., A behavioral comparison of three lesion-induced models of aggression in the rat, *Physiol. Behav.*, 5, 1970, 679–681.

Mos, J., M.R. Kruk, A.M. Van der Poel, W. Meelis, Aggressive behavior induced by electrical stimulation in the midbrain central gray of male rats, *Aggress. Behav.*, 8, 1982, 261–284.

Mos, J., J.H.C.M. Lammers, A.M. Van der Poel, B. Bermond, W. Meelis, M.R. Kruk, Effects of midbrain central gray lesions on spontaneous and electrically induced aggression in the rat, *Aggress. Behav.*, 9, 1983, 133–155.

Nakao, H., Emotional behavior produced by hypothalamic stimulation, *Amer. J. Physiol.*, 194, 1958, 411–418.

Oishi, R. and S. Ueki, Facilitation of muricide by dorsal norepinephrine bundle lesions in olfactory bulbectomized rats, *Pharmacol. Biochem. Behav.*, 8, 1978, 133–136.

Panksepp, J., Aggression elicited by electrical stimulation of the hypothalamus in albino rats, *Physiol. Behav.*, 6, 1971, 321–329.

Paxinos, G. and D. Bindra, Hypothalamic knife cuts: effects on eating, drinking, irritability, aggression, and copulation in the male rat, *J. Comp. Physiol. Psychol.*, 79, 1972, 219–229.

Pellis, S.M., D.P. O'Brien, V.C. Pellis, P. Teitelbaum, D.L. Wolgin, S. Kennedy, Escalation of feline predation along a gradient from avoidance through "play" to killing, *Behav. Neurosci.*, 102, 1988, 760–777.

Perachio, A.A. and M. Alexander, The neural bases of aggression and sexual behavior in the rhesus monkey, in R.C. Bourne, Ed., *The Rhesus Monkey*, Academic Press, New York, 1975, pp. 381–409.

Proshansky, E. and R. Bandler, Midbrain-hypothalamic interrelationships in the control of aggressive behavior, *Aggress. Behav.*, 1, 1975, 135–155.

Putkonen, P.T.S., Attack elicited by forebrain and hypothalamic stimulation in the chicken, *Experientia*, 22, 1966, 1–5.

Putkonen, P.T.S., Electrical stimulation of the avian brain: behavioral and autonomic reactions from the archistriatum, ventromedial forebrain and the diencephalon in the chicken, *Ann. Acad. Sci. Fenn.*, 130, 1967, 9–89.

Renfrew, J.W. The intensity function and reinforcing properties of brain stimulation that elicits attack, *Physiol. Behav.*, 4, 1969, 509–515.

Roberts, W.W. and R.J. Carey, Rewarding effect of performance of gnawing aroused by hypothalamic stimulation in the rat, *J. Comp. Physiol. Psychol.*, 59, 1965, 317–324.

Roberts, W.W. and H.O. Kiess, Motivational properties of hypothalamic aggression in cats, *J. Comp. Physiol. Psychol.*, 58, 1964, 187–193.

Roberts, W.W. and J. Nagel, First-order projections activated by stimulation of hypothalamic sites eliciting attack and flight in rats, *Behav. Neurosci.*, 110, 1996, 509–527.

Shaikh, M.B., J.A. Barrett, A. Siegel, The pathways mediating affective defense and quiet biting attack behavior from the midbrain central gray of the cat: an autoradiographic study, *Brain Res.*, 437, 1987, 9–25.

Shaikh, M.B., M. Brutus, A. Siegel, H.E. Siegel, Topographically organized midbrain modulation of predatory and defensive aggression in the cat, *Brain Res.*, 336, 1985, 308–312.

Sheard, M.H. and J.P. Flynn, Facilitation of attack behavior by stimulation of the midbrain of cats, *Brain Res.*, 4, 1967, 324–333.

Skultety, F.M., Stimulation of periaqueductal gray and hypothalamus, *Arch. Neurol.*, 8, 1963, 38–50.

Sledjeski, M.B. and J.P. Flynn, Poststimulus excitability at attack sites in the cat's hypothalamus, *Brain Res.*, 40, 1972, 516–522.

Smith, D.A., Increased perioral responsiveness: a possible explanation for the switching of behavior observed during lateral hypothalamic stimulation, *Physiol. Behav.*, 8, 1972, 617–621.

Sugarman, R.A. and L.S. Demski, Agonistic behavior elicited by electrical stimulation of the brain in Western collared lizards, *Crotaphytus Collaris, Brain Behav. Evol.*, 15, 1978, 446–469.

Thorne, M.B., M. Aaron, E.E. Latham, Olfactory system damage in rats and emotional, muricidal, and rat pup killing behavior, *Physiol. Psychol.*, 2, 1974, 157–163.

Van Hemel, P.E., Aggression as a reinforcer: operant behavior in the mouse-killing rat, *J. Exp. Anal. Behav.*, 17, 1972, 237–245.

Vanegas, H., W.E. Foote, J.P. Flynn, Hypothalamic influences upon activity of units of the visual cortex, *Yale J. Biol. Med.*, 1970, 191–201.

Vergnes, M. E. Kempf, G. Mack, Controle inhibiteur du comportement d'agression interspecifique du rat: systeme serotoninergique du raphe et afferences olfactives, *Brain Res.*, 70, 1974, 481–491.

Wolgin, D.L. and P. Teitelbaum, Role of activation and sensory stimuli in recovery from lateral hypothalamic damage in the cat, *J. Comp. Physiol. Psychol.*, 92, 1978, 474–500.

5 Limbic System I: Behavioral, Anatomical, and Physiological Considerations

In this chapter, results are summarized of studies attempting to identify the relationships of a group of forebrain regions in aggression and rage referred to as "limbic" structures. The word "limbic" means "border" in Latin, and thus, the structures comprising the limbic system are generally interposed between the overlying cerebral cortex and underlying diencephalon. These structures include phylogenetically older regions of the telencephalon and include the hippocampal formation, septal area, and amygdala. In considering the limbic system, it is also useful to include the prefrontal cortex and cingulate gyrus as components of this system because of their effects on visceral processes that parallel those of other limbic structures.

One of the defining properties of limbic structures is that they project directly or indirectly to the hypothalamus or midbrain periaqueductal gray (PAG). In so doing, limbic structures modulate the functions typically associated with the hypothalamus or PAG. The anatomical bases by which each of these limbic structures can regulate visceral processes will be considered in detail in the next chapter. While the major focus of the analysis considers how these limbic structures affect aggression and rage, it should also be noted that several of these structures also mediate other processes such as short-term memory and epileptogenic activity that are apparently unrelated to functions of the hypothalamus or PAG.

As part of an overview of this system, it is useful to mention several features concerning the input–output relationships of limbic structures, which perhaps can help to provide understanding of how these structures may modulate aggression and rage. As a general rule, limbic structures receive inputs from at least two sources: (1) either directly or indirectly from one or more sensory systems; and (2) monoaminergic fiber systems. Limbic neurons then project directly or indirectly to the hypothalamus or midbrain PAG. A likely explanation of this organization is that inputs to the hypothalamus and PAG from limbic structures serve to modulate their outputs, which are directed on somatic motor and autonomic neurons of the lower brainstem and spinal cord for the integration of the specific forms of emotional behavior comprising aggression and rage. To complete this overview, the possible feedback pathways that may come into play should also be noted. These include pathways from the hypothalamus and PAG that project to limbic structures (and cerebral cortex), as well as limbic structures that project to the cerebral cortex. In this manner, such feedback could serve to adjust the modulatory properties of limbic structures, especially with respect to their receipt of new information from the cortex and monoamine neurons, which could affect the reaction to emotional stimuli and the duration of the attack or rage response. Likewise, feedback to the cerebral cortex from the hypothalamus and limbic system could also provide the basis for the "patterning" mechanism governing the overall sensorimotor components of the expression of the attack or rage response that were considered in Chapter 4.

On the basis of experimental evidence, the discussion below considers the extent to which each of the limbic and related structures is involved in the regulation of aggression and/or rage behavior. In Chapter 6, evidence of limbic system involvement as determined from human studies and case histories will be further considered. This chapter will systematically examine each of the limbic structures by first identifying in brief their basic anatomical relationships and then their functional properties with respect to aggression and rage.

HIPPOCAMPAL FORMATION

Anatomical Considerations

The hippocampal formation consists of the cornu Ammonis, dentate gyrus, and subicular cortex. The cornu Ammonis refers to the hippocampus, which is a multi-layered primitive cortex whose primary cell type is the pyramidal cell. These cells are arranged in a C-shaped fashion, which is interlocked with a C-shaped arrangement of the dentate gyrus. The dendrites of pyramidal cells receive inputs from various sources. These include, in part, other regions of the hippocampus, entorhinal cortex, septal area, and brainstem monoamine neurons. Sensory information that reaches the hippocampus through oligo- or multi-synaptic pathways includes olfactory, visual, and auditory messages. The axons of pyramidal cells project outside the hippocampal formation to such regions as entorhinal cortex, septal area, and the contralateral hippocampus (Figure 5.1).

The dentate gyrus may also be viewed as primitive cortex, and has multiple layers as well whose primary neuron is the granule cell. Axons of the dentate gyrus generally do not project outside of the hippocampal formation. Instead, the axon of the granule cell is called a *mossy fiber* that projects to parts of the cornu Ammonis. The subicular cortex is an extension of the cornu Ammonis and is interposed between the cornu Ammonis and entorhinal cortex. Again, the principal cell type is a pyramidal cell whose axons project both to entorhinal cortex, diencephalon (i.e., mammillary bodies, medial hypothalamus, anteroventral thalamic nucleus) and septal area (Krayniak et al., 1979; Meibach and Siegel, 1975, 1977a; Siegel and Tassoni, 1971; Siegel et al., 1974, 1975) (Figure 5.2).

The key points to emphasize here are that (1) monoaminergic inputs provide the anatomical and functional bases underlying mood changes, which can thus affect or sculpture the hippocampal response to visual, auditory, olfactory stimuli and other cortical inputs; and (2) the hippocampal response to such inputs are reflected in terms of its direct or indirect outputs that modulate the functions of the hypothalamus.

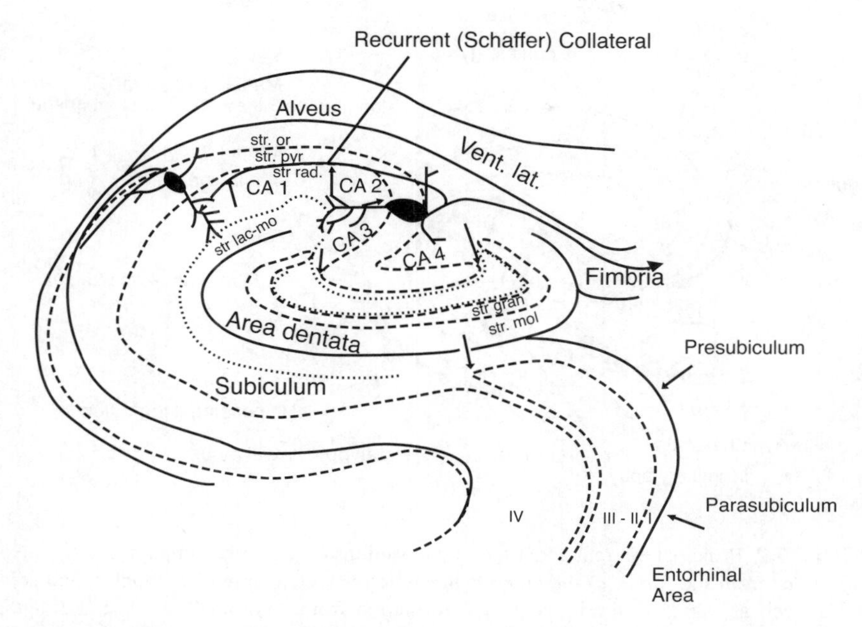

FIGURE 5.1 Horizontal section depicting the different layers of hippocampal formation. The layers of the hippocampus extending from the ventricular surface toward the dentate gyrus are as follows: alveus, stratum oriens (str. or.), pyramidal cell layer (str. pyr.), stratum radiatum (str. rad.), stratum lacunosum-moleculare (str. lac.-mol.) (of both hippocampus and dentate gyrus) and dentate gyrus (called area dentata). (Modified from M.B. Carpenter and J. Sutin, *Human Neuroanatomy*, 8th ed., Williams & Wilkins, Baltimore, 1983.)

The direct pathways are mediated from the subiculum to the mammillary bodies and medial hypothalamus, while the indirect pathways to the hypothalamus involve a synaptic relay in the septal area. (See discussion below concerning the afferent supply to the septal area for further details of this pathway.)

Relationship to Aggression and Rage

The notion that the hippocampal formation may play a role in emotional behavior received impetus from the early theoretical papers of Papez (1937) and MacLean (1949). During the 1950s, a number of studies were conducted attempting to link hippocampal functions with aggression and rage. Rothfield and Harman (1954) observed that following bilateral surgical interruption of the fornix in the cat, which was preceded by neocortical ablation, there was a lowering of the rage threshold in two cats. Several investigators also observed that hippocampal seizures following either electrolytic lesions, electrical or chemical stimulation have also been associated with the onset of rage responses (Green et al., 1957; Kaada et al., 1953; MacLean and Delgado, 1953). However, in these instances, there is difficulty in interpreting these data in terms of localization of function because of the nonspecificity of the seizure activity resulting from its spread to different regions of the forebrain.

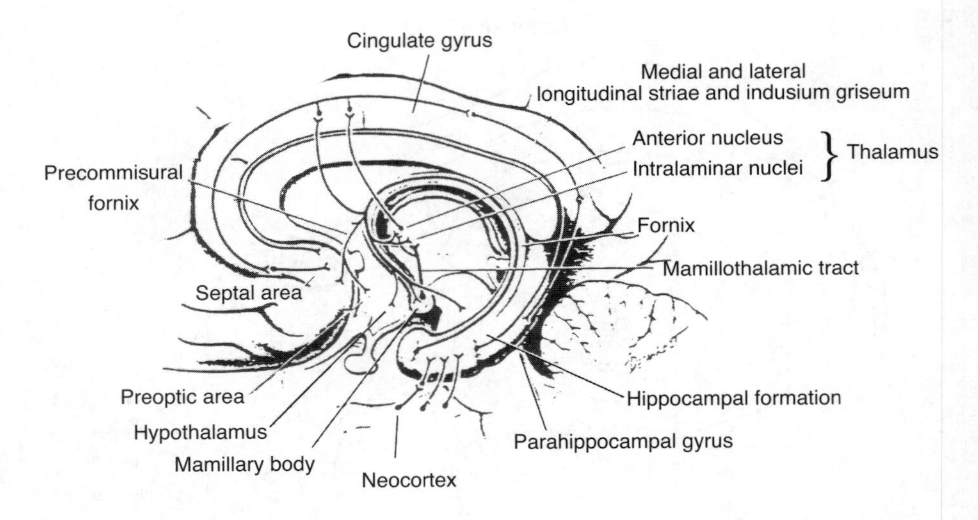

FIGURE 5.2 Principal efferent and afferent connections of the hippocampal formation. Note projections from the fornix to the anterior thalamic nucleus, mammillary bodies, and septal area as well as inputs to the hippocampal formation from entorhinal cortex and cingulate gyrus. (Modified from C.R. Noback and R.J. Demarest, *The Human Nervous System*, 2nd ed., McGraw-Hill, New York, 1975.)

In terms of the gross topography with respect to species such as rodents and felines, the hippocampus is situated in a manner that extends from its rostral tip that approaches the septal area caudally, passing immediately ventral to the corpus callosum parallel to the lateral ventricle, and then entering the temporal lobe where it runs ventrally and then rostrally, again parallel to the inferior horn of the lateral ventricle, ending in the temporal lobe just caudal to the amygdala. Thus, the position of a given region of hippocampal formation can be described in terms of its relationship to the septal or temporal pole. It is interesting to note that with reference to primates, including humans, a longitudinal axis is still present. However, while the primate hippocampal formation is homologous with that of lower forms, the position it occupies is slightly different due to significant expansion of the cerebral cortex in primate development. Accordingly, the longitudinal axis of the primate hippocampal formation is present exclusively in the temporal lobe, and extends from its rostral pole situated immediately behind the amygdala caudally to the posterior aspect of the temporal lobe (Siegel et al., 1975). Therefore, the hippocampal formation does not occupy a subcallosal position as it does in lower forms.

Because the hippocampal formation in the cat contains both a dorsal (septal) and ventral (temporal) aspect, Siegel and Flynn (1968) sought to systematically determine the role of each of these regions in modulating hypothalamically elicited predatory attack in the cat by using a dual stimulation paradigm described previously. The paradigm consisted of presenting paired trials of single (lateral hypothalamic-[attack site]) and dual (dorsal- or ventral hippocampal-[modulating] plus lateral hypothalamic) stimulation. The results of this study are shown in Figure 5.3, and

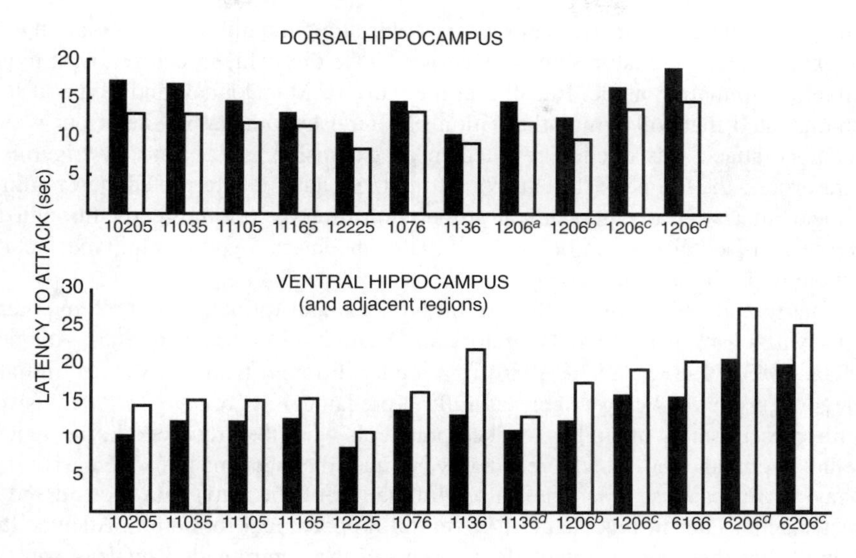

FIGURE 5.3 Upper bar graph indicates that concurrent stimulation of dorsal hippocampus and hypothalamus (filled bars) increases the latency to predatory attack relative to single stimulation of the lateral hypothalamus alone (striped bars). Lower bar graph indicates that concurrent stimulation of the ventral hippocampus (and adjoining regions) and hypothalamus (filled bars) facilitated the occurrence of predatory attack relative to single stimulation of the lateral hypothalamus alone (striped bars). Numbers along abscissa denote cases examined in this study. [a]Stimulation of ipsilateral hypothalamus and hippocampus (right side). [b]Current intensity applied through electrodes aimed for right hippocampus was raised to 0.30 mA from 0.20 mA, the intensity typically used. [c]Stimulation of hypothalamus and contralateral hippocampus. [d]Stimulation of ipsilateral hypothalamus and hippocampus (left). (From A. Siegel and J.P. Flynn, *Brain Res.*, 7, 1968, 252–267. With permission.)

illustrate the effects of dual stimulation on predatory attack. In brief, stimulation involving ventral hippocampal formation facilitated the occurrence of predatory attack, while stimulation involving the dorsal hippocampal formation suppressed this form of aggression. Thus, the striking feature of this experiment was that it demonstrated that different aspects of the hippocampal formation do not exert uniform effects on predatory attack, but instead, differentially modulate this form of aggression. An additional experiment also showed that lesions placed in the ventral hippocampal formation (and neighboring lateral nucleus of amygdala) suppressed predatory attack, while lesions of the dorsal hippocampal formation had little effect on the attack response. These data were interpreted to suggest that the ventral hippocampal formation tonically facilitates the occurrence of predatory attack, whereas the effects of the dorsal hippocampal formation on this response are phasic (i.e., occurring only during that period of time when this structure is activated).

Replication and extension of this finding were carried out later by Watson et al. (1983a). In this study, it was again demonstrated that the ventral and dorsal hippocampal formation differentially modulated predatory attack in the same manner as described above. In addition, it was further shown that ventral hippocampal stimulation altered the trigeminal sensory fields along the lipline in association with the expression of predatory attack over and above that which occurred with hypothalamic stimulation alone. Recall that the work of MacDonnell and Flynn (1966) demonstrated that following stimulation of lateral hypothalamic sites from which predatory attack was elicited resulted in an expansion of the sensory trigeminal fields around the lipline of the cat. Therefore, these authors interpreted their findings to mean that ventral hippocampal stimulation directly (or indirectly through an interneuron [see discussion below]) facilitates the lateral hypothalamic mechanism governing predatory attack behavior.

The relationship between the ventral hippocampal formation and the hypothalamus was examined in the laboratory of Adamec (Adamec and Stark-Adamec, 1983a, 1983b, 1983c, 1983d, 1986) in a totally different manner. Adamec initially categorized cats as being either naturally "(predatory) aggressive" or "defensive." In the present series of studies, evoked potentials were then recorded in the ventromedial hypothalamus (which presumably includes the mechanism for defensive rage behavior) in response to single pulse stimulation of the amygdala (considered in greater detail later in this chapter) or ventral hippocampal formation. Adamec discovered that the evoked potentials recorded in the ventromedial nucleus were of greater magnitude following amygdaloid stimulation and that smaller responses were recorded following ventral hippocampal stimulation. In addition, Adamec found that cats, which were more defensive, displayed reduced excitability with respect to evoked responses in the hippocampal formation to stimulation of the perforant pathway (i.e., a fiber bundle originating in entorhinal cortex that supplies the hippocampal formation) relative to cats, which were normally more predatory. It should be further noted that Adamec described similar relationships following seizure discharges emanating from the ventral hippocampal formation, thereby implicating epileptogenic activity in the modulation of aggression and rage. The role of seizure activity in different parts of the amygdala of experimental animals is further considered later in this chapter, and the effects of temporal lobe seizures on these processes in humans will be considered.

The results of two additional studies are relevant and should be mentioned here. In one early study conducted by Wasman and Flynn (1966), after-discharges were induced in the dorsal hippocampus of cats and their effects on predatory attack during the ictal (i.e., hippocampal stimulation) period. In brief, these investigators observed that seizure discharges within the dorsal hippocampus delayed the onset of the attack response, but did not completely block the expression of this response. These results were interpreted to mean that dorsal hippocampal seizures produced a generalized "motor" suppression of predatory attack similar to that observed when tested against other behavioral responses (e.g., learned responses and responses to nociceptive stimulation). These authors also concluded that hippocampal after-discharges did not prevent the animals from engaging in environmental contact with objects of interest.

While the previous study examined the effects of dorsal hippocampal after-discharges on predatory attack, it did not address the issue of how hippocampal seizures may affect aggression or rage during the "interictal" period. This question was considered in a study by Griffith et al. (1987). In this study, seizures in the dorsal hippocampus of cats were induced by unilateral microinjections of kainic acid into this region. These injections produced acute periods of intense seizure activity over the first few days, and were later followed by spontaneous complex partial seizures lasting up to 4 months. Here, these investigators examined the effects of defensive rage elicited from the medial hypothalamus and observed that rage response thresholds were lowered during the "interictal" period (when no seizures occurred). Moreover, mild stimulation was shown to induce "rage-like" responses in otherwise normal and affectionate animals. These findings thus provide additional evidence that activation of the dorsal hippocampus differentially affects predatory attack and defensive rage similar to the effects described below associated with the septal area and amygdala. One final point should be noted here: chronic seizure activity involving the hippocampal formation can bring about long-term changes in the propensity for the expression of defensive rage behavior. As discussed later in this chapter, similar findings were observed following the induction of chronic seizures of the amygdala in cats. The most likely explanation for this phenomenon is that chronic seizure activity involving a limbic structure causes, over an extended period of time, changes in the neurophysiological properties of the target neurons for such seizure activity. In this case, the most probable neuronal groups so affected would include extensive regions of the medial hypothalamus. This issue is also considered again later in this chapter.

These findings can be viewed as consistent with data provided above by Siegel and colleagues. This may be accomplished by integrating these two sets of studies. Here, it can be said that stimulation of the ventral hippocampal formation facilitates the occurrence of predatory attack, and that greater excitation of this region of hippocampus occurs in association with predatory relative to defensive behavior. Moreover, stimulation of the ventral hippocampal formation results in smaller evoked potentials in the ventromedial hypothalamus than sites in the amygdala that are typically associated with suppression of predatory attack and facilitation of defensive behavior. In relating the sets of data obtained from each of these laboratories, it should be recalled that the medial and lateral hypothalamus are linked by reciprocal inhibitory connections. Thus, a reduced evoked potential observed in the ventromedial nucleus following ventral hippocampal stimulation obtained from an aggressive cat would be predicted. This conclusion is based on the principle of "disinhibition." What this means is that a reduction in excitation of the ventromedial nucleus will result in greater excitation in the lateral hypothalamus — the primary region of the brain associated with predation. This is indeed due to a reduction in excitation of the inhibitory neurons passing from the medial to lateral hypothalamus, which thus releases the lateral hypothalamus from inhibition by the medial hypothalamus. (See Chapter 3, Figure 3.12, for description of the anatomical and functional relationship between the medial and lateral hypothalamus.)

SEPTAL AREA

Anatomical Considerations

The histological appearance of the septal area is relatively similar in a variety of species such as rodents, felines, and lower primates. In brief, the septal area consists of different groups of cell groups, several of which are described in Figure 5.4. This figure illustrates the loci of the principal nuclei shown in a frontal section taken through a midlevel of the septal area of the rat. The dorsal aspect of the septal area contains the lateral septal nucleus, which occupies the lateral two-thirds of this region. The region medial to the lateral septal nucleus within the dorsal half of the septal nucleus contains the medial septal nucleus. The medial septal nucleus becomes continuous with the vertical and horizontal limbs of the diagonal band of Broca, which are situated in the ventral aspect of the septal area (Andy and Stephan, 1964; Krayniak et al., 1980; Meibach and Siegel, 1977b). It should be noted that in humans, the septal area is essentially located in a ventral position. The dorsal aspect in humans is limited to the septum pellucidum and, consequently, there are few neurons present in this aspect.

Concerning the inputs and outputs of the septal area, the hippocampal formation constitutes the primary afferent supply to the septal area (Krayniak et al., 1979; Meibach and Siegel, 1977a; Siegel and Tassoni, 1971; Watson et al., 1983b). This projection arises from different fields of the cornu Ammonis and subiculum and project topographically along the longitudinal axis of hippocampal formation through the precommissural fornix to the lateral septal area. In this manner, neurons arising close to the septal pole of the hippocampal formation project to the medial aspect of the lateral septal nucleus, while progressively more distal aspects of hippocampal formation that approach its temporal pole, project to progressively more lateral aspects of the lateral septal nucleus (Meibach and Siegel, 1977a; Siegel and Tassoni, 1971; Siegel et al., 1974, 1975).

In turn, the septal area projects principally to different nuclei of the hypothalamus. The primary target within the hypothalamus of septal efferent fibers is the perifornical region and parts of the medial hypothalamus, including the mammillary bodies (Krayniak et al., 1980; Meibach and Siegel, 1977b; Stoddard-Apter and MacDonnell, 1980). The projection from the septal area can reach the hypothalamus from two groups of nuclei — the lateral septal nucleus and nuclei of the diagonal band (Krayniak et al., 1980). The key point to remember here is that the anatomical relationships of the septal area strongly suggest that one of its primary functions is to serve as a relay nucleus by which the hippocampal formation can indirectly modulate hypothalamic processes (Figure 5.5). Studies describing the role of the septal area in regulating aggression and rage are summarized below.

Relationship to Aggression and Rage

Over the past 4 decades, many studies focused on the role of the septal area in emotional behavior, including aggression and rage. Typical of the approaches used in research in this field, the earlier studies conducted sought to identify how lesions and ablations of the septal area altered the expression of rage and aggression in

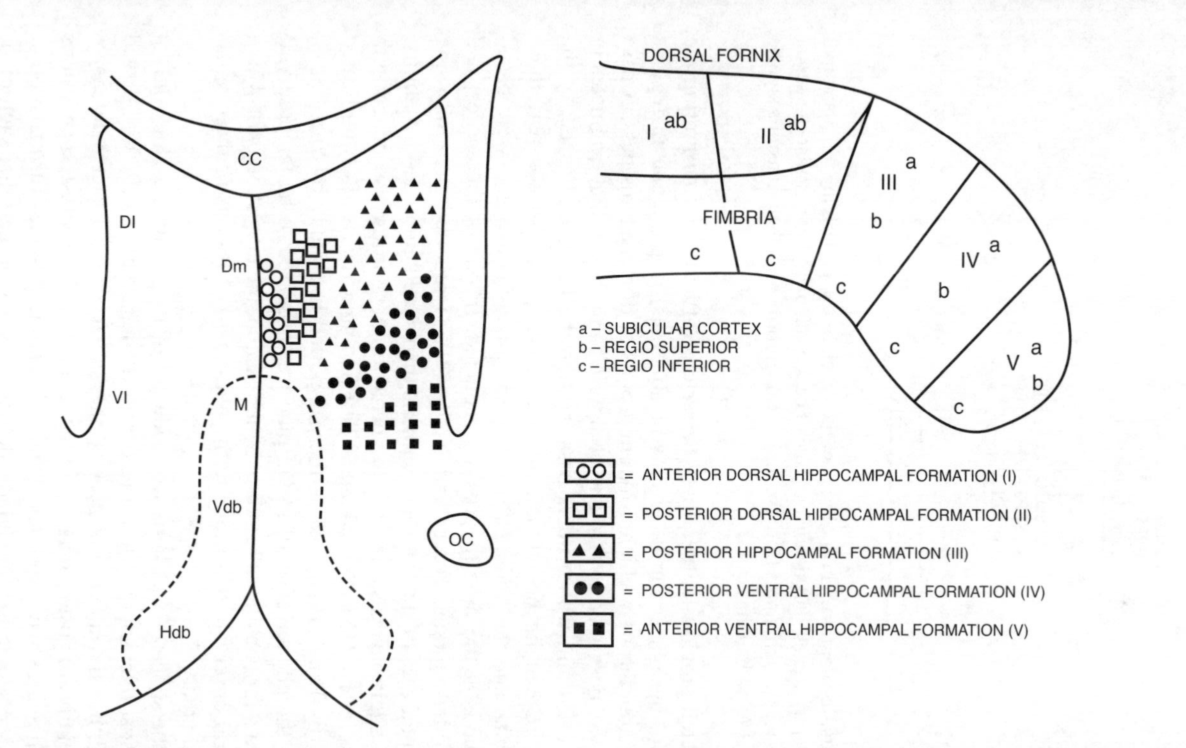

FIGURE 5.4 Schematic diagram summarizing the locations of the major cell groups of the septal area and the topographical organization of the fornix projections to the lateral septal nucleus. The dorsal fornix and fimbria, divided into five zones, represent the positions of fibers that arise from five different levels along the longitudinal axis of the hippocampal formation. a, subicular cortex; b, regio superior; c, regio inferior. (From R. Meibach and A. Siegel, *Brain Res.*, 119, 1977b, 1–20. With permission.)

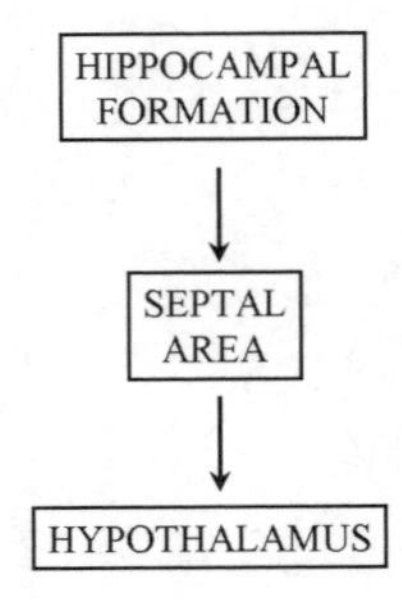

FIGURE 5.5 Schematic diagram illustrating the likely circuitry underlying hippocampal modulation of predatory attack and defensive rage. The key element here is that the septal area, which receives significant inputs from the hippocampal formation, then transmits signals mainly to the medial hypothalamus.

rodents. As indicated previously, there may be serious difficulty in interpreting the data resulting from the placement of lesions or ablations, in particular in regions such as the septal area through which different groups of fiber systems pass. A case in point is a pathway described by Troiano et al. (1978) in the cat that originates in the infra-limbic cortex and whose fibers pass through the far medial aspect of the septal area to terminate in the medial thalamus. Nevertheless, for the most part, these studies used rodents, and the findings, which are relatively consistent, were based on available methods at the time they were conducted. In one of the earliest and most frequently cited studies, Brady and Nauta (1953) reported that following ablation of the septal area of the rat, these animals displayed heightened irritability, emotionality, and aggressiveness. In fact, the behavioral syndrome associated with this procedure was referred to as "septal rage." This finding was essentially replicated in a related study by King (1958) and further supported in a study by Yutzy et al. (1967). Although the phenomenon of septal rage was observed to be temporary, it provided an impetus for considerable subsequent research, the largest body of which involved behavioral observations following the placement of lesions or direct ablations of the septal area.

It is beyond the scope of this book to review all of these studies. Thus, only a selection of the most relevant ones are considered. Blanchard et al. (1979) observed that lesions involving the anterior aspect of the septal area enhanced defensive behavior in rats relative to posterior aspect of the septal area. In contrast, such lesions had little effect on aggressive responses. Thus, this finding provides further support for the view that the septal area modulates defensive behavior, and that such modulation is associated to a greater extent with the anterior aspect of this structure. Gotsick and Marshall (1972) also observed the presence of the septal rage syndrome following septal lesions in rats. However, they noted that the time course of this syndrome was a function of the extent of handling. For example, if the animals were handled once a day, they returned to normal within 6 days. When they were handled 12 times on the first postoperative day, the syndrome persisted for a longer period of time. This data demonstrated the importance of behavioral manipulations in regulating the septal rage syndrome as well as the interaction between neural and behavioral processes. In contrast to the findings of Blanchard and colleagues, Lathem and Thorne (1974) reported that septal lesions enhanced aggressive behavior as measured by muricide, but that such effects were strain related. Here, the authors reported that the primary effects were observed following lesions of Long-Evans strain as opposed to Sprague Dawley rats. The difference in findings between these two studies could possibly be accounted for on the basis of strain differences since Blanchard and colleagues employed the Wistar strain. This finding would further suggest that strain differences imply variations in the neural organization and processing of information from the septal area in each of these strains. However, it is not possible to assess what these differences might constitute. That enhancement of muricide as a function of septal lesions is a temporary phenomenon is provided from a study by Miczek and Grossman (1972), and suggests that perhaps dissipation of attack may reflect some kind of neural reorganization that takes place over the first few weeks following surgery.

Other approaches have been employed to determine the role of the rodent septal area in aggression and rage. Here, several studies are cited. In one study, Albert and Richmond (1976) microinjected the anesthetic lidocaine into the lateral septal nucleus and observed an increase in reactivity and aggression (as measured by responses to a sharp tap on the back, placement of a gloved hand near the snout, gentle prodding or grasping of the body, tail capture, vocalization, and biting). Microinjections placed in the region of the nucleus accumbens or adjoining regions produced similar results, while no effects followed microinjections into the caudate nucleus. This finding, which parallels those resulting from lesion studies described above, led these authors to suggest that septal lesions interfere with neural pathway mediating aggression that pass from the hippocampal commissure through the lateral septal area into the anterior aspect of the septal area and adjoining basal forebrain. It should be noted that the one pathway that would correspond to such a trajectory is the precommissural fornix, which, as noted above, originates in the hippocampal formation. In a second approach, Bawden and Racine (1979) attempted to determine whether kindling of the septal area (or amygdala) could affect aggressive reactions in rats. This approach was employed as a means of comparing the effects described by Adamec and colleagues (see discussion below) who demonstrated modulating effects of amygdaloid and hippocampal kindling on aggression and rage in the cat. In brief, Bawden and Racine reported that kindled seizures in the septal area or amygdala did not induce or inhibit muricide or ranicide, or modulate intraspecific (defensive) aggression. In contrast, stimulation of the septal area of the rat at current levels below that which could elicit seizures suppressed muricidal behavior (Potegal et al., 1980). It should be pointed out that similar findings were observed with septal stimulation in cats on predatory attack.

Fewer studies have been conducted in the cat. The studies described below specifically used electrical stimulation of different aspects of the septal area to determine the role of this region on aggression and rage. One of the first studies that systematically examined the role of the septal area in feline aggression was carried out by Siegel and Skog (1970). The paradigm for this experiment was identical to that described above for the analysis of hippocampal modulation of attack behavior in the cat. Here, dual stimulation of mainly the lateral septal area suppressed predatory attack behavior elicited from the lateral hypothalamus. Several sites were also tested against defensive rage and flight behavior elicited from the medial hypothalamus. It was observed that septal stimulation facilitated these responses.

That activation of wide regions of the septal area suppresses predatory attack and facilitates defensive rage received support from several additional studies. In one study, Brutus et al. (1984) replicated the findings of Siegel and Skog by demonstrating again that stimulation of wide regions of the septal area suppressed predatory attack behavior (Figure 5.6). These authors further showed that septal area inhibition might be linked to sensory mechanisms associated with trigeminal reflexes activated during hypothalamic stimulation. Specifically, stimulation of septal area sites shown to suppress predatory attack also decreased the lateral extent of the effective sensory fields of the lipline established during hypothalamic stimulation. In a further aspect of this study, ^{14}C-deoxyglucose was injected into the animal, and stimulation was applied to septal area sites, which suppressed predatory attack while the cats were anesthetized with pentobarbital. Deoxyglucose autoradiography revealed that a descending pathway labeled by this procedure included fibers that passed to the medial hypothalamus and adjoining perifornical region (Figure 5.7). (A more extensive analysis of the functional pathways of the limbic system activated by electrical stimulation is presented in Chapter 6.)

This finding suggests that the lateral septal area and adjoining nuclei project directly to sites in the hypothalamus from which defensive rage is elicited. Likewise, in a more extensive series of experiments concerning defensive rage in the cat, MacDonnell and Stoddard-Apter (1978) and Stoddard-Apter and MacDonnell (1980) showed that stimulation of sites located in either the medial septal nucleus or in the medial aspect of the lateral septal nucleus facilitated this form of aggression. These authors also used autoradiographic tracing procedures to identify the descending pathways from septal area sites that facilitated defensive rage to the hypothalamus, and observed that these fibers project directly to the medial hypothalamus.

Collectively, results of the studies conducted in the cat indicate that stimulation of the lateral septal nucleus and adjoining regions of the medial septal nucleus suppresses predatory attack and facilitates defensive rage behavior. These results further suggest that these effects are mediated directly through the medial hypothalamus. On the basis of studies described previously concerning the reciprocal inhibitory relationship of the medial to lateral hypothalamus, it is proposed that the differential effects of septal stimulation on predatory attack and defensive rage can be understood simply as follows: activation of the septal area produces direct excitation of neurons of the medial hypothalamus, which serves to facilitate defensive rage, and through an inhibitory interneuron projecting from the medial to lateral hypothalamus, inhibits neuronal activity in the lateral hypothalamus, thus causing suppression of predatory attack behavior. This principle will be discussed again

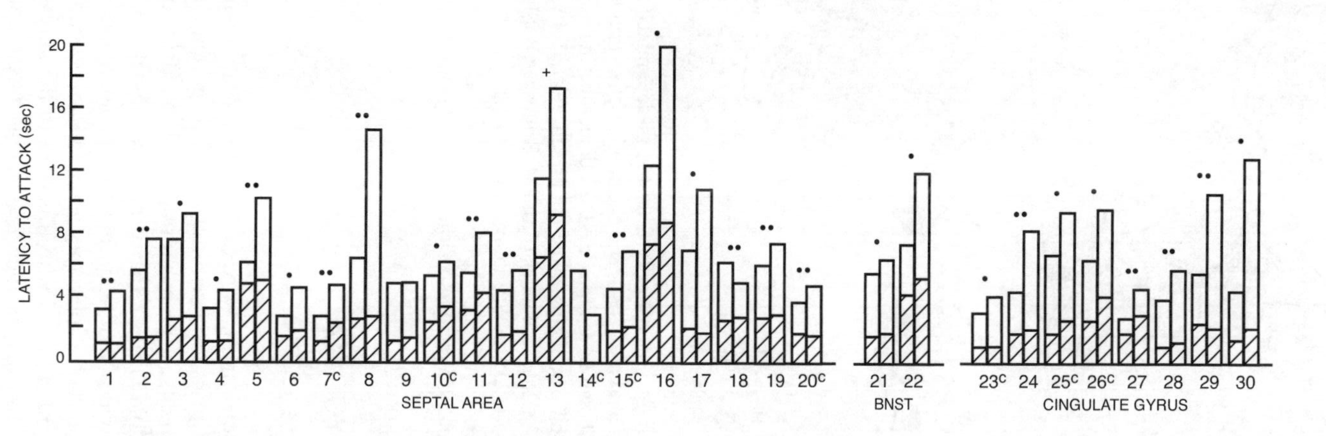

FIGURE 5.6 Bar graphs indicate that concurrent stimulation of the hypothalamus with the septal area, BNST, or cingulate gyrus alters the latency to attack (open bars), whereas this stimulation has no effect upon latencies to initial movement of the forepaws (crosshatched bars). For each pair of histograms, the first bar corresponds to the mean attack latency for stimulation of the hypothalamus alone; the second bar represents the mean attack latency for dual stimulation. Numbers centered below pairs of bar graphs identify the case numbers. The "C" superscripts indicate that during concurrent stimulation, the modulatory site was contralateral to the hypothalamic attack site. In all other cases, the modulatory site was ipsilateral to the attack site. + $p < 0.10$; * $p < 0.05$; ** $p < 0.02$. (From M. Brutus, M.B. Shaikh, A. Siegel, H.E. Siegel, *Brain Res.*, 310, 1984, 235–248. With permission.)

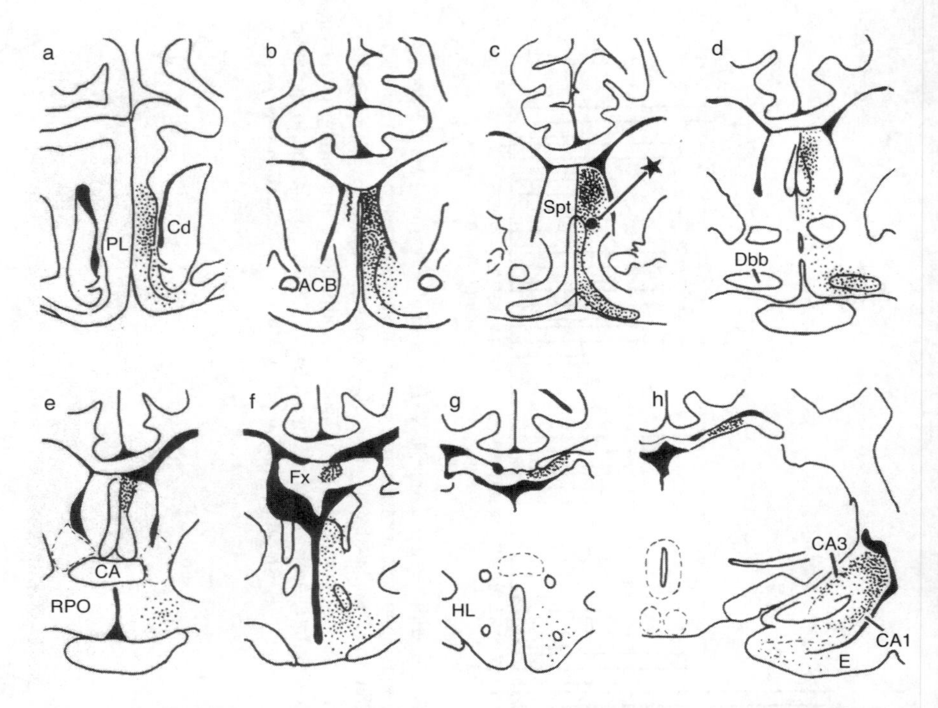

FIGURE 5.7 Diagrams illustrate the structures metabolically activated as determined from ^{14}C-2-DG autoradiographs taken from an individual cat. The location of the stimulating electrode is depicted with a closed circle shown in section "c." Dots depict the intensity of uptake of label in regions activated by septal stimulation. Note the presence of label within the anterior third of the medial and perifornical hypothalamus. (From M. Brutus, M.B. Shaikh, A. Siegel, H.E. Siegel, *Brain Res.*, 310, 1984, 235–248. With permission.)

when I attempt to account for the differential effects of the amygdala on these two forms of aggression.

AMYGDALA

Anatomical Considerations

The amygdala consists of different nuclei and is surrounded on its ventral surface by a primitive cortical mantle referred to as the pyriform lobe. This cortex can be subdivided into the following components: a prepyriform area located at rostral levels of amygdala and a periamygdaloid cortex situated at more caudal levels of the amygdala. The pyriform cortex provides significant input to the amygdala from regions that mediate sensory information such as olfactory and temporal neocortices. For this reason, the pyriform cortex is often linked with functions of the amygdala. The principal nuclear groups of the amygdala include the medial, basal, central, lateral, and anterior nuclei (Figure 5.8). However, many investigators have chosen to simplify this system functionally by dividing it into two regions — a corticomedial

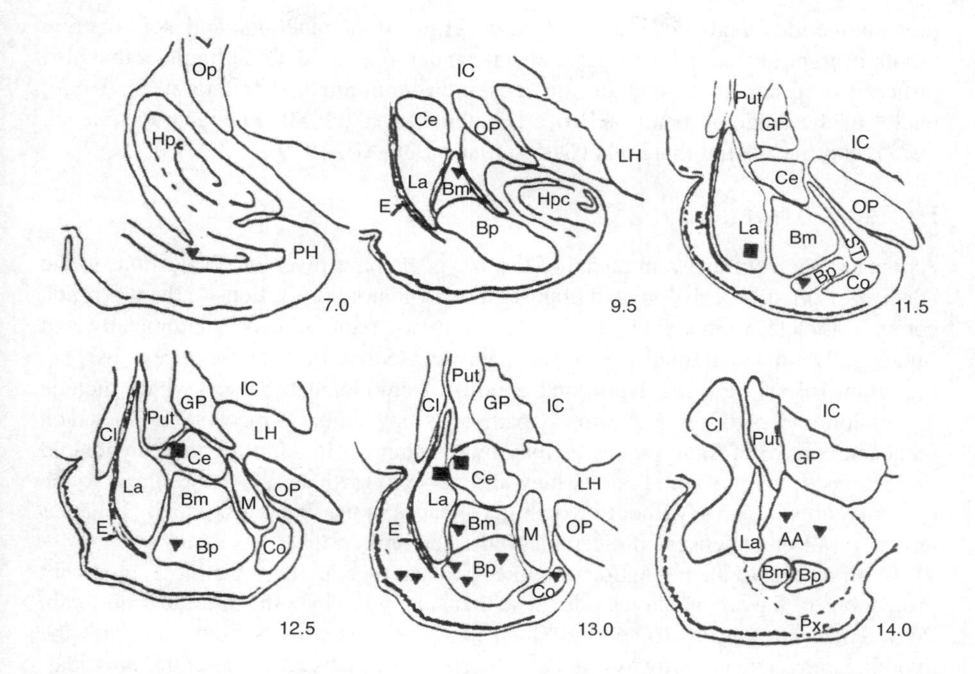

FIGURE 5.8 Electrode sites in the amygdala and surrounding pyriform cortex that modulated predatory attack and the "effective sensory fields." Filled triangles, suppression of predatory attack and constriction of "effective sensory fields"; filled squares, facilitation of predatory attack and expansion of "effective sensory fields." The number at the bottom right corner indicates the frontal plane of the section in stereotaxic coordinates. AA, anterior amygdala; Bm, basal nucleus (pars magnocellular); Bp, basal nucleus (pars parvocellular); Ce, central nucleus; Cl, claustrum; Co, cortical nucleus; E, external capsule; GP, globus pallidus; Hpc, hippocampus; IC, internal capsule; La, lateral nucleus; LH, lateral hypothalamus; M, medial nucleus; Op, optic tract; PH, parahippocampal gyrus; Put, pyriform cortex. (From C.H. Block, A. Siegel, H.M. Edinger, *Brain Res.*, 197, 1980a, 39–55. With permission.)

group that includes structures situated within the medial and dorsal aspects, and a basolateral group that encompasses the remaining nuclei.

Afferent Connections

Consistent with other limbic structures, the amygdala receives afferents from sensory and monoaminergic systems (LeDoux, 1996; Price et al., 1987; Swanson and Petrovich, 1998). Olfactory inputs project directly from the olfactory bulb to the amygdala or indirectly via a synapse in the pyriform cortex. Other sensory inputs to the amygdala include tertiary auditory and taste afferent fibers, which arise from the temporal neocortex or medial geniculate nucleus and parabrachial or solitary nucleus, respectively. In addition to the temporal neocortex, the prefrontal cortex has also been shown to have significant input into different nuclei of the amygdala. (See discussion below for further discussion of the relationship between the prefrontal cortex and amygdala.) Noradrenergic, dopaminergic, and serotonergic inputs from brainstem neurons have also been demonstrated. Other regions that also project to the amygdala include the substantia innominata (Femano et al., 1983), nuclei of the diagonal band of Broca (Krayniak et al., 1980; Meibach and Siegel, 1977), and the medial thalamus (Ottersen and Ben-Ari, 1979).

Efferent Connections

Axons arising from different nuclei of the amygdala have divergent projections to the telencephalon, diencephalon, and brainstem. Amygdaloid projections to the prefrontal cortex complete a circuit which enables these two regions to be anatomically and functionally related through reciprocal pathways. Since both of these regions play important roles in the regulation and control of emotional behavior, which include aggression and rage, these reciprocal pathways may reflect a mechanism by which feedback control of these processes may be maintained. In addition, the amygdaloid projections to the prefrontal cortex may also serve to mediate olfactory signals to the cerebral cortex directly without involving a synapse in the thalamus, which is characteristic of all other sensory projections to the cerebral cortex.

With respect to the regulation of aggression and rage, perhaps the most significant projections of the amygdala include those that supply the hypothalamus and midbrain PAG. There are various routes by which the amygdala can communicate with the hypothalamus. Direct pathways include the stria terminalis and the ventral amygdalofugal pathway. The stria terminalis arises mainly from neurons situated in the corticomedial group of nuclei and the fibers project to the rostro-caudal extent of the preoptico-medial hypothalamus. In contrast, the ventral amygdalofugal pathway arises principally from the basolateral complex of amygdaloid nuclei and supplies the lateral hypothalamus and midbrain PAG. Thus, by virtue of these two pathways, the corticomedial and basolateral nuclear complexes of the amygdala can directly regulate the activity of neurons in the medial and lateral hypothalamus, respectively. The amygdala can also modulate hypothalamic indirectly. It may do so by its projections to the bed nucleus of the stria terminalis, which in turn projects to the intermediate regions of the hypothalamus proximal to the fornix as well as to the PAG.

Relationship to Aggression and Rage

Of all the limbic structures, the amygdala has received the most attention concerning its possible relationship to aggressive behavior. In fact, one can easily argue that the literature has clearly established that the amygdala and adjoining pyriform cortex do play highly significant roles in the regulation of aggression and rage across species, including humans. The evidence spans four decades of research and involves a variety of methodologies, such as the effects of regional brain ablation, localized lesions, electrical and chemical stimulation, and seizure activity whose focus includes the amygdala. However, a number of these studies and especially the earlier ones were highly contradictory. Because of the very large numbers of studies that have been conducted over this period of time, the discussion below does not attempt to review them all, but summarizes representative experiments that I believe have had an impact on the field.

As an illustration of the conflicting findings of the early studies, both Kluver and Bucy (1939) and Schreiner and Kling (1953) placed amygdaloid lesions in monkeys and cats, respectively, and each reported that such lesions produced taming effects in these species. In contrast, Bard and Mountcastle (1948) observed that bilateral ablation of the amygdala of the cat produced ferocious behavior. Similarly, other investigators have also reported conflicting data following the placement of lesions most frequently conducted in cat or rat. For example, Shealy and Peele (1957) elicited nondirected rage behavior by amygdaloid stimulation, which also included salivation, sniffing, and gagging responses. They then placed lesions at these sites where stimulation was applied. The lesions were found to block the behavioral responses observed from stimulation. Anatomical localization was difficult to assess because of the variability in the locus and general sizes of the lesions. There is one additional difficulty in interpreting the findings of this study. On the basis of the descriptions of the behavioral responses associated with amygdaloid stimulation, it was likely that stimulation-induced seizure activity resulting in the appearance of rage behavior was secondary to seizure activity, and thus did not constitute "true" rage relative to the response patterns associated with medial hypothalamic or dorsal PAG stimulation. Nakao (1960) also observed that bilateral amygdaloid lesions involving mainly the basal complex in cats reduced or eliminated rage and other emotional responses elicited by the presence of a barking dog. In contrast, a more recent study by Zagrodzka et al. (1998) demonstrated that lesions involving the central nucleus of amygdala in the cat increased defensive behavior to humans.

In two related studies, Zagrodzka and Fonberg (1977, 1978) examined the effects of ventromedial amygdaloid lesions on predatory attack in the cat. The results of both of these studies revealed that such lesions abolished predatory attack while having little effect on food intake. This finding suggests the specificity of effects of amygdaloid lesions on aggression relative to other hypothalamic processes such as feeding. In addition, the results of this and other studies described to this point indicate that lesions of the amygdala appeared to have differential effects on defensive rage and predatory attack. Further discussion of these differential effects as well as the implications is considered in the section concerning the effects of amygdaloid stimulation on aggression and rage.

Studies conducted in rats have also revealed varying results, which may relate to both the model (and paradigm) employed as well as the locus of the amygdaloid lesion. For example, Allikmets and Ditrikh (1965) found that lesions placed in the basal amygdaloid complex blocked the expression of rage typically elicited by such perturbations as touching of its muzzle or back or lifting the rat by its tail. Using intermale aggression as a model, Vochteloo and Koolhaas (1977) demonstrated that aggression observed by this paradigm could also be blocked when lesions are placed in the medial amygdala. However, these investigators qualified their conclusions about the effects of such amygdaloid lesions by indicating that blockade only occurred when the animals were exposed to at least four aggressive interactions. This finding would suggest the existence of an interaction between brain mechanisms and environmental exposure in amygdaloid modulation of (at least certain forms of) aggression.

Investigators have also examined the effects of amygdaloid lesions on mouse killing in rats. Lesions placed in different nuclei of the amygdala appear to have produced differential effects on this form of aggression. For example, mouse killing is induced or facilitated following lesions placed in either the medial (Karli et al., 1977; Vergnes, 1975) or central (Eclancher et al., 1975) nucleus or stria terminalis (Karli et al., 1977). In contrast, lesions placed in the lateral amygdala had the effect of reducing mouse killing in rats (Vergnes, 1976). In attempting to understand the nature of this form of aggression, it should be noted that it has been suggested that this model may be the equivalent of predatory attack in the cat. However, it may also be argued that the functional homology between predation in the cat and mouse killing in the rat is at best questionable, because the ethological objectives in each situation likely differ. Specifically, in the cat the ultimate goal of predation is the acquisition of food, while such an objective is not likely the objective in mouse killing. Thus, any attempt to equate this form of aggression with predation in the cat should be viewed with caution.

Studies conducted using electrical stimulation have been very useful in clarifying the roles of different nuclear groups of the amygdala in aggression and rage. In particular, Egger and Flynn (1963) published a seminal paper that contributed significantly in clarifying some of the apparent contradictory evidence generated from experiments employing lesions or ablations. In their study, Egger and Flynn used dual stimulation of specific amygdaloid nuclei and hypothalamic sites from which predatory attack behavior was elicited in the cat. Results of their study demonstrated that stimulation of the lateral amygdala, which included the lateral nucleus and lateral aspect of the basal complex, facilitated the occurrence of predatory attack, while stimulation of more medial aspects of the amygdala, which included the medial nucleus and medial aspects of the basal complex, suppressed predatory attack. Thus, this study was the first to demonstrate that different nuclei of the amygdala differentially modulate predatory attack behavior. In so doing, this study provided a possible explanation for the divergent findings observed in studies employing lesions. On the basis of the results obtained by Egger and Flynn, the opposing effects observed by different investigators described above may be accounted for on the basis that lesions associated with facilitation of aggression primarily affected different nuclei than those associated with suppression of aggression. In the discussion below, evidence is provided that suggests another level of complexity — namely, that the effects of perturbation of amygdaloid nuclei on aggression, either by stimulation or by lesions, are dependent on the nature of the aggressive response considered.

A number of other studies that built on the work of Egger and Flynn have served to further clarify the relationships of various regions of the amygdala in the regulation of aggression and rage. Two studies examined various properties of amygdaloid modulation of hypothalamically elicited predatory attack in the cat. In one study, Block et al. (1980a) initially demonstrated in a fashion similar to the findings of Egger and Flynn that dual stimulation of separate regions of amygdala differentially modulated predatory attack elicited from the lateral hypothalamus of the cat (Figure 5.8). More recent studies by Han et al. (1996a, 1996b) replicated and extended the findings of Block et al., in part, by focusing on inhibitory role of the amygdala on predatory attack. Collectively, from the results of these studies, stimulation at sites

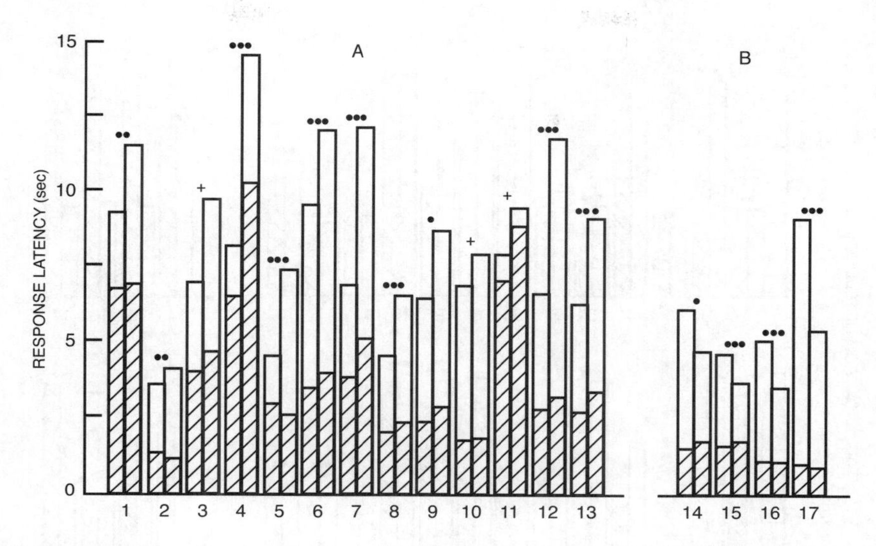

FIGURE 5.9 Bar graphs indicate that concomitant stimulation of the amygdala and hypothalamus alter latencies to predatory attack (open bar), whereas this stimulation has no effect on latencies to initial movement of the forepaws (hatched bar). For each pair of histograms, the first bar corresponds to the mean latency for single stimulation of the hypothalamus alone; the second bar represents the mean attack latency for dual stimulation. Numbers centered below pairs of bar graphs identify the individual cats used in the study. + $p < 0.10$; * $p < 0.05$; ** $p < 0.01$; *** $p < 0.001$. (A), inhibitory sites; (B), facilitatory sites. (From C.H. Block, A. Siegel, H.M. Edinger, *Brain Res.*, 197, 1980a, 39–55. With permission.)

located in the basal complex, medial aspect of amygdala, including the medial nucleus, or pyriform cortex suppressed predatory attack, while stimulation at sites situated in the central or lateral nuclei facilitated the occurrence of predatory attack (Figure 5.9). It would seem that such differential effects on predatory attack may be best accounted for in terms of the efferent pathways from the amygdala to the hypothalamus and PAG. As already noted, the outputs from the medial and basal amygdala provide the origins of the stria terminalis that supply the medial hypothalamus. In contrast, the lateral and central amygdala provide the origins of the amygdalofugal pathways, including those that supply the PAG. The second aspect of the study by Block et al. demonstrated that the sensory trigeminal fields along the lipline of the cat were altered as a function of stimulation of sites in the amygdala that suppressed or facilitated predatory attack. Similar to the findings observed with the hippocampal formation and septal area, stimulation of inhibitory sites in the amygdala reduced the size of the effective sensory fields along the lipline, while stimulation of amygdaloid sites that facilitated predatory attack increased the size of the effective sensory fields (Figure 5.10). Again, for reasons elaborated previously regarding the effects of stimulation of other limbic structures, these findings would also suggest that the effects of amygdaloid stimulation on predatory attack were mediated directly on the mechanism related to the expression of this form of aggression. Additional evidence in support of this notion is discussed below.

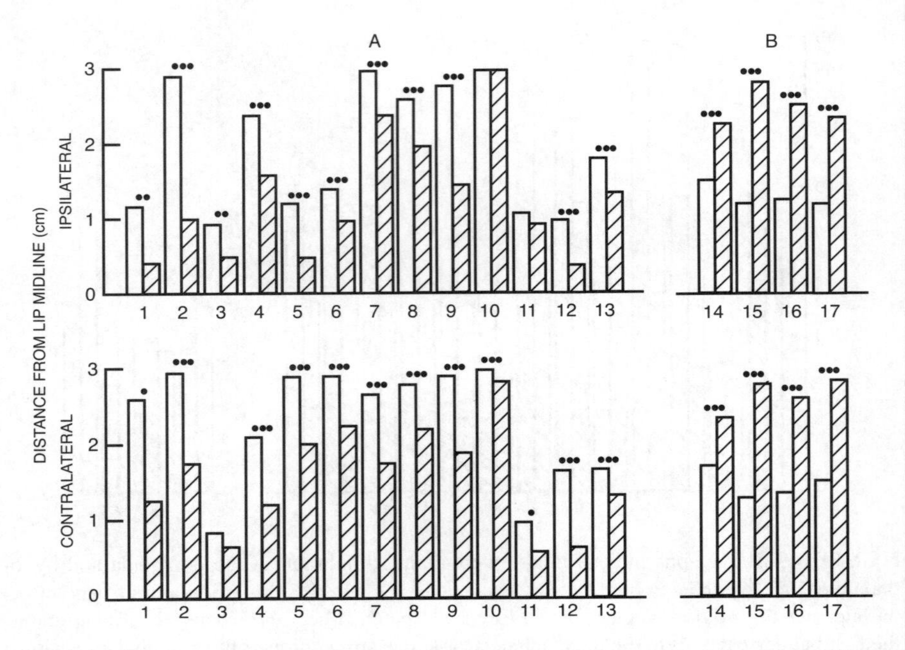

FIGURE 5.10 Bar graphs indicate that concomitant stimulation of the amygdala and hypothalamus (hatched bar) alters the size of the "effective sensory field" established during single stimulation of the hypothalamus (open bar). (A) Constriction following stimulation of inhibitory sites. (B) Expansion following stimulation of facilitatory sites for both ipsilateral and contralateral sensory fields. Numbers centered below pairs of bar graphs identify the individual cats used in the study. * $p < 0.05$; ** $p < 0.01$, *** $p < 0.001$. (From C.H. Block, A. Siegel, H.M. Edinger, *Brain Res.*, 197, 1980a, 39–55. With permission.)

Consider now the effects of stimulation of these regions of amygdala on defensive rage behavior elicited from the medial hypothalamus or PAG. Here, at least four studies are relevant. The first, conducted by Stoddard-Apter and MacDonnell (1980), was part of the same study employed by these authors for the study of the anatomical projections of septal sites that facilitated defensive rage. These investigators initially identified sites in the medial amygdala (with the aid of a cannula-electrode), which upon stimulation facilitated defensive rage in the cat elicited by medial hypothalamic stimulation. Then the pathways from these sites were determined using a radioactive tracer, which was microinjected through the cannula-electrode. The major pathway labeled was shown to be the stria terminalis in which the pathways were traced principally to the bed nucleus of the stria terminalis and medial hypothalamus. Thus, this paper provided the first piece of evidence that the medial amygdala facilitates defensive rage by acting through a monosynaptic projection to sites in the medial hypothalamus associated with this form of aggression.

Additional lines of evidence linking the amygdala with facilitation of defensive rage in the cat were provided in a study by Shaikh et al. (1993). While aspects of this and several other related studies are described in more detail in Chapter 7, other features of this study basically replicated the findings of Stoddard-Apter and MacDonnell. These include the observations that stimulation of the medial nucleus of amygdala facilitates defensive rage elicited from the medial hypothalamus, and that when using retrograde tracing methods, amygdaloid projections to sites in the medial hypothalamus from which defensive rage had been elicited arise mainly from the medial nucleus and adjoining aspects of the basal nuclear complex. (See Chapter 7 for further details on the loci of amygdaloid nuclei that project to sites in the hypothalamus and PAG from rage and aggression are elicited.)

The studies described above focused on the facilitatory effects of amygdaloid stimulation on defensive rage behavior. In a separate study, Shaikh et al. (1991) showed that stimulation of central or lateral nucleus of amygdala powerfully suppressed defensive rage behavior in the cat elicited from stimulation of the dorsolateral PAG. Thus, similar to the findings described above with respect to amygdaloid modulation of predatory attack, here too, defensive rage behavior is also differentially modulated from different nuclei of the amygdala as well. It should be noted, however, that the directionality of modulation by amygdaloid nuclei for predatory attack appears to be the opposite to that for defensive rage behavior. Specifically, stimulation of the central or lateral nuclei of the amygdala facilitates predatory attack, but suppresses defensive rage. In contrast, stimulation of the medial nucleus and adjoining regions of the basal nuclear complex suppresses predatory attack, but facilitates defensive rage. Further evidence in support of the differential effects of specific amygdaloid nuclei on predatory attack and defensive rage behavior in the cat in association with seizure activity is provided below from the experiments of Brutus et al. (1986).

Two other studies are worth noting here that involve amygdaloid stimulation and aggression conducted by Potegal et al. (1996a, 1996b). In one study, brief high-frequency stimulation was applied to the corticomedial amygdala of hamsters and the effects of such stimulation on aggression measured at different times (i.e., 1 to 90 minutes) following stimulation, using the resident–intruder design. These authors observed that amygdaloid (priming) stimulation facilitated aggressive responses in the stimulated (resident) hamster by lowering attack latencies, the maximal effect occurring 10 minutes following stimulation. In the second study, these authors observed that allowing a hamster a single "priming" attack on a conspecific (intruder) increased a transient increase in aggression by reducing the latency response for attack on a second intruder. Of particular interest was the fact that priming stimulation significantly enhanced *c-fos* expression in the corticomedial amygdala. A careful analysis revealed that these increases in *c-fos* expression were specifically associated with aggressive reactions, and not with arousal or motor functions when measured against locomotor activity. Thus, this study showed that in a species other than cat (i.e., hamster), activation of the corticomedial amygdala is associated with facilitation of a defensive form of aggression. Therefore, this finding indicates that the corticomedial amygdala in both the hamster and cat functions in similar ways.

EFFECTS OF SEIZURES

Seizures Induced by Electrical Stimulation of Amygdala

In one early study, Alonso-de Florida and Delgado (1958) stimulated the amygdala of cats for 0.5 seconds every 5 seconds for 1 hour daily for 1 to 15 days. These authors then reported that such stimulation increased playful and contact activities as well as increased or decreased rage responses when placed in a colony of cats. While the authors suggested that the directionality of the effects of seizures on rage behavior is related to the anatomical locus, they did not specify the precise nature of this relationship other than to indicate that sites where seizures were initiated included the corticomedial and basolateral components of amygdala. One significant feature of this study is that it was the first to link seizures of the amygdala to modulation of rage behavior. The authors also noted that the seizure paradigm resulted in the onset of spontaneous seizures, thus suggesting the presence of "kindling-like" phenomenon. The studies described below have clarified and extended this original finding.

The study by Brutus et al. (1986) differed from previous experiments described above in that it sought to determine the effects of seizure discharges generated from different nuclei of the amygdala on defensive rage and predatory attack elicited from the hypothalamus of the cat. Initially, these authors identified sites in the amygdala and adjoining pyriform cortex from which subseizure stimulation modulated one or both of these forms of aggression. Then, current thresholds for elicitation of either (or both) defensive rage and predatory attack were identified over a period of 1 to 2 weeks. After these "baseline" thresholds were determined, three to five seizures per day were initiated by stimulation of an amygdaloid modulating site for an additional period of 1 to 4 weeks, which was then followed by a "post-seizure" period of 1 to 2 weeks when no seizures were induced.

The results of this study revealed that seizures induced at sites in the medial amygdala and pyriform cortex associated with facilitation of defensive rage at subseizure levels of stimulation lowered the threshold for elicitation of defensive rage elicited from hypothalamic stimulation alone. In a parallel manner, seizures induced at sites in the central or lateral nuclei of amygdala associated with suppression of defensive rage at subseizure levels of stimulation elevated the threshold for elicitation of defensive rage. In the post-seizure period, threshold responses for elicitation of defensive rage returned to baseline levels, thus demonstrating the potent effects of seizure discharges at functionally identifiable sites in the amygdala on defensive rage, as well as the fact that these seizures had not produced long-term functional damage to the stimulated region (Figure 5.11). Several cats were also tested for the effects of seizure induction at sites that modulated predatory attack behavior. In one cat, stimulation-induced seizures at a site in the basolateral nucleus (which facilitated predatory attack at subseizure levels of stimulation) resulted in an elevation of the threshold for elicitation of this form of aggression. In a second cat, pyriform cortex stimulation at subseizure levels suppressed predatory attack. Seizures induced at this site resulted in a lowering of response thresholds. This latter case was of particular interest since subseizure stimulation had been shown to

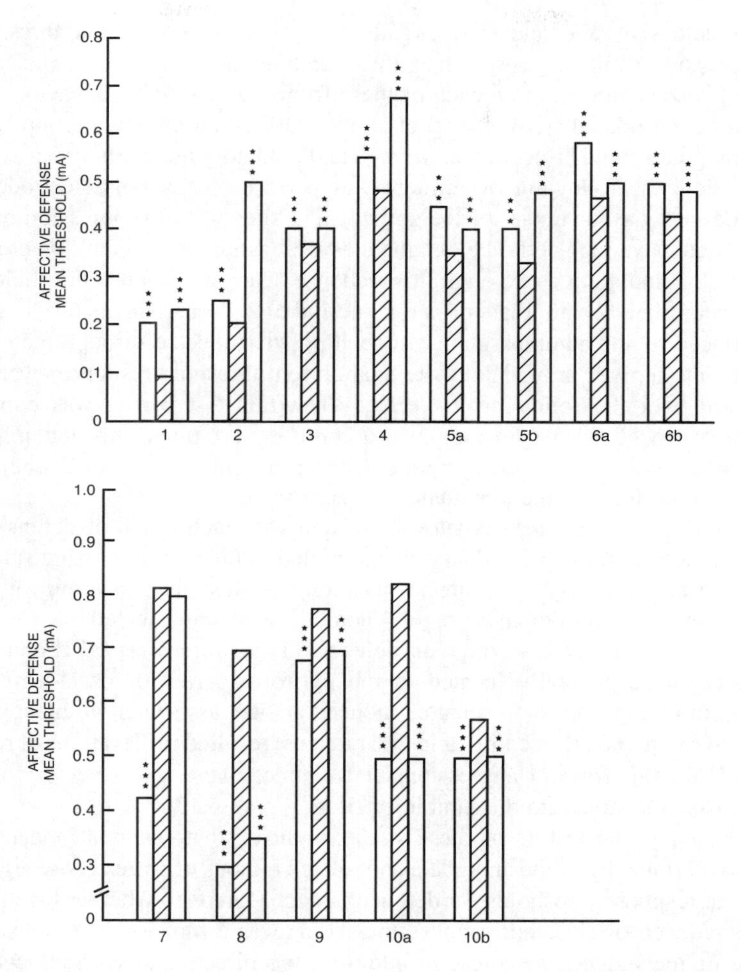

FIGURE 5.11 (A) Bar graphs indicate that current thresholds for defensive rage are lowered (hatched bar) when seizures are induced at sites from which response facilitation had been noted from preliminary experiments. (B) In contrast, bar graphs indicate that defensive rage thresholds are elevated (hatched bar) following seizure induction at sites from which suppression of this response had been previously determined. Note recovery of post-seizure threshold values (shown as the third [open] bar for each three-bar series) toward pre-seizure baseline levels, shown as the first (open) bar for each series. First open bar — mean baseline response threshold determined prior to induction of seizures; middle (hatched) bar — mean response threshold obtained during last 3 days in which seizures were induced; last open bar — mean response threshold obtained during recovery in the post-seizure period; asterisks above first open bar in each three-bar series indicate t-test probabilities comparing defensive rage during pre-seizure sessions and seizure induction sessions; asterisks above third bar indicate t-test probabilities between seizure-induction and post-seizure sessions. Numbers along abscissa indicate the cases examined. Letters a and b indicate different test sites within the same animal. *** $p < 0.001$. NS, not significant. (From C.H. Block, A. Siegel, H.M. Edinger, *Brain Res.*, 197, 1980a, 39–55. With permission.)

facilitate defensive rage, and seizures induced at this site lowered the threshold for the expression of this response. Thus, these findings indicate the specificity of the effects of seizure induction on each of these forms of aggressive behavior.

In a second related study, Siegel et al. (1986) focused on how chronic seizures of the amygdala could affect defensive rage and predatory attack. In order to answer this question, the following experiment was performed. Cannula-electrodes were implanted into various nuclei of the amygdala at sites where suppression or facilitation of defensive rage could be obtained, as well as into the medial hypothalamus from which stimulation could elicit this response. Current response thresholds were initially determined on alternate days for a period of 2 weeks. Then, chronic seizures were induced by the administration of cobalt powder injected through the cannula-electrode into a given amygdaloid site, and current threshold responses were again determined over the following 2 weeks. When this test period was completed, lidocaine hydrochloride (Xylocaine) was administered through an Alzet® minipump into the seizure focus over a 7-day period in order to significantly reduce or eliminate the excitatory effects of the amygdaloid seizure focus.

Chronic seizures induced at sites associated with facilitation of defensive rage (as determined from the preliminary dual stimulation procedure of using subseizure stimulation at the amygdaloid site in question) resulted in a lowering of current threshold responses for defensive rage. When lidocaine was injected into these sites, the effects of the chronic seizures on defensive rage were reversed (Figure 5.12). These sites were generally located in the pyriform cortex or basal amygdaloid complex. In contrast, chronic seizures induced at sites associated with suppression of defensive rage (i.e., the central or lateral nucleus) resulted in elevations of response thresholds for this form of aggression. These effects were also reversed following injections of lidocaine into the inhibitory sites (Figure 5.13).

Collectively, the results of these studies indicate that seizures induced either acutely or chronically of the amygdala and related regions of cortex cause significant changes in response thresholds for defensive rage. Moreover, whether the effects of seizures reduced or elevated response thresholds was a function of the anatomical location of the induced seizures. In addition, the directionality of effects of the seizures is also linked to the form of aggression considered. As indicated above, seizures generated from the lateral or central nucleus of amygdala are associated with lowering of current thresholds for predatory attack and with elevations of response thresholds for defensive rage. The opposite effects are observed following seizures induced at more medial sites within the amygdala as well as pyriform cortex.

Effects of Amygdaloid Seizures on Spontaneously Elicited Attack

The discussion above considered how seizures of different aspects of the amygdala could affect the expression of aggression and rage elicited by electrical stimulation of the hypothalamus. The question may be asked here how seizure activity involving the temporal lobe could affect these forms of aggression when they occur spontaneously. This question was addressed in an extensive series of elegant studies by Adamec and colleagues.

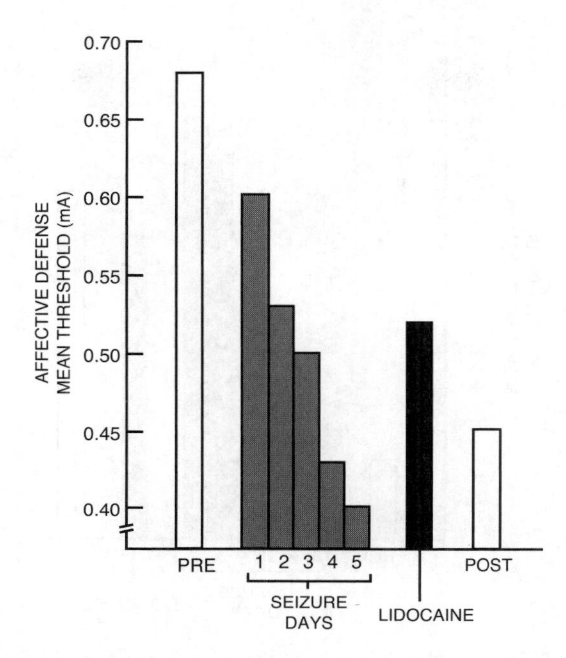

FIGURE 5.12 Bar graphs indicate a reduction in defensive rage thresholds following establishment of a chronic amygdaloid seizure focus. This finding indicates that chronic seizures induced at sites within the amygdala (as determined from preliminary experiments which demonstrated that subseizure stimulation of these sites facilitated defensive rage) also result in a lowering of the current threshold for elicitation of this response. This finding is thus parallel to the results described in Figure 5.11, which demonstrated similar effects following administration of acute seizures. Note that lidocaine, which blocks nerve conduction and which also acts as an anticonvulsant, reverses the effects of cobalt-induced seizures when microinjected into the region of the seizure focus. This figure further shows that current threshold responses are lowered even further over time in the post-seizure period in the absence of lidocaine administration. (From A. Siegel, M. Brutus, M.B. Shaikh, H. Edinger, *Int. J. Neurol.*, 19–20, 1985–1986, 59–73. With permission.)

In an initial set of studies, Adamec observed that cats that spontaneously and vigorously attacked rats would have a greater tendency to approach novel stimuli, but would display weaker responses to threatening stimuli than cats that did not spontaneously attack rats (Adamec, 1975a, 1975b, 1978; Adamec and Stark-Adamec, 1983a, 1983b, 1983c, 1983d, 1985, 1986). Adamec then discovered that after-discharge thresholds from stimulation of the basolateral amygdala were higher in cats that spontaneously attacked rats compared to those which did not spontaneously attack rats. In addition, cats, which displayed the lowest threshold for basal amygdaloid after-discharges, also had the weakest tendencies for the expression of predatory attack. However, these animals were far more sensitive to threat stimuli in the environment. These observations are consistent with the findings of Brutus et al. (1986) described above, which suggest the presence of a neurophysiological

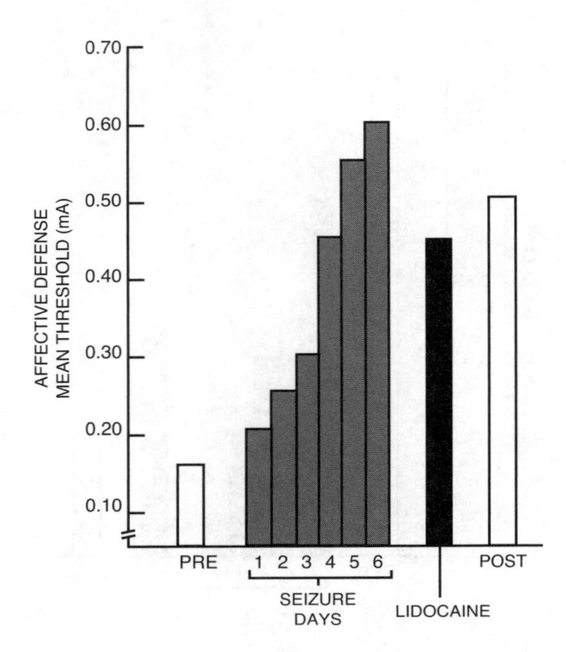

FIGURE 5.13 Bar graphs indicate an elevation in current threshold responses for defensive rage following establishment of a chronic seizure focus with cobalt in the amygdala. Note that lidocaine administration reversed the effects of cobalt seizures, but that the response threshold tended to become elevated in the absence of lidocaine. (From A. Siegel, M. Brutus, M.B. Shaikh, H. Edinger, *Int. J. Neurol.*, 19–20, 1985–1986, 59–73. With permission.)

mechanism within parts of the amygdala relating epileptiform activity with aggressive behavior. Other experiments by Adamec, which are described below, provided further support for this notion.

In one set of studies, Adamec and Stark-Adamec (1983b, 1983c) reported that after-discharges involving other parts of the temporal lobe (i.e., ventral hippocampus) enhanced defensive responses in cats while increasing the latency for biting attack. The question may be raised here why after-discharges elicited from the ventral hippocampus should suppress predatory attack and facilitate defensive responses when subseizure stimulation of this region was reported to produce the opposite results (Siegel and Flynn, 1968; Watson et al., 1983a). Adamec provided an answer to this apparent contradiction by indicating that as a result of seizure discharges induced in the hippocampal formation, there is spread of neural activity to regions of the amygdala such as the corticomedial and adjoining parts of the basal complex that normally facilitate defensive responses and suppress predatory attack. This conclusion was based on the additional observation that seizures induced in the ventral hippocampus in cats also resulted in an increase in the evoked response recorded in the ventromedial hypothalamus of the same animals in response to basal amygdaloid stimulation.

In another set of studies, Adamec and Stark-Adamec (1986) showed that the amplitudes of evoked responses in the ventromedial nucleus were larger following amygdaloid stimulation in cats that were more defensive than the evoked responses in cats that displayed more aggressiveness to a prey object. In a later study reported by Adamec (1990), stimulation of amygdaloid sites that were shown to lower the size of evoked potentials in the medial hypothalamus were also associated with both a reduction in the extent of defensive behavior as well as increases in predatory attack expressed by these animals.

Collectively, the findings of Adamec suggest the following conclusions:

1. The likelihood that a cat will be more defensive or predatory is linked to the threshold for elicitation of seizure discharges in specific regions of the amygdala.
2. The size of the evoked potential recorded in the ventromedial hypothalamus following basal amygdaloid stimulation is related to the degree to which the cat is likely to display defensive behavior, indicating the presence of an underlying neurophysiological mechanism.
3. Long-term changes in defensive behavior may be mediated, in part, by seizure activity initiated in the ventral hippocampus, which is then transmitted to the basal amygdala, thus influencing the neural mechanism governing activity in the ventromedial hypothalamus.
4. The temporal lobe mechanism modulating the expression of defensive behavior in a given direction has an opposing effect on predatory behavior.

Another important point to note here is that the differential effects of amygdaloid excitation on defensive and predatory attack are entirely consistent with the studies described previously in this chapter that involved direct subseizure stimulation of amygdaloid nuclei, as well as distinct regions of the medial and lateral hypothalamus as described in previous chapters. Moreover, the findings clearly reflect the similarity of the effects associated with the amygdaloid stimulation on these two forms of aggression when these responses are elicited either by electrical brain stimulation of the hypothalamus or by natural "spontaneous" occurrence.

PREFRONTAL CORTEX AND CINGULATE GYRUS

Background

The prefrontal cortex and (anterior) cingulate gyrus are considered together because they both share similar anatomical and functional properties. These include the facts that: (1) both regions are characterized mainly by an agranular cortex; (2) both regions project heavily to the mediodorsal thalamic nucleus; and (3) both regions suppress predatory attack behavior. Details concerning these facts are presented below together with a brief review of the literature. While it is beyond the scope of this chapter to consider the many important functions other than aggression and rage, suffice it to say that considerable research has been conducted on the prefrontal cortex, linking it to such processes as emotional behavior and related learning (Pujic et al., 1993; Sato, 1971), delayed responding (Wilcott, 1977), and schizophrenia (Winn, 1994). Thus, the prefrontal cortex may be viewed as a highly complex structure with respect to both its anatomical connections with other regions of the brain and to functions that it mediates.

Anatomical Considerations

A unique feature of the prefrontal cortex is that it receives inputs from all other regions of the cerebral cortex and, in addition, it also receives major inputs from the mediodorsal thalamic nucleus and amygdala (Bacon, 1973; Divac et al., 1978). With respect to the efferent projections of the prefrontal cortex, the primary target regions include the mediodorsal thalamic nucleus, amygdala, and midbrain PAG (Alexander and Fuster, 1973; An et al., 1998; Siegel et al., 1974, 1975, 1977). Other investigators have also described connections from the prefrontal cortex directly to the hypothalamus (Öngür et al., 1998), which may be inhibitory (Kita and Oomura, 1982). From the descriptions provided above, one can see that there are various ways by which pathways originating from the prefrontal cortex may enter the hypothalamus. Nevertheless, the overwhelming numbers of fibers that pass to the diencephalon terminate in the mediodorsal thalamic nucleus (Figure 5.14). Thus, it is reasonable to conclude that these projections are the most likely ones over which the primary modulating effects of the prefrontal cortex are mediated. In fact, this issue is addressed directly in several of the experiments described below (Siegel et al., 1972, 1977).

As noted above, a study of the projection pathways of the cingulate gyrus revealed that its anterior aspect is highly similar to the prefrontal cortex with respect to its projection patterns (Siegel et al., 1973a). Namely, the primary thalamic projection target of this region of cortex is the mediodorsal nucleus. In contrast, more posterior aspects of the cingulate gyrus project to the anteroventral thalamic nucleus.

Relationship to Aggression and Rage

The historical basis for the study of the role of the prefrontal cortex in emotional behavior presumably had its origin with the case reports of a New England railroad construction worker, Phineas Gage (Harlow, 1848, 1868). A heavy iron bar that penetrated his skull struck him, inflicting massive damage of the frontal region of the brain, in part, by destroying many of the anatomical connections between the prefrontal cortex and other regions of the cerebrum as well as with subcortical structures. Thus, the accident resulted in Gage becoming the first prefrontal lobotomy. Following his accident, Gage displayed marked personality changes. Several of these included child-like behavior in intellectual abilities and passions of a strong man. One of the implications of this observation is that the intact prefrontal cortex normally inhibits various forms of emotional behavior, and that removal of such inputs from the prefrontal cortex results in a release of its inhibitory influences. On the other hand, the display of temper tantrums by chimpanzees as a result of making incorrect choices was reduced or eliminated following the cutting of the fiber tracts associated with the prefrontal cortex (see studies cited in Fulton, 1952). In addition, after Moniz introduced prefrontal lobotomy in humans in 1936, Freeman and Watts (1942) reported favorable results from this procedure in treating aggressive

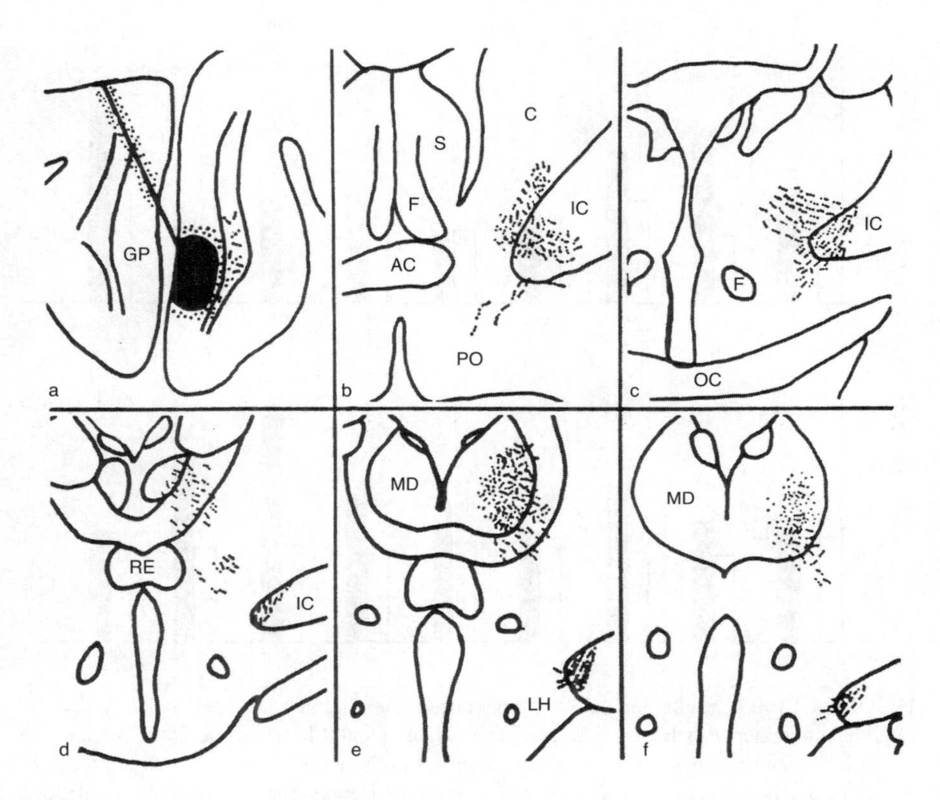

FIGURE 5.14 Course of degeneration following a lesion of the medial aspect of the prefrontal cortex in cat. Note the heavy projection to the lateral segment of the mediodorsal thalamic nucleus and the paucity of degeneration in the hypothalamus. AC, anterior commissure; C, caudate nucleus; F, fornix; GP, gyrus proreus; IC, internal capsule; LH, lateral hypothalamus; MD, mediodorsal thalamic nucleus; OC, optic chiasm; PO, proptic area; RE, nucleus reuniens; S, septal area. (From A. Siegel, M. Brutus, M.B. Shaikh, H. Edinger, *Int. J. Neurol.*, 19–20, 1985–1986, 59–73. With permission.)

schizophrenic patients. In spite of these positive reports, significant ethical as well as treatment concerns about the continued use of prefrontal lobotomies for the treatment of mental disorders were raised over the subsequent two decades. For these reasons coupled with the advent of noninvasive drug therapy for the treatment of severe emotional and aggressive disturbances, the prefrontal lobotomy procedure was ultimately discontinued. Nevertheless, the issues raised here concerning the nature of how the prefrontal cortex affects emotional behavior are significant ones that needed to be addressed experimentally. In the section below, results of several studies concerning this issue are summarized.

One set of studies conducted over a period of several years focused on the method of dual stimulation of the medial and lateral aspects of the prefrontal cortex, and lateral and medial regions of the hypothalamus from which predatory attack and defensive rage, respectively, could be elicited (Siegel et al., 1974, 1975). It was noted that stimulation of either the medial wall or lateral (orbital) aspect of the

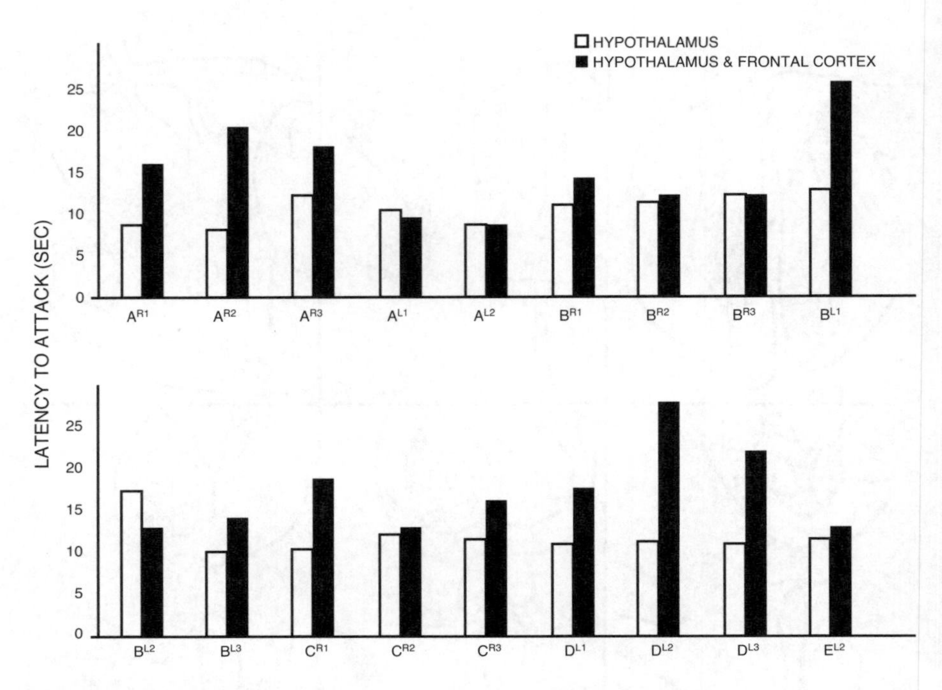

FIGURE 5.15 Bar graphs indicate that concurrent stimulation of the prefrontal cortex and hypothalamus increases latencies to predatory attack; capital letters centered below pairs of bar graphs identify cats used in this study. Superscripts R and L denote right and left side of brain, respectively; superscripts 1, 2, and 3 refer to frontal planes 26.0, 24.5, and 23.0, respectively, in stereotaxic coordinates. (From A. Siegel, H. Edinger, R. Troiano, *Exp. Neurol.*, 38, 1973, 202–217. With permission.)

prefrontal cortex caused profound suppression of predatory attack and defensive rage (Figure 5.15). However, such stimulation had little or no effect on flight behavior elicited from neighboring regions of the hypothalamus, indicating that the suppressive effects of prefrontal cortical stimulation were specific to aggression and rage, and did not reflect a nonspecific motor suppression. In addition, the effects of bilateral stimulation of the prefrontal cortex on spontaneously elicited predatory attack were determined in several cats. It was noted that likewise, stimulation blocked spontaneously elicited predatory attack. However, when the same stimulation parameters were applied against hypothalamically elicited flight behavior or on simple responses, such as the corneal reflex, righting reflex, or vocalization, there was little or no change in any of these responses. It should be pointed out that, while the prefrontal cortex is the only structure studied where stimulation had similar (suppressive) effects on both predatory attack and defensive rage, the suppressive effects of stimulation did not extend to other responses studied, thus allowing these authors to conclude that the suppressive effects of the prefrontal cortex were specifically directed against aggressive responses but not on other forms of behavior tested.

A basic question remaining here concerns how the prefrontal cortex mediates its effects on aggression and rage. This issue was addressed in several studies and is briefly noted above. In one study (Siegel et al., 1974), a small lesion was placed at the tip of an electrode site from which prefrontal cortical stimulation had powerfully suppressed predatory attack, and the degenerating axons from this site were traced to the diencephalon. As shown in Figure 5.14, the major distribution of fibers from this region of cortex was directed principally through the internal capsule to the lateral half of the mediodorsal thalamic nucleus.

In a subsequent study (Siegel et al., 1977), the hypothesis that the suppressive effects of the prefrontal cortex are mediated primarily to the hypothalamus through the mediodorsal thalamic nucleus was tested directly. In this study, the suppressive effects of prefrontal cortical stimulation on hypothalamically elicited predatory attack were initially demonstrated. Then, a lesion was placed in the mediodorsal thalamic nucleus and the effects of prefrontal cortical stimulation on predatory attack were reassessed. The results showed that lesions of the anterior half of the mediodorsal nucleus blocked the suppressive effects of the prefrontal cortex, but that lesions of the posterior half of the mediodorsal nucleus but which spared the anterior aspect of this nucleus had little effect on suppression. Apparently, the locus of the lesion was of importance because it was previously shown that the prefrontal cortical fibers, which project to the mediodorsal nucleus, supply mainly the anterior half of this nucleus. Thus, lesions of the anterior aspect of the mediodorsal nucleus presumably were more effective in blocking the suppressive effects of the prefrontal cortex because such lesions had a much greater effect in disrupting these afferent (prefrontal) fibers than lesions of the posterior half of the mediodorsal nucleus.

With respect to the cingulate gyrus, a study by Siegel and Chabora (1971) used dual stimulation procedures to demonstrate that, similar to the prefrontal cortex, stimulation of the anterior cingulate gyrus suppresses predatory attack, while stimulation of more posterior aspects of this structure have little or no effect on this form of aggression. Based on the anatomical and brain stimulation findings, it may be concluded that the suppressive effects of the anterior cingulate gyrus are mediated through the mediodorsal nucleus to the lateral hypothalamus. Intuitively, it is understandable that the posterior cingulate gyrus has little effect on predatory attack because its primary thalamic projection target — the anteroventral thalamic nucleus — is not known to project to the hypothalamus or have any effect on predatory attack (MacDonnell and Flynn, 1967).

A further question was raised from these studies: How do signals from the prefrontal cortex and anterior cingulate gyrus reach the hypothalamus after they terminate in the mediodorsal thalamic nucleus? This issue was addressed in several earlier studies (Siegel et al., 1972, 1973b). These studies were based on the tracing of degenerating axons resulting from the placement of small lesions in the mediodorsal thalamic nucleus and at different levels of the midline thalamus. The results, summarized in Figure 5.16, indicate that the pathway from the mediodorsal nucleus to the lateral hypothalamus is multisynaptic. It includes synapses in both caudal and rostral aspects of the midline thalamus where the final neuron in this circuit arises from the rostral aspect of the nucleus reuniens, and terminates mainly in the lateral hypothalamus.

An analysis of the role of the prefrontal region of the rat has also been carried out by de Bruin (1983, 1990). These studies revealed several significant findings. In

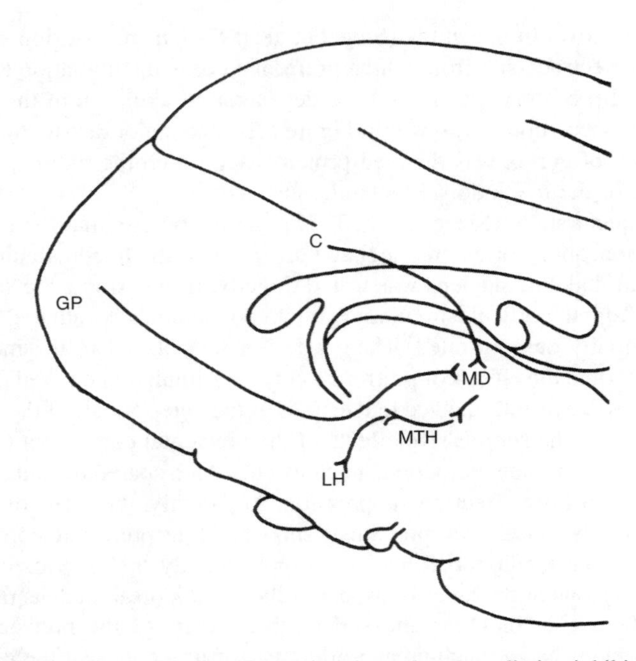

FIGURE 5.16 Schematic diagram illustrating the pathway mediating inhibition of predatory attack from the prefrontal cortex (GP) and anterior cingulate gyrus (C) to the lateral hypothalamus (LH). This pathway involves synaptic relays in the mediodorsal thalamic nucleus (MD) and posterior and anterior midline thalamus (MTH). (From A. Siegel and H. Edinger, Neural control of aggression and rage behavior, in P.J. Morgane and J. Panksepp, Eds., *Handbook of the Hypothalamus*, vol. 3, Marcel Dekker, New York, 1981, pp. 203–240. With permission.)

one experiment, thermal lesions were placed bilaterally in the ventrolateral aspect of the prefrontal region of the rat and dyadic interactions with respect to aggression were determined. The most clearly observed changes in the experimental animals included significant increases in duration of lateral threats and keeping down the opponent. Other aspects of the behavioral sequences examined showed few differences between experimental and control groups, leading de Bruin to suggest that the orbitofrontal lesions affect primarily the mechanism governing aggression and not other features such as social interactions.

A number of hypotheses could be proposed to account for the observed findings. The hypothesis proposed by de Bruin was that enhanced aggressiveness following the lesions was due to reduced levels of dopamine present in the prefrontal cortex, which normally receives dopaminergic inputs from the ventral tegmental area. In order to test this hypothesis, 6-hydroxydopamine (6-OHDA), which depletes dopamine, was bilaterally microinjected into the prefrontal cortex of other groups of rats and the effects of injection of this neurotoxin on dyadic encounters and dopamine levels in the prefrontal cortex were assessed. The results revealed that, following administration of 6-OHDA, there were significant increases in the duration of both lateral threats and keeping the opponent down relative to control animals. In addition, immunocytochemical staining of the brain tissue suggested that dopamine levels were reduced in the prefrontal regions where microinjections of 6-OHDA were placed. Overall, these findings are consistent with the data described above in the cat and indicate that the prefrontal cortex normally provides significant inhibition of aggressive reactions in several species. Moreover, these findings further suggest that dopaminergic inputs to the prefrontal cortex are necessary for the prefrontal cortex to act in its "normal" manner by inhibiting the mechanism governing aggression, which is presumably maintained at the level of the hypothalamus or PAG.

RELATED STRUCTURES OF LIMBIC FOREBRAIN: SUBSTANTIA INNOMINATA AND BED NUCLEUS OF STRIA TERMINALIS

In addition to the limbic structures already considered, two additional nuclear groups — the bed nucleus of the stria terminalis (BNST) and substantia innominata — are reviewed here because of their anatomical connections with limbic structures, hypothalamus, or PAG, and because they have been shown to modulate aggression and rage.

Bed Nucleus of Stria Terminalis

Anatomical Considerations

The BNST consists of a band of neurons that are located approximately at the level of the anterior commissure just below the lateral ventricle, and at the confluence of the ventromedial and ventrolateral aspects of the caudate nucleus and septal area, respectively (see Chapter 6). That the BNST receives significant inputs mainly from the central nucleus of amygdala (and from other amygdaloid nuclei as well) is well documented from both anatomical (Beltramino et al., 1996; Petrovich and Swanson, 1997; Sun et al., 1991) and electrophysiological (Dalsass and Siegel, 1987) studies. In turn, the BNST projects heavily to the region of the ventromedial hypothalamus immediately below the fornix as determined from functional 2-DG analysis (Watson et al., 1983c) and electrophysiology (Dalsass and Siegel, 1987), and to the PAG and related regions of the brainstem as determined from anatomical tracing methods (Holstege et al., 1985). Thus, the input–output relationships of the BNST endow it with the capacity to receive information concerning visceral, autonomic, and emotional functions and to influence the primary regions of the diencephalon and brainstem mediating rage and aggression. The results of an experiment described below provide evidence of BNST involvement in modulating rage and aggression.

Relationship to Aggression and Rage

In this study, a dual stimulation paradigm involved concurrent stimulation of the BNST and a medial or lateral hypothalamic site from which defensive rage or predatory attack could be elicited, respectively (Shaikh et al., 1986). As shown in Figure 5.17, stimulation of the BNST facilitated the occurrence of defensive rage

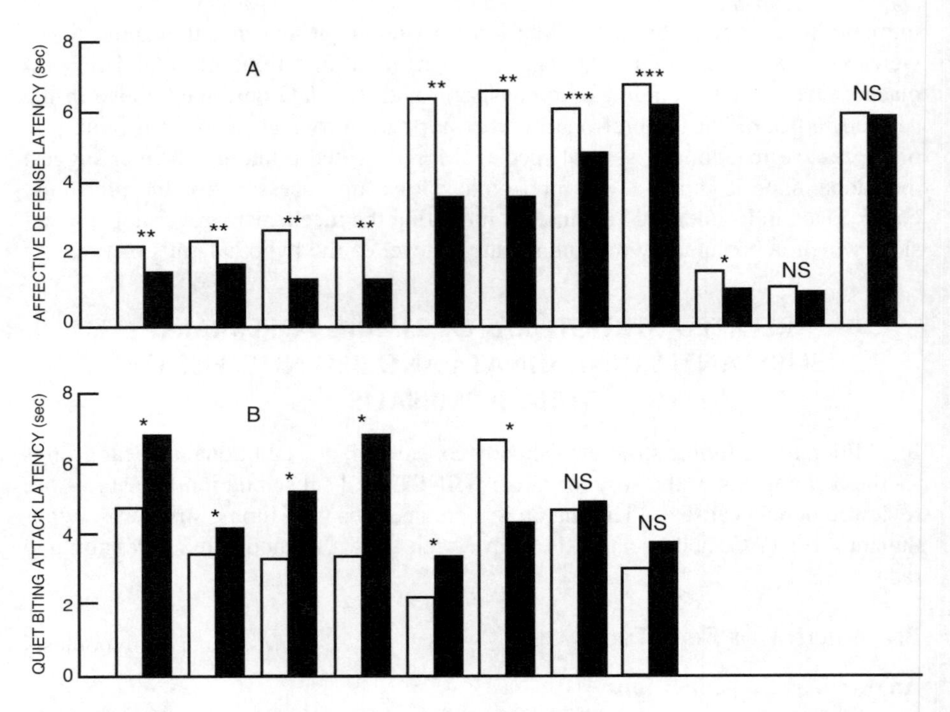

FIGURE 5.17 Bar graphs indicate that concurrent stimulation of the hypothalamus and BNST alters latencies for hypothalamically elicited defensive rage (panel A) and predatory attack (panel B). For each pair of histograms, the first (hatched) bar corresponds to the mean response latency following single stimulation of the hypothalamus alone; the second (cross-hatched) bar represents the mean response latency following dual stimulation of the hypothalamus and BNST. * $p < 0.05$; ** $p < 0.01$, *** $p < 0.001$. NS, not significant. (From M.B. Shaikh, M. Brutus, A. Siegel, H.E. Siegel, *Brain Res. Bull.*, 16, 1986, 179–182. With permission.)

in 9 of 11 sites tested, but suppressed predatory attack in 6 of 8 sites tested. These findings parallel those described earlier in this chapter concerning the effects of stimulation of the medial amygdala and related regions of the amygdala, which give rise to neurons forming the stria terminalis. Thus, one possible way of interpreting the findings from BNST stimulation is that this structure represents a relay nucleus of selective regions of amygdala that give rise to the stria terminalis.

Substantia Innominata

Anatomical Considerations

The substantia innominata is a region of the basal forebrain whose medial border is the lateral hypothalamus and preoptic region, and whose lateral border is the anterior third of amygdala. Dorsal to the substantia innominata lies the internal capsule and striatum, and its ventral aspect forms the floor of the forebrain. It consists of groups of nuclei and fiber tracts. The majority of cells is amorphous in appearance and does not form distinct nuclear groups. One group of cells, however, is large and quite prominent. It is situated near the base of the brain and is called the basal nucleus of Meynert. One characteristic of the basal nucleus and adjacent neurons is that it has ubiquitous projections to wide regions of the cerebral cortex, hippocampal formation, and in particular, the amygdala (Femano et al., 1979). With respect to this chapter, it is of special interest to note that the medial aspect of the substantia innominata proximal to the preoptic region contains descending projections that are highly similar to the preoptic region. In brief, such fibers enter the medial forebrain bundle and descend in this pathway, terminating at different levels throughout the lateral hypothalamus (Troiano and Siegel, 1978; Watson et al., 1983c). It is reasonable to suggest that this region of the substantia innominata may serve as an interneuron for transmission of signals from the lateral aspect of amygdala to the lateral hypothalamus.

Relationship to Aggression

As indicated above, the substantia innominata likely serves as a relay nucleus of certain cell groups of the amygdala such as the central and lateral nuclei, which have been shown to facilitate predatory attack. On this basis, it is reasonable to assume that the regions of substantia innominata receiving inputs from these regions of amygdala would likewise facilitate this form of feline aggression. This hypothesis was directly tested in a study by Block et al. (1980b). In this study, stimulating electrodes were placed into both the lateral and medial aspects of the substantia innominata as well as the lateral hypothalamus from which predatory attack could be elicited. The methods employed were identical to the analysis of the modulating effects of the amygdaloid nuclei (Block et al., 1980a) and hippocampal formation (Watson et al., 1983a) on aggression and rage described earlier in the chapter.

The results indicate that stimulation of the medial half of the substantia innominata facilitated predatory attack and expanded the size of the trigeminal sensory fields along the lipline activated by sensory stimulation. On the other hand, stimulation of the lateral aspect of the substantia innominata suppressed predatory attack, and caused a constriction of the same sensory fields. These findings indicate that the substantia innominata plays an important role in modulating predatory attack, and it appears to do so, at least in part, by acting on the trigeminal sensory fields that control jaw closing and opening reflexes associated with predation. There are nevertheless a number of issues that remain unanswered from these studies:

How does the lateral aspect of the substantia innominata suppress predatory attack?

To what extent are different regions of the substantia innominata critical for amygdaloid modulation of predatory attack?

How (if at all) would the substantia innominata modulate defensive rage behavior?

Answers to these questions would certainly help to clarify our understanding of the mechanisms by which the substantia innominata modulates aggressive reactions.

LIMBIC SYSTEM AND HUMAN AGGRESSION

While most studies dealing with limbic system functions have been conducted in animals, there has also been an increasing number of clinical reports detailing the relationship between limbic system dysfunctions and emotional disorders, including rage and aggression. Some of these reports are summarized briefly below.

The aim of most studies has been uncovering the relationship between dysfunctions associated with the temporal lobe and its underlying structures (i.e., hippocampal formation and amygdala) and violence and aggression. In general, many of these individuals displayed marked outbursts of anger and violence to what are considered innocuous statements by another individual. In addition, violent episodes may even occur in such individuals in the absence of any clear-cut identifiable threatening stimuli. Because of the nature of these characteristic responses, such displays have been referred to as the "episodic dyscontrol syndrome" by Monroe (1978). To site several examples of this syndrome, Martinius (1983) provided a report of a 14-year-old boy who brutally murdered an 8-year-old boy following an argument, and who was found to have a tumor bordering on the hippocampal formation. This author further reported that a tumor was situated in a similar region of hippocampal formation in Charles Whitman, the individual who killed 16 people and wounded 32 others from a tower at the University of Texas in Austin.

Other studies provided further support for the association between temporal lobe dysfunction and violence and aggression. Serafetinides (1965) described the results of a study involving 100 medial temporal lobe patients, 36% of whom displayed overt physical aggressiveness. Following temporal lobectomy, long-term follow-up studies resulted in significant reductions in aggressivity. Parallel findings were also reported in association with surgery involving the amygdala and adjoining periamygdaloid cortex (Hood et al., 1983). In more recent times, with the advent of magnetic resonance imaging techniques, Tonkonogy and Geller (1992) was able to demonstrate that lesions localized to the anterior inferior aspect of the temporal lobe in 5 of 14 patients were associated with violent episodes of rage behavior.

Thus, these studies point to the likelihood that structures situated deep within the temporal lobe — namely, the hippocampal formation and amygdala — including adjoining regions of the periamygdaloid, entorhinal, and prepyrifom cortices, play significant roles in the regulation of rage and aggression in a manner similar to that described above for these structures in regulating aggression in animals. It should also be noted that other regions of the brain associated with the limbic system also contribute significantly to the organization and control of human aggression. Perhaps the most important region studied here is the hypothalamus. A number of studies have described the effects of tumors (Alpers, 1940; Reeves and Plum, 1969), hematomas (Berkovic et al., 1988), or cholinergic receptor dysfunction (Devinsky et al., 1992) in the hypothalamus on aggressive behavior. These studies demonstrated that such disturbances of the hypothalamus resulted in displays of violence, rage, and aggression. These behaviors included stabbing, biting of the victim, and murder. These and related studies are summarized in Table 5.1.

TABLE 5.1
Violence in Humans Resulting from Dysfunction of Temporal Lobe and Hypothalamus

Citation	Type of Dysfunction	Behavioral Manifestations	Comment
Serafetinides, 1965	TLE, medial temporal sclerosis with tumors	Character disorders with aggressive outbursts	Reduction in aggression following temporal lobectomy
Malamud, 1967	Temporal lobe tumor	Marked increase in rage and violence with little provocation	Tumors in nine patients displaying rage were located in temporal lobe
Ounsted, 1969	TLE	Hyperkinetic syndrome with rage outbursts	Positive correlation between age, early onset, and occurrence of rage
Sweet et al., 1969	Temporal lobe tumor	Two patients: one tried to kill his family; the other physically attacked family members	Symptoms had disappeared after removal of tumor
Mark and Sweet, 1974	TLE	Violence induced with little provocation; patient stabbed stranger when he was accidentally bumped	Amygdaloid lesion reduced rage behavior
Falconer, 1973	TLE	Interictal aggression	Seven of 12 patients improved after unilateral temporal lobectomy
Hermann et al., 1980	TLE	Psychopathic deviation and hypomania	Chronological age is inversely correlated with aggression scores
Delgado-Escueta et al., 1981	TLE	Aggressive behavior associated with epileptic episodes	Aggression is not directed during seizure
Hood et al., 1983	TLE involving amygdala and periamygdala cortex	Increased aggression and irritable behavior	Amygdalectomy markedly reduced rage behavior
Martinius, 1983	Temporal lobe tumor	Killed 16 people and wounded 32 before police shot him	Malignant tumor adjacent to lateral amygdala
Taylor, 1983	TLE	Epileptic episodes and behavioral problems	Benefits of temporal lobectomy
Vaernet, 1983	TLE	Hyperaggressive behavior	Temporal lobe resection reduced aggression
Birmaher et al., 1993	Low serotonin	Recurrent aggression behavior and impulsivity	Decreased serotonergic activity is associated with aggression; platelet IB is inversely correlated to degree of aggression

TABLE 5.2 (CONTINUED)
Violence in Humans Resulting from Dysfunction of Temporal Lobe and Hypothalamus

Citation	Type of Dysfunction	Behavioral Manifestations	Comment
Alpers, 1940	Anterior hypothalamic tumor	Hyperaggressivity, patient flew into rage with little provocation	Sudden onset of rage associated with development of tumor
Reeves and Plum, 1969	Tumor of ventro-medial hypothalamus	Person had low threshold for violent and aggressive behavior; hit and bit examiner	Behavior paralleled responses seen in animals with lesions of medial hypothalamus, that is, rage and obesity
Berkovic et al., 1988	Hypothalamic hematoma	Uncontrolled violent rage lasting for hours, that is, assaulted teacher and parents with a knife	Author concluded that response paralleled that seen in animal models
Tonkonogy and Geller, 1992	Craniopharyngiomas with lesions of medial hypothalamus	Intermittent explosive rage disorders	Hypothalamic lesions played major role in development of rage
Devinsky et al., 1992	Hypothalamic cholinergic receptor dysfunction	Violent responses with little provocation, that is, multiple stabbing, killing, and biting of victims	Termination of exposure to cholinesterase inhibitor reduced violence
De Reuck, 1983	Cerebral infarct	Convulsive episodes	Convulsive disorder accompanying stroke is considered sign of temporal lobe herniation and brainstem compression
Gedye, 1989	Frontal lobe seizure	Involuntary aggression	Similarity to Tourette's syndrome
Morell, 1993	Genetic mutation on monoamine oxidase-A gene	Periodic aggressive outbursts	Implications of diet or drugs in counteracting mutation effects

TLE, temporal lobe epilepsy.

Source: From K. Schubert and A. Siegel, *Int. J. Group Tensions*, 24, 1994, 237–265. With permission.

SUMMARY

In this chapter, I have provided evidence in both animals and humans of the roles of different components of the limbic system and related nuclei on aggression and rage behavior. The conclusions that may be drawn follow: (1) the regions shown to facilitate predatory attack include the ventral hippocampus, lateral aspects of amygdala, and medial half of the substantia innominata; (2) the regions shown to suppress predatory attack appear to be more widespread and include the prefrontal cortex, dorsal hippocampus, considerable parts of the septal area, lateral aspect of the substantia innominata, and medial amygdala, pyriform cortex, and BNST; (3) a number of regions have been shown to differentially modulate aggression and rage, the most notable of which include the amygdala, BNST, and septal area. Research has shown that sites within these structures that facilitate predatory attack also suppress defensive rage, while sites that suppress predatory attack facilitate the occurrence of defensive rage. The likely mechanism for this mechanism, discussed previously, suggests that direct or indirect inputs from these structures to the medial or lateral hypothalamus excite neurons governing defensive rage and predatory attack, respectively, and that such excitation further causes a suppression of the other form of attack by virtue of reciprocal inhibitory connections between the medial and lateral hypothalamus; and (4) limbic modulation of aggression and rage is mediated by direct or indirect connections from these regions to the hypothalamus or midbrain PAG. These functional anatomical relationships are further considered in Chapter 6.

REFERENCES

Adamec, R.E., The behavioral bases of prolonged suppression of predatory attack in cats, *Aggress. Behav.*, 1, 1975a, 297–314.

Adamec, R.E., The neural basis of prolonged suppression of predatory attack. I. Naturally occurring physiological differences in the limbic systems of killer and non-killer cats, *Aggress. Behav.*, 1, 1975b, 315–330.

Adamec, R.E., Normal and abnormal limbic system mechanisms of emotive biasing, in K.E. Livingston and O. Hornykiewicz, Eds., *Limbic Mechanisms*, Plenum Press, New York, 1978, pp. 405–455.

Adamec, R.E., Role of the amygdala and medial hypothalamus in spontaneous feline aggression and defense, *Aggress. Behav.*, 16, 1990, 207–222.

Adamec, R.E. and C. Stark-Adamec, Limbic control of aggression in the cat, *Prog. Neuro-Psychopharmacol. Biol. Psychiatry*, 7, 1983a, 505–512.

Adamec, R.E. and C. Stark-Adamec, Limbic kindling and animal behavior — implications for human psychopathology associated with complex partial seizures, *Biol. Psychiatry*, 18, 1983b, 269–293.

Adamec, R.E. and C. Stark-Adamec, Partial kindling and emotional bias in the cat: lasting aftereffects of partial kindling of the ventral hippocampus: I. Behavioral changes, *Behav. Neural Biol.*, 38, 1983c, 205–222.

Adamec, R.E. and C. Stark-Adamec, Partial kindling and emotional bias in the cat: lasting aftereffects of partial kindling of the ventral hippocampus: II. Physiological changes, *Behav. Neural Biol.*, 38, 1983d, 223–239.

Adamec, R.E. and C. Stark-Adamec, Kindling and interictal behavior: an animal model of personality change, *Psychiatr. J. (Ottawa)*, 10, 1985, 220–230.

Adamec, R.E. and C. Stark-Adamec, Partial kindling and behavioural change — some rules governing behavioural outcome of repeated limbic seizures, *Kindling* 3, 1986, 195–211.

Albert, D.J. and S.E. Richmond, Hyperreactivity and aggressiveness following infusion of local anesthetic into the lateral septum or surrounding structures, *Behav. Biol.*, 18, 1976, 211–226.

Alexander, G.E. and J.M. Fuster, Effects of cooling prefrontal cortex on cell firing in the nucleus medialis dorsalis, *Brain Res.*, 61, 1973, 93–105.

Allikmets, L. Kh. and M.E. Ditrikh, Effects of lesions of limbic system on emotional reactions and conditioned reflexes in rats, Zh. Vyssh. Nerv. *Deyatel'nosti imen. I.P. Pavlova*, 15, 1965, T1003–T1007.

Alonso-de Florida, F. and J.M.R. Delgado, Lasting behavioral and EEG changes in cats induced by prolonged stimulation of amygdala, *Am. J. Physiol.*, 193, 1958, 223–229.

Alpers, B.J., Personality and emotional disorders associated with hypothalamic lesions, *Res. Publ. Assoc. Nerv. Ment. Dis.*, 20, 1940, 725–752.

An, X., R. Bandler, D. Öngür, J.L. Price, Prefrontal cortical projections to longitudinal columns in the midbrain periaqueductal gray in macaque monkeys, *J. Comp. Neurol.*, 401, 1998, 455–479.

Andy, O.J. and H. Stephan, *The Septum of the Cat*, Charles C. Thomas, Springfield, Ill., 1964.

Bacon, S.J., A.J.N. Headlam, P.L.A. Gabbott, A.D. Smith, Amygdala input to medial prefrontal cortex, mPFC, in the rat: a light and electron microscope study, *Brain Res.*, 720, 1996, 211–219.

Bard, P. and V.B. Mountcastle, Some forebrain mechanisms involved in expression of rage with special reference to suppression of angry behavior, *Assoc. Res. Nerv. Ment. Dis.*, 27, 1948, 362–399.

Bawden, H.N. and R.J. Racine, Effects of bilateral kindling or bilateral sub-threshold stimulation of the amygdala or septum on muricide, ranacide, intraspecific aggression and passive avoidance in the rat, *Physiol. Behav.*, 22, 1979, 115–123.

Beltramino, C.A., M.S. Forbes, D.J. Swanson, G.F. Alheid, L. Heimer, Amygdaloid input to transiently tyrosine hydroxylase immunoreactive neurons in the bed nucleus of the stria terminalis of the rat, *Brain Res.*, 706, 1996, 37–46.

Berkovic, S., F. Andermann, F. Melanson, R. Ethier, W. Feindell, P. Gloor, Hypothalamic hematomas and ictal laughter: evolution of a characteristic epileptic syndrome and diagnostic value of magnetic resonance imaging, *Ann. Neurol.*, 23: 429–439, 1988.

Birmaher, B., M. Stanley, L. Greenhill, J. Twomey, A. Gavrilescu, H. Rabinovich, Platelet imipramine binding in children and adolescents with impulsive behavior, *J. Am. Acad. Child Adolesc. Psychiatr.*, 29, 1993, 914–918.

Blanchard, D.C., R.J. Blanchard, E.M.C. Lee, S. Nakamura, Defensive behaviors in rats following septal and septal-amygdala lesions, *J. Comp. Physiol. Psychol.*, 93, 1979, 378–390.

Block, C.H., A. Siegel, H.M. Edinger, Effects of electrical stimulation of the amygdala upon trigeminal sensory fields established during hypothalamically-elicited quiet biting attack in the cat, *Brain Res.*, 197, 1980a, 39–55.

Block, C.H., A. Siegel, H.M. Edinger, Effects of stimulation of the substantia innominata upon attack behavior elicited from the hypothalamus in the cat, *Brain Res.*, 197, 1980b, 57–74.

Bobo, E.G. and M. Bonvallet, Amygdala and masserteric reflex. I. Facilitation, inhibition and diphasic modifications of the reflex, induced by localized amygdaloid stimulation, *Electroencephalogr. Clin. Neurophysiol.*, 39, 1975, 329–339.

Bonvallet, M. and E.G. Bobo, Amygdala and masseteric reflex, II. Mechanism of the diphasic modifications of the reflex elicited from the "defence reaction area": role of the spinal trigeminal nucleus, Pars oralis, *Electroencephalogr. Clin. Neurophysiol.*, 39, 1975, 341–352.

Brady, J.V. and W.J.H. Nauta, Subcortical mechanisms in emotional behavior: affective changes following septal forebrain lesions in the albino rat, *J. Comp. Physiol. Psychol.*, 46, 1953, 339–346.

Brutus, M., M.B. Shaikh, A. Siegel, H. Edinger, Effects of experimental temporal lobe seizures upon hypothalamically elicited aggressive behavior in the cat, *Brain Res.*, 366, 1986, 53–63.

Brutus, M., M.B. Shaikh, A. Siegel, H.E. Siegel, An analysis of the mechanisms underlying septal area control of hypothalamically elicited aggression in the cat, *Brain Res.*, 310, 1984, 235–248.

Carpenter, M.B. and J. Sutin, *Human Neuroanatomy*, 8th ed., Williams & Wilkins, Baltimore, 1983.

Dalsass, M. and A. Siegel, The bed nucleus of the stria terminalis: electrophysiological properties and responses to amygdaloid and hypothalamic stimulation, *Brain Res.*, 425, 1987, 346–350.

de Bruin, J.P.C., Orbital prefrontal cortex, dopamine, and social agonistic behavior of male Long Evans rats, *Aggress. Behav.*, 16, 1990, 231–248.

de Bruin, J.P.C., H.G.M. Van Oyen, N. Van De Poll, Behavioural changes following lesions of the orbital prefrontal cortex in male rats, *Behav. Brain Res.*, 10, 1983, 209–232.

Delgado-Escueta, A.V., R.H. Mattson, L. King, E.S.Goldensohn, H. Spiegel, J.Madsen, P. Crandall, F. Dreifuss, R.J. Porter, *New Eng. J. Med.*, 305, 1981, 711–716.

De Reuck, J.L., Neuropathology of Epilepsy Resulting from Cerebrovascular Disorders, presented at Advances in Epileptology, XIVth Epilepsy International Symposium, M. Parsonage et al., Eds., Raven Press, New York, 1983, pp. 95–98.

Devinsky, O., J. Kernan, D.M. Bear, Aggressive behavior following exposure to cholinesterase inhibitors, *J. Neuropsychiatry*, 4, 1992, 189–194.

Divac, I., A. Kosmal, A. Bjorklund, O. Lindvall, Subcortical projections to the prefrontal cortex in the rat as revealed by the horseradish peroxidase technique, *Neuroscience*, 3, 1978, 785–796.

Eclancher, F., P. Schmitt, P. Karli, Effets de lesions precoces de l'amygdale sur le developpement de l'agressivite interspecifique du rat, *Physiol. Behav.*, 14, 1975, 277–283.

Egger, M.D. and J.P. Flynn, Effects of electrical stimulation of the amygdala on hypothalamically elicited attack behavior in cats, *J. Neurophysiol.*, 26, 1963, 705–720.

Falconer, M.A., Reversibility by temporal-lobe resection of the behavioral abnormalities of temporal-lobe epilepsy, *N. Engl. J. Med.*, 289, 1973, 451–455.

Femano, P.A., H.M. Edinger, A. Siegel, Evidence of a potent excitatory influence from substantia innominata on basolateral amygdaloid units: a comparison with insular-temporal cortex and lateral olfactory tract stimulation, *Brain Res.*, 177, 1979, 361–366.

Femano, P.A., H.M. Edinger, A. Siegel, The effects of stimulation of substantia innominata and sensory receiving areas of the forebrain upon the activity of neurons within the amygdala of the anesthetized cat, *Brain Res.*, 269, 1983, 119–132.

Freeman, W. and J.W. Watts, *Psychosurgery, Intelligence, Emotion and Social Behavior Following Prefrontal Lobotomy for Mental Disorders*, Charles C. Thomas, Springfield, Ill., 1942.

Fulton, J.F., *The Frontal Lobes and Human Behavior*, The Sherrington Lectures, No. 11, Charles C. Thomas, Springfield, Ill., 1952.

Gedye, A., Episodic rage and aggression attributed to frontal lobe seizures, *J. Ment. Defic. Res.*, 33, 1989, 369–379.

Gotsick, J.E. and R.C. Marshall, Time course of the septal rage syndrome, *Physiol. Behav.*, 9, 1972, 685–687.

Green, J.D., C.D. Clemente, J. de Groot, Experimentally induced epilepsy in the cat with injury of cornu ammonis, *Arch. Neurol., Psychiatry*, 78, 1957, 259–263.

Griffith, N., J. Engel, R. Bandler, Ictal and enduring interictal disturbances in emotional behaviour in an animal model of temporal lobe epilepsy, *Brain Res.*, 400, 1987, 360–364.

Han, Y., M.B. Shaikh, A. Siegel, Medial amygdaloid suppression of predatory attack behavior in the cat: I. Role of a substance P pathway from the medial amygdala to the medial hypothalamus, *Brain Res.*, 716, 1996a, 59–71.

Han, Y., M.B. Shaikh, A. Siegel, Medial amygdaloid suppression of predatory attack behavior in the cat: II. Role of a GABAergic pathway from the medial to the lateral hypothalamus, *Brain Res.*, 716, 1996b, 72–83.

Harlow, J.M., Passage of an iron rod through the head., *Boston Med. Surg. J.*, 39, 1848, 389–393.

Harlow, J.M., Recovery from the passage of an iron bar through the head, *Publ. Mass. Med. Soc.*, 2, 1868, 327–346.

Hermann, B.P., M.S. Schwartz, S. Whitman, W.E. Karnes, Aggression and epilepsy: seizure-type comparisons and high-risk variables, *Epilepsia*, 22, 1980, 691–698.

Holstege, G., L. Meiners, K. Tan, Projections of the bed nucleus of the stria terminalis to the mesencephalon, pons, and medulla oblongata in the cat, *Exp. Brain Res.*, 58, 1985, 379–391.

Hood, T.W., J. Siegfried, H.G. Wieser, The role of stereotactic amygdalotomy in the treatment of temporal lobe epilepsy associated with behavioral disorders, *Appl. Neurophysiol.*, 46, 1983, 19–25.

Kaada, B.R., J. Jansen Jr., P. Andersen, Stimulation of the hippocampus and medial cortical areas in unanesthetized cats, *Neurology* 3, 1953, 844–857.

Karli, P., M. Vergnes, F. Eclancher, C. Penot, Involvement of amygdala in inhibitory control over aggression in the rat: a synopsis, *Aggress. Behav.*, 3, 1977, 157–162.

King, F.A., Effects of septal and amygdaloid lesions on emotional behavior and conditioned avoidance responses in the rat, *J. Nerv. Ment. Dis.*, 126, 1958, 57–63.

Kita, H. and Y. Oomura, Evidence for a glycinergic cortico-lateral hypothalamic inhibitory pathway in the rat, *Brain Res.*, 235, 1982, 131–136.

Kluver, H. and P.C. Bucy, Preliminary analysis of functions of the temporal lobes in monkeys, *Arch. Neurol., Psychiatry*, 42, 1939, 979–1000.

Krayniak, P.F., A. Siegel, R.C. Meibach, D. Fruchtman, M. Scrimenti, Origin of the fornix system in the squirrel monkey, *Brain Res.*, 160, 1979, 401–411.

Krayniak, P.F., S. Weiner, A. Siegel, An analysis of the efferent connections of the septal area in the cat, *Brain Res.*, 189, 1980, 15–29.

Latham, E.E. and B.M. Thorne, Septal damage and muricide: effects of strain and handling, *Physiol. Behav.*, 12, 1974, 521–526.

LeDoux, J.E., *The Emotional Brain*, Simon & Schuster, New York, 1996.

MacDonnell, M.F. and J.P. Flynn, Control of sensory fields by stimulation of hypothalamus, *Science*, 152, 1966, 1406–1408.

MacDonnell, M.F. and J.P. Flynn, Attack elicited by stimulation of the thalamus and adjacent structures of cats, *Behaviour*, 31, 1967, 185–202.

MacDonnell, M.F. and S. Stoddard-Apter, Effects of medial septal stimulation on hypothalamically-elicited intraspecific attack and associated hissing in cats, *Physiol. Behav.*, 21, 1978, 679–683.

MacLean, P.D., Psychosomatic disease and the "visceral brain," *Psychosom. Med.*, 11, 1949, 338–353.

MacLean, P.D. and J.M.R. Delgado, Electrical and chemical stimulation of frontotemporal portion of limbic system in the waking animal, *EEG Clin. Neurophysiol.*, 5, 1953, 91–100.

Malamud, N., Psychiatric disorder with intracranial tumors of limbic system, *Arch. Neurol.*, 17, 1967, 113–123.

Mark, V.H. and W.H. Sweet, The role of limbic brain dysfunction in aggression, *Res. Publ. Assoc. Nerv. Ment. Dis.*, 152, 1974, 186–200.

Martinius, J., Homicide of an aggressive adolescent boy with right temporal lesion: a case report, *Neurosci. Biobehav. Rev.*, 7, 1983, 419–422.

Meibach, R. and A. Siegel, The origin of fornix fibers which project to the mammillary bodies in the rat: a horseradish peroxidase study, *Brain Res.*, 88, 1975, 1–5.

Meibach, R. and A. Siegel, Efferent connections of the hippocampal formation in the rat, *Brain Res.*, 124, 1977a, 197–224.

Meibach, R. and A. Siegel, Efferent connections of the septal area in the rat: an analysis utilizing retrograde and anterograde transport methods, *Brain Res.*, 119, 1977b, 1–20.

Miczek, K.A. and S.P. Grossman, Effects of septal lesions on inter- and intraspecies aggression in rats, *J. Comp. Physiol. Psychol.*, 79, 1972, 37–45.

Monroe, R.R., *Brain Dysfunction in Aggressive Criminals*, Lexington Books, Lexington, Mass., 1978.

Morell, V., Evidence found for a possible "aggression gene", *Science*, 260, 1993, 1722–1723.

Nakao, H., Hypothalamic emotional reactivity after amygdaloid lesions in cats. *Folia Psychiat. Neurol. Jap.*, 14, 1960, 357–366.

Noback, C.R. and R.J. Demarest, *The Human Nervous System*, 2nd ed., McGraw-Hill, New York, 1975.

Ottersen, O.P. and Y. Ben-Ari, Afferent connections to the amygdaloid complex of the rat and cat: I. Projections from the thalamus, *J. Comp. Neurol.*, 187, 1979, 401–424.

Ounsted, C., Aggression and epilepsy rage in children with temporal lobe epilepsy, *J. Psychosom. Res.*, 13, 1969, 237–242.

Öngür, D., X. An, J.L. Price, Prefrontal cortical projections to the hypothalamus in macaque monkeys, *J. Comp. Neurol.*, 401, 1998, 480–505.

Papez, J.W., A proposed mechanism of emotion, *Arch. Neurol. Psychiatry*, 38, 1937, 725–743.

Petrovich, G.D. and L.W. Swanson, Projections from the lateral part of the central amygdalar nucleus to the postulated fear conditioning circuit, *Brain Res.*, 763, 1997, 247–254.

Potegal, M., C.F. Ferris, M. Hebert, J. Meyerhoff, L. Skaredoff, Attack priming in female Syrian golden hamsters is associated with a c-fos-coupled process within the corticomedial amygdala, *Neuroscience*, 75, 1996, 869–880.

Potegal, M., J.L. Gibbons, M. Glusman, Inhibition of muricide by septal stimulation in rats, *Physiol. Behav.*, 24, 1980, 863–867.

Potegal, M., M. Hebert, M. DeCoster, J.L. Meyerhoff, Brief, high-frequency stimulation of the corticomedial amygdala induces a delayed and prolonged increase of aggressiveness in male syrian golden hamsters, *Behav. Neurosci.*, 110, 1996a, 401–412.

Potegal, M., C.F. Ferris, M. Hebert, J. Meyerhoff, L. Skaredoff, Attack priming in female Syrian golden hamsters is associated with a c-fos-coupled process within the corticomedial amygdala, *Neuroscience*, 75, 1996b, 869–880.

Price, J.L., F.T. Russchen, D.G. Amaral, The limbic region II: The amygdaloid complex, in A. Bjorklund and T. Hokfelt, Eds., *Handbook of Chemical Neuroanatomy*, vol. 5, Elsevier, Amsterdam, 1987, pp. 279–388.

Pujic, Z., I. Matsumoto, P.A. Wilce, Extinction of emotional learning: contribution of medial prefrontal cortex, *Neurosci. Lett.*, 162, 1993, 67–70.

Reeves, A.G. and F. Plum, Hyperphagia, rage, and dementia accompanying a ventromedial hypothalamic neoplasm, *Arch. Neurol.*, 20, 1969, 616–624.

Rothfield, L. and P.J. Harman, On the relation of the hippocampal-fornix system to the control of rage responses in cats, *J. Comp. Neurol.*, 101, 1954, 265–282.

Sato, M., Prefrontal cortex and emotional behaviors, *Folia Psychiatr. Neurol. Jpn.*, 25, 1971, 69–78.

Schubert, K. and A. Siegel, What animal studies have taught us about the neurobiology of violence, *Int. J. Group Tensions*, 24, 1994, 237–265.

Schreiner, L. and A. Kling, Behavioral changes following rhinencephalic injury in cat, *J. Neurophysiol.*, 16, 1953, 643–659.

Serafetinides, E.A., Aggressiveness in temporal lobe epileptics and its relation to cerebral dysfunction and environmental factors, *Epilepsia*, 6, 1965, 33–42.

Shaikh, M.B., M. Brutus, A. Siegel, H.E. Siegel, Regulation of feline aggression by the bed nucleus of stria terminalis, *Brain Res. Bull.*, 16, 1986, 179–182.

Shaikh, M.B., C.-L. Lu, A. Siegel, Affective defense behavior elicited from the feline midbrain periaqueductal gray is regulated by mu- and delta-opioid receptors, *Brain Res.*, 557, 1991, 344–348.

Shaikh, M.B., A. Steinberg, A. Siegel, Evidence that substance P is utilized in medial amygdaloid facilitation of defensive rage behavior in the cat, *Brain Res.*, 625, 1993, 283–294.

Shealy, C. and T. Peele, Studies on amygdaloid nucleus of cat, *J. Neurophysiol.*, 20, 1957, 125–139.

Siegel, A., M. Brutus, M.B. Shaikh, H. Edinger, Effects of temporal lobe epileptiform activity upon aggressive behavior in the cat, *Int. J. Neurol.*, 19–20, 1985–1986, 59–73.

Siegel, A. and J. Chabora, Effects of electrical stimulation of the cingulate gyrus upon attack behavior elicited from the hypothalamus in the cat, *Brain Res.*, 32, 1971, 169–177.

Siegel, A. and H. Edinger, Neural control of aggression and rage behavior, in P.J. Morgane and J. Panksepp, Eds., *Handbook of the Hypothalamus*, vol. 3, Marcel Dekker, New York, 1981, pp. 203–240.

Siegel, A., H. Edinger, M. Dotto, Effects of electrical stimulation of the lateral aspect of the prefrontal cortex upon attack behavior in cats, *Brain Res.*, 93, 1975, 473–484.

Siegel, A., H. Edinger, A. Koo, Suppression of attack behavior in the cat by the prefrontal cortex: role of the mediodorsal thalamic nucleus, *Brain Res.*, 127, 1977, 185–190.

Siegel, A., H. Edinger, H. Lowenthal, Effects of electrical stimulation of the medial aspect of the prefrontal cortex upon attack behavior in cats, *Brain Res.*, 66, 1974, 467–479.

Siegel, A., H. Edinger, S. Ohgami, The topographical organization of the hippocampal projection to the septal area: a comparative neuroanatomical analysis in the gerbil, rat, rabbit and cat, *J. Comp. Neurol.*, 157, 1974, 359–378.

Siegel, A., H. Edinger, R. Troiano, The pathway from the mediodorsal nucleus of thalamus to the hypothalamus in the cat, *Exp. Neurol.*, 38, 1973a, 202–217.

Siegel, A. and J.P. Flynn, Differential effects of electrical stimulation and lesions of the hippocampus and adjacent regions upon attack behavior in cats, *Brain Res.*, 7, 1968, 252–267.

Siegel, A., J.P. Flynn, R. Bandler, Thalamic sites and pathways related to elicited attack, *Brain Behav. Evol.*, 6, 1972, 542–555.

Siegel, A., S. Ohgami, H. Edinger, The topographical organization of the hippocampal projection to the septal area in the squirrel monkey, *Brain Res.*, 99, 1975, 247–260.

Siegel, A. and D. Skog, Effects of electrical stimulation of the septum upon attack behavior elicited from the hypothalamus in the cat, *Brain Res.*, 23, 1970, 371–380.

Siegel, A. and J.P. Tassoni, Differential efferent projections from the ventral and dorsal hippocampus of the cat, *Brain Behav. Evol.*, 4, 1971, 185–200.

Siegel, A., R. Troiano, A. Royce, Differential efferent projections of the anterior and posterior cingulate gyrus to the thalamus in the cat, *Exp. Neurol.*, 38, 1973b, 192–201.

Stoddard-Apter, S.L. and M.F. MacDonnell, Septal and amygdalar efferents to the hypothalamus which facilitate hypothalamically elicited intraspecific aggression and associated hissing in the cat: an autoradiographic study, *Brain Res.*, 193, 1980, 19–32.

Sun, N., L. Roberts, M.D. Cassell, Rat central amygdaloid nucleus projections to the bed nucleus of the stria terminalis, *Brain Res. Bull.*, 27, 1991, 1–12.

Swanson, L.W. and G.D. Petrovich, What is the amygdala? *Trends Neurosci.*, 21, 1998, 323–331.

Sweet, W.H., F. Ervin, V.H. Mark, The relationship of violent behaviour to focal cerebral disease, in S. Garratini and E.B. Sigg, Eds., *Aggressive Behavior*, Excerta Medica Foundation, Amsterdam, 1969, pp. 336–352.

Taylor, D.C., Child Behavioural Problems and Temporal Lobe Epilepsy, Proceedings, Advances in Epileptology, XVth Epilepsy International Symposium, 1983, pp. 243–247.

Tonkonogy, J.M. and J.L. Geller, Hypothalamic lesions and intermittent explosive disorder, *J Neuropsychiat. Clin. Neurosci.*, 4, 1992, 45–50.

Troiano, R., C. Boehme, and A. Siegel, A corthicothalamic projection involving the midline septum in the cat, *Brain Res.*, 139, 1978, 348–353.

Troiano, R. and A. Siegel, Efferent connection of the basal forebrain in the cat: II. The substantia innominata, *Exp. Neurol.*, 61, 1978, 185–197.

Vaernet, K., Temporal Lobotomy in Children and Young Adults, Proceedings, Advances in Epileptology, XVth Epilepsy International Symposium, M. Parsonage et al., Eds., Raven Press, New York, 1983, 255–261.

Vergnes, M., Controle amygdalien de comportements d'agression chez le rat, *Physiol. Behav.*, 17, 1976, 439–444.

Vergnes, M., Declenchement de reactions d'agression interspecifique apres lesion amygdalienne chez le rat, *Physiol. Behav.*, 17, 1975, 221–226.

Vochteloo, J.D. and J.M. Koolhaas, Medial amygdala lesions in male rats reduce aggressive behavior: interference with experience, *Physiol. Behav.*, 41, 1987, 99–102.

Wasman, M. and J.P. Flynn, Directed attack behavior during hippocampal seizures, *Arch. Neurol.*, 14, 1966, 408–414.

Watson, R.E., H. Edinger, A. Siegel, An analysis of the mechanisms underlying hippocampal control of hypothalamically-elicited aggression in the cat, *Brain Res.*, 269, 1983a, 327–345.

Watson, R.E., H.E. Edinger, A. Siegel, A 14C-2-deoxyglucose analysis of the functional neural pathways of the limbic forebrain in the rat: III. The hippocampal formation, *Brain Res. Brain Res. Rev.*, 5, 1983b, 133–176.

Watson, R.E., R. Troiano, J.J. Poulakos, S. Weiner, C.H. Block, A. Siegel, A (14C)2-deoxyglucose analysis of the functional neural pathways of the limbic forebrain in the rat: I. The amygdala, *Brain Res. Brain Res. Rev.*, 5, 1983c, 1–44.

Weiss, S.R.B., R.M. Post, P.W. Gold, G. Chrousos, T.L. Sullivan, D. Walker, A. Pert, CRF-induced seizures and behavior: interaction with amygdala kindling, *Brain Res.*, 372, 1986, 345–351.

Wilcott, R., Electrical stimulation in the prefrontal cortex and delayed response in the cat, *Neuropsychologia*, 15, 1977, 115–121.

Winn, P., Schizophrenia research moves to the prefrontal cortex, *Trends Neurosci.*, 17, 1994, 265–268.

Yutzy, D.A., D.R. Meyer, P.M. Meyer, Effects of simultaneous septal and neo- or limbic-cortical lesions upon emotionality in the rat., *Brain Res.*, 5, 1967, 452–458.

Zagrodzka, J., C.E. Hedberg, G.L. Mann, A.R. Morrison, Contrasting expressions of aggressive behavior released by lesions of the central nucleus of the amygdala during wakefulness and rapid eye movement sleep without atonia in cats, *Behav. Neurosci.*, 112, 1998, 589–602.

6 Limbic System II: Functional Neuroanatomy

Chapter 5 focused on the anatomical, physiological, and behavioral properties of limbic structures with respect to their roles in the regulation of aggression and rage behavior. A large number of the studies that contributed to our knowledge linking the limbic system to aggression and rage were based on the application of mainly electrical, and to a lesser extent, chemical stimulation of these structures. Accordingly, a series of studies was conducted utilizing microstimulation of different regions of the limbic system in conjunction with (^{14}C)2-deoxyglucose (2-DG) autoradiography. As pointed out in earlier chapters, a close relationship exists between functional activity and incorporation of 2-DG into neurons. For this reason, this method was used to identify the regions of the brain that are demonstrably activated following discrete stimulation of different sites within each of the limbic structures. This information was deemed valuable in providing insight into the anatomical substrates involved in limbic modulation of aggression and rage behavior.

This chapter describes the results of mapping studies of the regions that are metabolically activated in the rat brain following electrical stimulation of the different components of the limbic system. Limbic structures from which stimulation was applied and described below include the hippocampal formation, septal area, amygdala, and prefrontal cortex. It should be pointed out that this method was not intended to replace neuroanatomical methods. Instead, the results should be viewed as reflecting a means by which the major regions activated as a result of stimulation of a given site within the limbic system can be identified.

METHODS USED

The methods employed for these studies involving different components of the limbic system were basically identical and were described in Chapter 3. Details concerning this methodology and related issues related to limbic system anatomy and function can be found in Brutus et al. (1984), Meibach and Siegel (1977), Siegel and Edinger (1976), Siegel et al. (1974, 1975), Siegel and Tassoni (1971), and Watson et al. (1982, 1983a, 1983b, 1985). In brief, adult rats were anesthetized with ketamine or pentobarbital, and a femoral vein was cannulated in which the patency of the cannula was maintained with heparinized isotonic saline. The paradigm consisted of electrical stimulation, which was delivered continuously for periods of 30 seconds on and 30 seconds off for both 5 minutes preceding the injection of 2-DG and for 45 minutes following the injection of 2-DG. Current values were generally maintained at approximately 0.2 mA at 60 Hz as a balanced biphasic pulse with 1 millisecond per half-cycle duration. The findings presented below represent principally the results of stimulation at current intensities that did not induce any seizure activity. At the completion of the 45-minute stimulation procedure, the animal was sacrificed, the brain was removed, and quickly frozen. Brain sections were cut and mounted on coverslips, which were then serially mounted on cardboard. X-ray films were then placed over each cardboard array of sections with the emulsion side apposed to the tissue and sealed in a light-tight case in the darkroom. Brain sections were exposed for a period of 4 weeks, and the x-rays were then developed. The optical densities present on the x-rays were a direct function of the relative levels of 2-DG uptake among the structures in the brain, which, in turn, is a reflection of the degree of metabolic activation in these structures.

STUDY RESULTS

Hippocampal Formation

Principal Findings

For analysis of the hippocampal formation, stimulation was applied in separate animals at different sites along its longitudinal axis (Figure 6.1). Consider first the distribution of 2-DG labeled regions following stimulation of the *dorsal hippocampus* of the rat, an example of which is shown in Figure 6.2. A principal finding was that stimulation of the dorsal hippocampal formation resulted in bilateral metabolic activation of only the dorsal hippocampus and a restricted region of the dorsomedial aspect of the lateral septal nucleus (Figure 6.3). When stimulation was applied at more posterior levels of the dorsal hippocampus, a similar pattern of labeling in these regions was also observed. Further, it did not matter whether the stimulating electrode was placed in the regio inferior (CA3) or regio superior (CA1) of the dorsal hippocampus, since similar patterns of labeling resulted from stimulation of each of these regions.

Stimulation of sites at progressively more distal to the septal pole of the hippocampus (i.e., which corresponds to sites located progressively more proximal to the temporal pole) resulted in activation patterns that were localized to progressively more ventrolateral aspects within the septal area (Figure 6.4). This functional topographical organization of hippocampal outputs to the septal area corresponds quite closely to the anatomical findings described in Chapter 5 (Meibach and Siegel, 1977; Siegel and Edinger, 1976; Siegel et al., 1974, 1975; Siegel and Tassoni, 1971). It is of further interest to note that when stimulation was applied to the subicular cortex of either the dorsal or ventral hippocampal formation, prominent labeling within the mammillary bodies was noted. In addition, stimulation of the ventral subicular cortex resulted in activation of the medial corticohypothalamic tract and its target region in the ventromedial hypothalamus extending from the suprachiasmatic nucleus to the arcuate region (Figure 6.4). Of further interest, stimulation of this region also produced activation of the stria terminalis and amygdalo-hippocampal area, as well as the medial, cortical, and basolateral amygdaloid nuclei. A summary of these findings is shown in Figure 6.5.

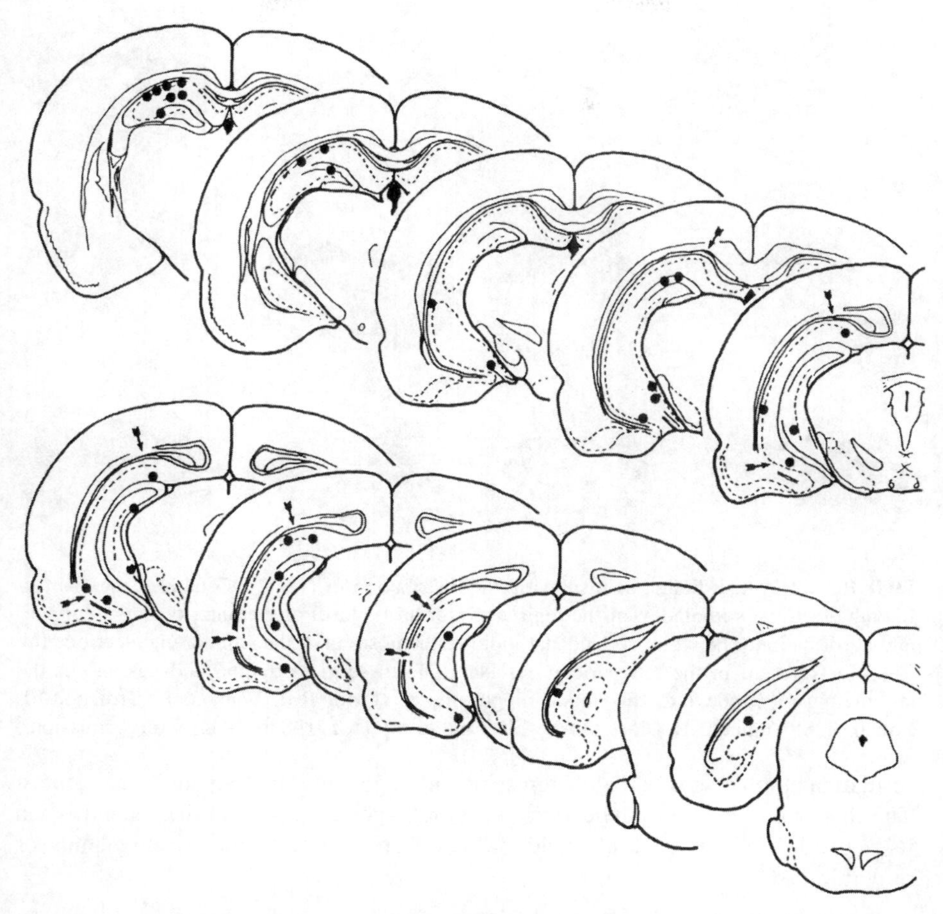

FIGURE 6.1 Map of representative sites electrically stimulated in the hippocampal formation. Not all sites stimulated within the regio superior of the dorsal hippocampal formation are depicted due to the considerable degree of overlap. Arrows delimit border between the cornu Ammonis and subicular cortex in this and all subsequent line diagrams.

Conclusions

The results described above suggest the following conclusions. First, a primary target of the hippocampal formation is the septal area. The medial-to-lateral zones of the septal area activated are determined by the loci of the sites along the longitudinal axis of the hippocampal formation extending from the septal pole to its temporal limit. Thus, this finding suggests that a most likely route by which the hippocampal formation modulates aggression and rage is via its connections with the septal area. Second, activation of the mammillary bodies occurs following stimulation of the dorsal and ventral subicular cortices. However, specific activation of the ventromedial hypothalamus and nuclei of the amygdala takes place following stimulation of the ventral subicular cortex. This would further suggest that the ventral hippocampal

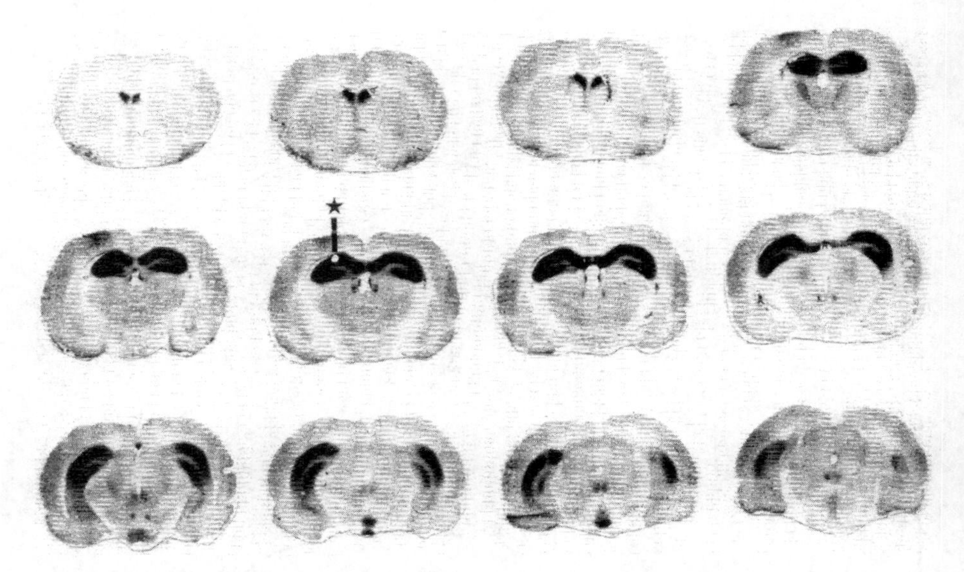

FIGURE 6.2 Autoradiographs displaying the distribution of (^{14}C)2-DG labeled regions following electrical stimulation of the regio superior at the level of the anterior dorsal hippocampal formation. The white dot and star indicate the position of the stimulating electrode tip. The label present in the left parietal cortex is due to surgical trauma and passage of the stimulating electrode into the dorsal hippocampus. (From R.E. Watson, R. Troiano, J.J. Poulakos, S. Weiner, C.H. Block, A. Siegel, *Brain Res. Rev.*, 5, 1983b, 1–44. With permission.)

formation could also modulate aggression and rage either by two additional routes. The first is by its direct connections with the hypothalamus, and the second is via its activation of amygdaloid nuclei, which then project to the hypothalamus or midbrain PAG.

Septal Area

Principal Findings

For this study, stimulation was applied within different zones of the lateral septal area, medial septal nucleus, septofimbrial nucleus, and vertical and horizontal limbs of the diagonal band of Broca (Figure 6.6). Consider first the regions activated following stimulation of the intermediolateral aspect of the lateral septal nucleus (Figure 6.7). Two major patterns of activation should be noted. First, stimulation of the septal area activated wide regions of both the dorsal and ventral hippocampal formation bilaterally. The pattern of activation included all of the CA fields with minimal involvement of the subicular cortex. Additional label was also seen in the ipsilateral basolateral nucleus of amygdala. A second pattern involved activation of the medial septal nucleus, vertical and horizontal limbs of the diagonal band of Broca, preoptic region, and ventral and medial hypothalamus, including the mamillary bodies. Similar patterns of labeling (with the exception of the amygdala) were seen after stimulation was applied to the septofimbrial nucleus (Figure 6.8). The

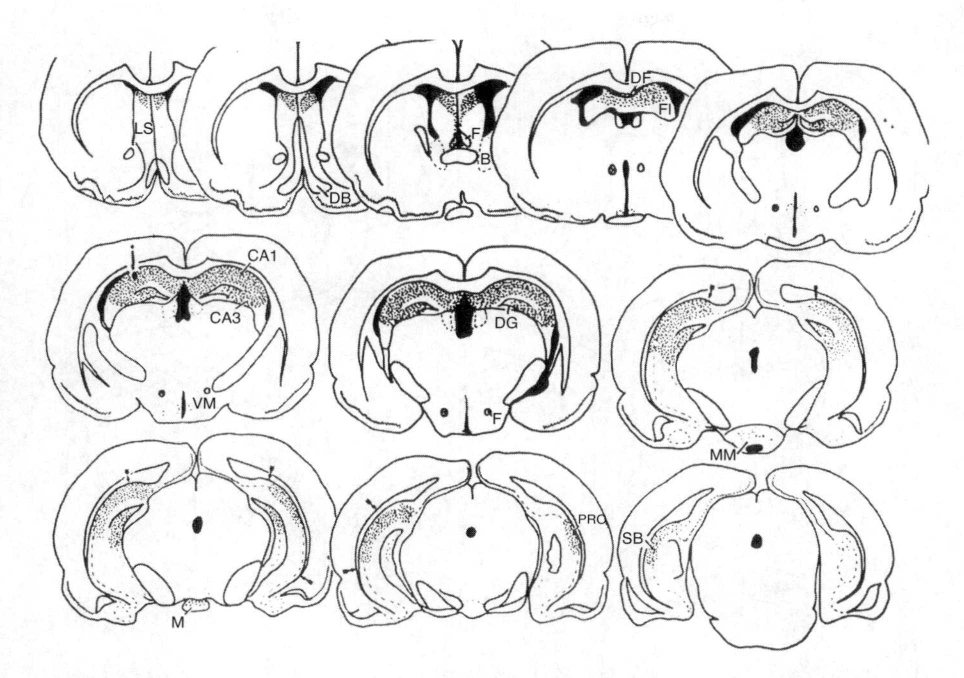

FIGURE 6.3 Distribution of ^{14}C-2DG labeled regions following electrical stimulation of the regio superior at the level of the anterior dorsal hippocampal formation. The label shown in this and subsequent line diagrams depicts only regions metabolically activated as a result of the applied stimulus and does not include regions normally associated with augmented labeling in control brains. (From R.E. Watson, R. Troiano, J.J. Poulakos, S. Weiner, C.H. Block, A. Siegel, *Brain Res. Rev.*, 5, 1983b, 1–44. With permission.)

patterns of labeling observed following stimulation of the medial septal nucleus, vertical limb of the diagonal band of Broca and horizontal limb of the diagonal band were highly similar (Figures 6.9 and 6.10). Here, the basic sites of activation included the medial forebrain bundle and lateral hypothalamus.

Conclusions

Several important conclusions concerning functions of the septal area are evident from these findings. As noted above, the first is that the lateral septal area serves as a relay of the hippocampal formation to the hypothalamus. This relay can be either through direct projections to the hypothalamus, or indirectly, via projections from the lateral septal nucleus to the vertical and horizontal limbs of the diagonal band of Broca. The second is that the septal area can directly activate wide areas of the hippocampal formation. In this manner, the hippocampal formation and septal area may be viewed as components of a single system. Namely, the hippocampal formation can modulate the functions of the hypothalamus via its connections with the septal area, and, in turn, the projections back to the hippocampal formation provide a "feedback" mechanism for regulation of this circuit.

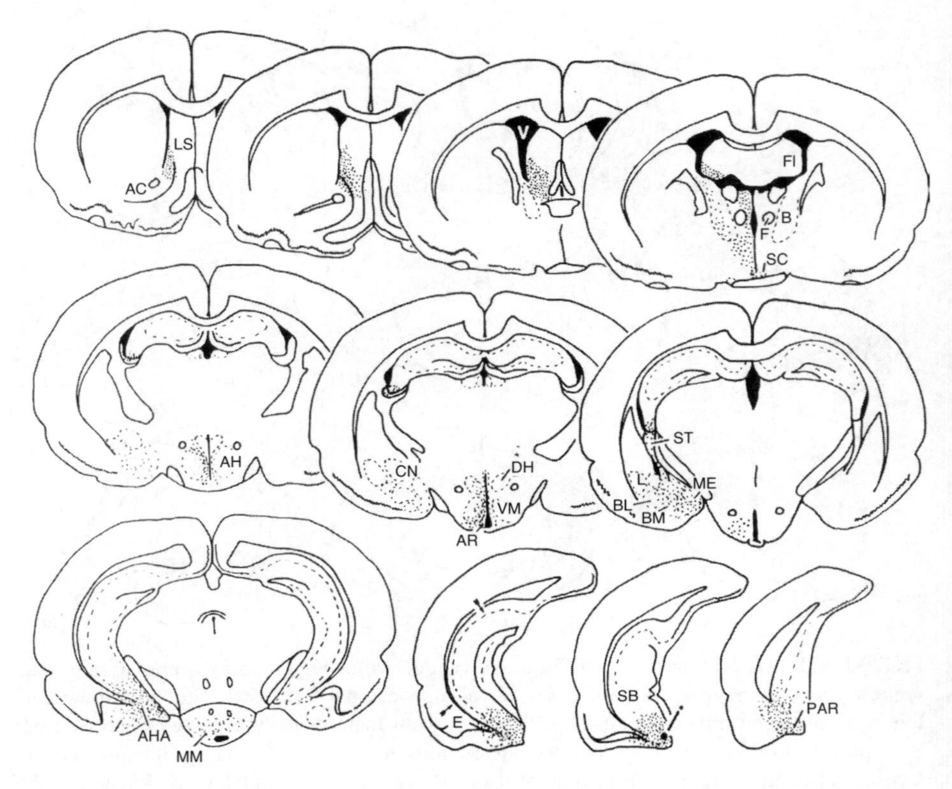

FIGURE 6.4 Distribution of label following electrical stimulation of the presubiculum at the level of the ventral hippocampal formation. Note that label is present over the ipsilateral forebrain and is extensive over the ventrolateral aspect of the lateral septal nucleus. Clearly defined label is also present over the bed nucleus of the stria terminalis, medial corticohypothalamic tract, and ventromedial hypothalamus. There is also diffuse label over the medial half of the amygdala. (From R.E. Watson, R. Troiano, J.J. Poulakos, S. Weiner, C.H. Block, A. Siegel, *Brain Res. Rev.*, 5, 1983b, 1–44. With permission.)

Amygdala

Principal Findings

Stimulation was applied at sites within the principal nuclei of the amygdala as well as the adjoining pyriform cortex (Figure 6.11).

Basomedial Nucleus

Following stimulation of the basomedial nucleus, activation was seen in both the stria terminalis and ventral amygdalofugal pathways. One band of activation extended through the substantia innominata into the bed nucleus of the stria terminalis and adjoining septal area. A second band of activation could be seen passing caudally in an intermediolateral position within the ventral aspect of the hypothalamus to the level of

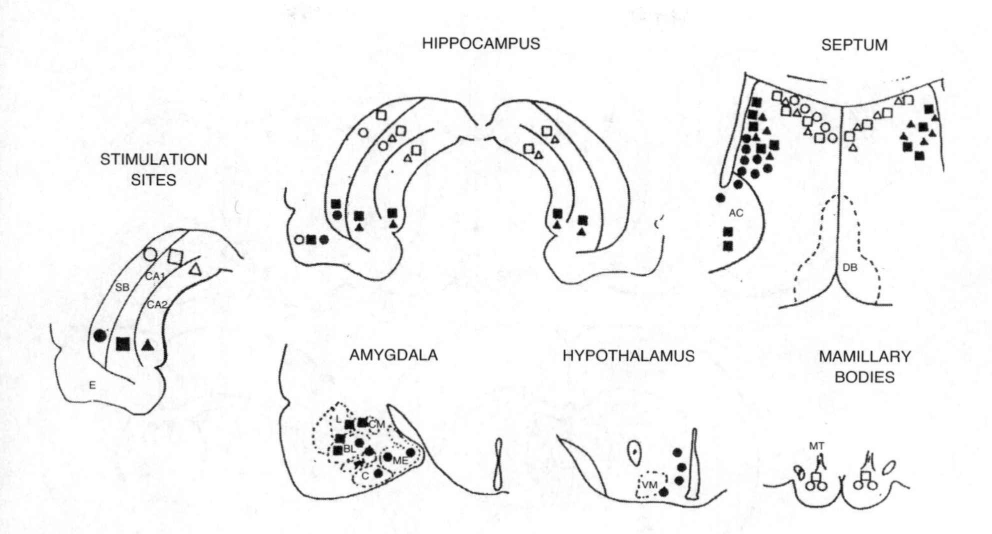

FIGURE 6.5 Summary diagram depicting the organization of systems activated following electrical stimulation of different components of the hippocampal formation. Stimulation is applied to the hippocampal formation as shown in the panel labeled "stimulation sites." The distribution of the symbols in the subsequent figures depicts the distribution of ^{14}C-2DG following stimulation of each hippocampal component. Open circles, dorsal subicular cortex; open squares, dorsal CA1 field; open triangles, dorsal CA3 field; filled circles, ventral subicular cortex; filled squares, ventral CA1 field; filled triangles, ventral CA3 field. (From R.E. Watson, R. Troiano, J.J. Poulakos, S. Weiner, C.H. Block, A. Siegel, *Brain Res. Rev.*, 5, 1983b, 1–44. With permission.)

the mammillary bodies. A third band of activation extended to the pyriform cortex at the level where stimulation was applied and more caudally in the entorhinal cortex and ventral and posterior aspects of the subicular cortex (Figure 6.12).

Cortical Nucleus

Stimulation of the cortical amygdaloid nucleus resulted in activation patterns that were not significantly different than those described for the basomedial nucleus. In brief, one band of labeling included the substantia innominata, bed nucleus of the stria terminalis, and lateral septal nucleus. A second band involved the stria terminalis and ventromedial hypothalamus; and a third band of activation extended to the pyriform cortex and adjoining entorhinal and subicular cortices.

Medial Nucleus

Stimulation of the medial nucleus produced the most intense labeling of the stria terminalis, bed nucleus of the stria terminalis, lateral septal nucleus, and ventromedial nucleus, and adjacent portions of the ventromedial hypothalamus. Additional labeling was also noted in the adjoining cortical nucleus, entorhinal cortex, and ventral posterior aspect of the subicular cortex.

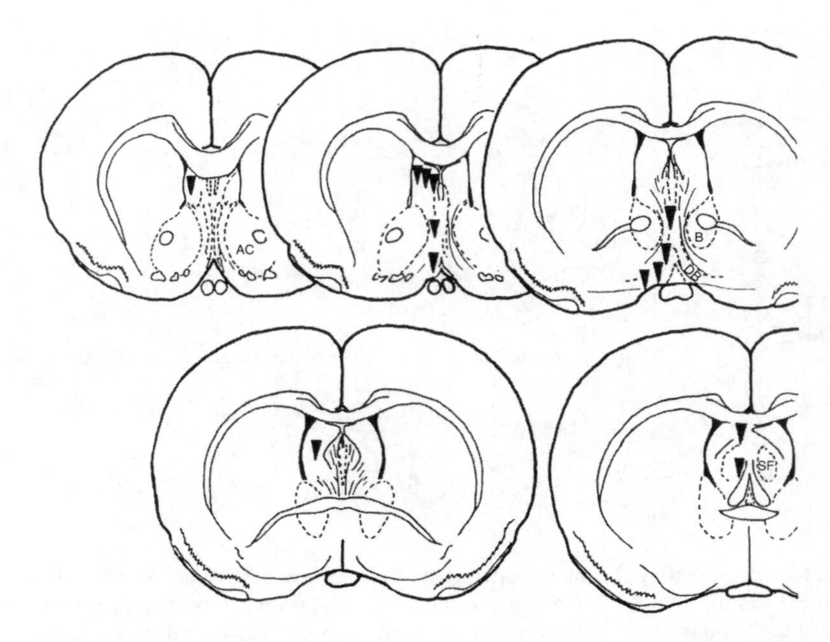

FIGURE 6.6 Map of sites electrically stimulated within the septal area. Not all stimulation sites within the lateral septal nucleus are shown due to considerable degree of overlap. Regions analyzed within the septal region include the lateral septal nucleus, medial septal nucleus, septofimbrial nucleus, and region of the diagonal band. (From R.E. Watson, R. Troiano, J.J. Poulakos, S. Weiner, C.H. Block, A. Siegel, *Brain Res. Rev.*, 5, 1983b, 1–44. With permission.)

Amygdalo-Hippocampal Area

The amygdalo-hippocampal area constitutes the transitional region between the posterior aspect of the amygdala and rostral pole of the ventral hippocampal formation. Stimulation of this group of neurons, shown in Figure 6.13, resulted in moderate labeling of the bed nucleus of the stria terminalis and weaker labeling of the lateral septal area and dorsomedial hypothalamus. However, stimulation did activate other nuclei of the amygdala, the adjoining subicular cortex, and associated CA1 field of the hippocampus.

Bed Nucleus of Stria Terminalis

The bed nucleus of the stria terminalis is considered a primary relay (to the hypothalamus and brainstem) of several nuclei of the amygdala, such as the central nucleus, basal nuclear complex, and medial amygdala (including the cortical nucleus). Thus, one purpose of this study was to identify the primary regions activated following stimulation of this nucleus. The results revealed intense labeling of the dorsomedial and ventromedial hypothalamus as well as most nuclei of the amygdala, including the amygdalo-hippocampal area and adjoining subicular cortex (Figure 6.14). It should be pointed out that the bed nucleus is known to project to brainstem structures such

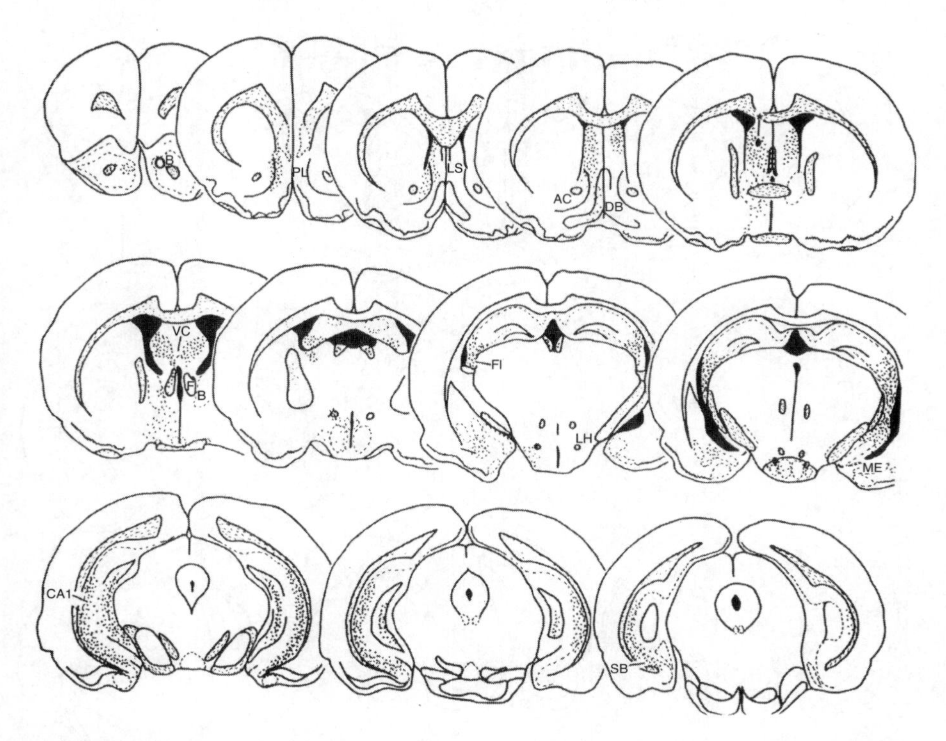

FIGURE 6.7 Distribution of label following electrical stimulation of the intermediolateral region of the lateral septal nucleus. Note the relatively dense, bilateral pattern of labeling over all CA fields and ventral subiculum, which is presumably the result of activation of fibers of passage that arise from the diagonal band region. Observe also the bilateral distribution of label within the lateral septal nucleus, as well as the presence of label distributed ipsilaterally to the diagonal band region, nucleus accumbens, basolateral nucleus of amygdala, prelimbic cortex, and olfactory bulb. Descending label can be seen within the bed nucleus of the stria terminalis, preoptic region, lateral hypothalamus, and mammillary bodies. (From R.E. Watson, R. Troiano, J.J. Poulakos, S. Weiner, C.H. Block, A. Siegel, *Brain Res. Rev.*, 5, 1983b, 1–44. With permission.)

as the PAG and tegmentum of the lower pons, but that in this study, the methods were not sufficiently sensitive to permit identification of such pathways.

Anterior Amygdaloid Area

The anterior amygdaloid area constitutes the most rostral extent of the amygdala. Surprisingly, stimulation of this region produced little activation of the forebrain with the exception of other amygdaloid nuclei and adjoining pyriform cortex (Figure 6.15).

FIGURE 6.8 Distribution of label following electrical stimulation of the septofimbrial nucleus. Note the intense, bilateral labeling of the hippocampal CA fields and subicular cortex. Observe also the bilateral labeling over the lateral septal nucleus, nucleus reuniens and anteromedial nucleus of the thalamus. Ipsilateral label is also present over the nucleus accumbens, region of the diagonal band, prelimbic cortex, and adjacent part of the caudate nucleus. Descending levels of activation are seen in the lateral preoptico-hypothalamus and mammillary bodies via the medial forebrain bundle and habenular complex via the stria medullaris. Additional label is also shown in the interpeduncular nucleus of the ventral midbrain. (From R.E. Watson, R. Troiano, J.J. Poulakos, S. Weiner, C.H. Block, A. Siegel, *Brain Res. Rev.*, 5, 1983b, 1–44. With permission.)

Lateral Nucleus

Stimulation of the lateral nucleus produced activation in the bed nucleus of the stria terminalis, nucleus accumbens, substantia innominata, anterior lateral hypothalamus, other amygdaloid nuclei, neostriatum, and pyriform cortex (Figure 6.16). Note that labeling could not be detected in any aspect of the medial hypothalamus.

Central Nucleus

Stimulation of the central nucleus produced a more restricted pattern of activation limited mainly to the substantia innominata, bed nucleus of the stria terminalis, and far lateral hypothalamus (Figure 6.17).

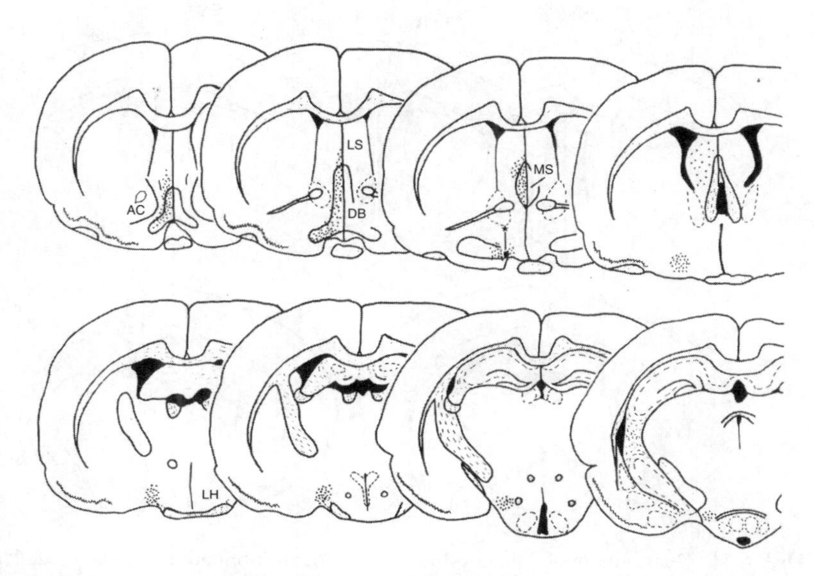

FIGURE 6.9 Distribution of label following electrical stimulation of the horizontal limb of the diagonal band. Label passes downstream in the medial forebrain bundle to the level of the mammillary bodies. Other label is diffuse and can be identified with the medial and lateral septal nuclei. Little or no label could be identified within the hippocampal formation. (From R.E. Watson, R. Troiano, J.J. Poulakos, S. Weiner, C.H. Block, A. Siegel, *Brain Res. Rev.*, 5, 1983b, 1–44. With permission.)

Basolateral Nucleus

Stimulation of the basolateral nucleus resulted in activation patterns within both the stria terminalis and ventral amygdalofugal pathways. Strial activation was associated with intense labeling in the bed nucleus of the stria terminalis and labeling in the lateral septal nucleus, while ventral amygdalofugal pathway activation with labeling in the substantia innominata and nucleus accumbens. Little or no labeling was detected in the hypothalamus, subicular cortex, or hippocampus (Figure 6.18).

Prepyriform Cortex

Stimulation of the prepyriform cortex resulted in labeling of the posterior aspect of the olfactory bulb, substantia innominata, rostral part of the lateral septal nucleus, and infralimbic and prelimbic cortices. More intense labeling extended to most amygdaloid nuclei except the central nucleus (Figure 6.19).

Entorhinal Cortex

Stimulation of the entorhinal cortex also produced limited patterns of labeling. Within the amygdala, labeling was restricted to the lateral and basolateral nuclei. Labeling was clearly seen in the posterior aspects of the subicular cortex and small amounts in the lateral septal nucleus (Figure 6.20).

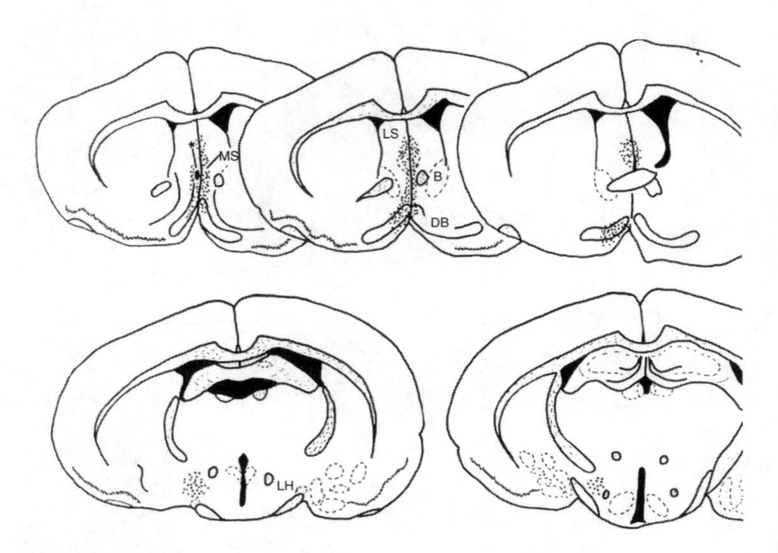

FIGURE 6.10 Distribution of label following electrical stimulation of the medial septal nucleus. The pattern of labeling in this instance parallels that observed from stimulation of the horizontal limb of the diagonal band (Figure 6.9) and that occurring from stimulation of the vertical limb of the diagonal band (not shown in this chapter). Label is generally noted over the diagonal band region and medial forebrain bundle. Where stimulation was applied at either the medial septal nucleus or diagonal band region, it was most surprising to note a marked absence of label within the hippocampal formation in spite of the fact that these groups of neurons form the origin of the septal-hippocampal projection system. (From R.E. Watson, R. Troiano, J.J. Poulakos, S. Weiner, C.H. Block, A. Siegel, *Brain Res. Rev.*, 5, 1983b, 1–44. With permission.)

Substantia Innominata

Stimulation of the substantia innominata resulted in labeling of two regions — the lateral and basolateral amygdaloid nuclei, and the rostral-to-caudal extent of the lateral hypothalamus (Figure 6.21).

Conclusions

The 2-DG findings following stimulation of nuclei of the amygdala and associated regions indicate their functional (but not necessarily their anatomical) relationship with different regions of the hypothalamus. As shown in the summary diagram (based on the 2-DG studies), stimulation of nuclear regions of the amygdala differentially activated areas within the hypothalamus. In particular, stimulation of the medial and cortical nuclei labeled fibers supplying the medial hypothalamus, including the ventromedial nucleus. Stimulation of the amygdalo-hippocampal area activated the preoptic region. In contrast, activation of parts of the ventrolateral hypothalamus was associated with stimulation of the basomedial nucleus. However, since these fibers could actually be followed into the midbrain, it was difficult to unequivocally discern whether the label in the lateral hypothalamus reflected the presence of axon

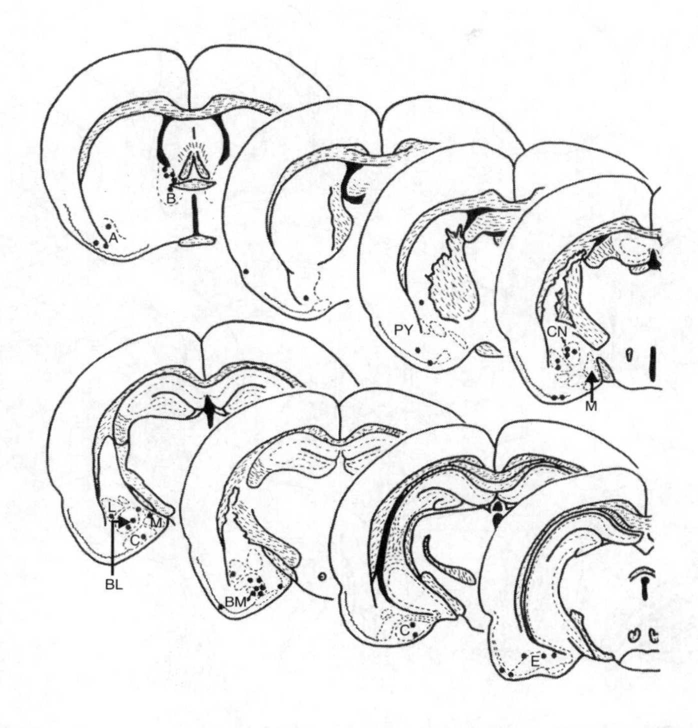

FIGURE 6.11 Map of sites electrically stimulated within the amygdaloid nuclear complex, adjacent cortical regions, bed nucleus of the stria terminalis, and substantia innominata. (From R.E. Watson, R. Troiano, J.J. Poulakos, S. Weiner, C.H. Block, A. Siegel, *Brain Res. Rev.*, 5, 1983b, 1–44. With permission.)

terminals or fibers of passage. Stimulation of other regions of the amygdala did not show direct labeling of any region of hypothalamus. Of considerable importance were the observations that stimulation of the amygdaloid nuclei also activates the bed nucleus of the stria terminalis or substantia innominata. *Since these structures project heavily to the hypothalamus and brainstem tegmentum, they may be viewed as key relay nuclei of the amygdala in that the amygdala is enabled to communicate both directly and indirectly with both the hypothalamus and brainstem.* Again, it should be pointed out that while the overall patterns of labeling appear to follow the anatomical projections of the amygdaloid nuclei, some differences are also present such as the absence of appearance of 2-DG labeling in the PAG following stimulation of the central, lateral, or basolateral nuclei, which do, in fact, receive projections from these amygdaloid nuclei.

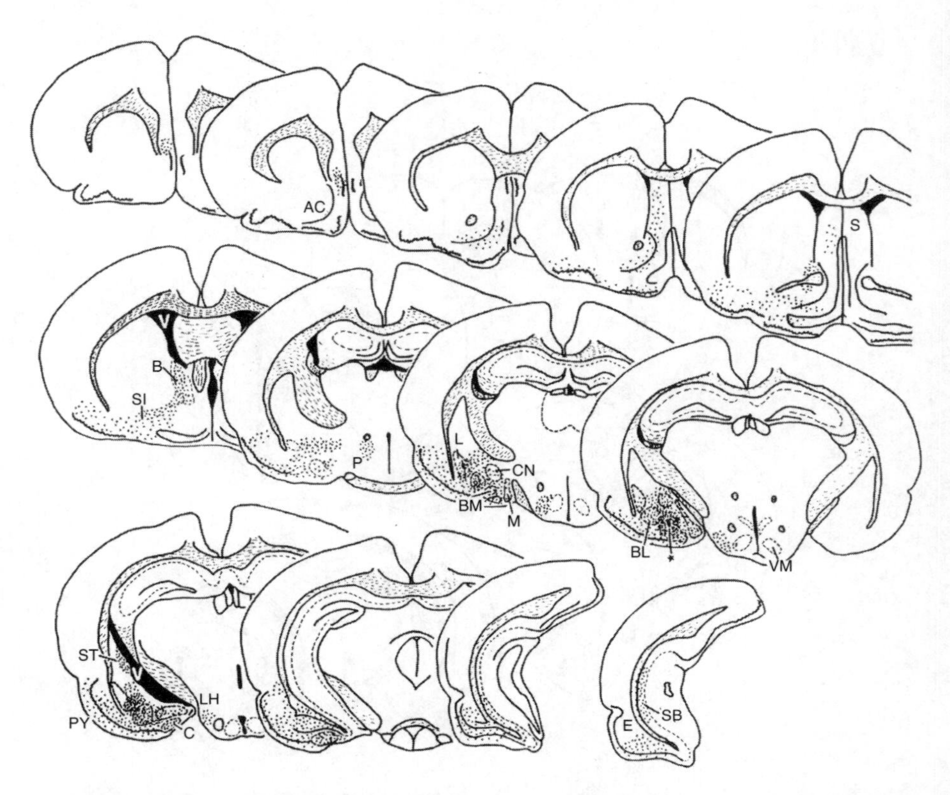

FIGURE 6.12 Distribution of label following electrical stimulation of the basomedial nucleus of the amygdala. Note label within the ventral amygdalofugal pathway and stria terminalis. Label from the ventral amygdalofugal pathway is present in the substantia innominata and ventrolateral hypothalamus. Label from the stria terminalis is present in the bed nucleus. Label presumably attributable to both pathways is present in the nucleus accumbens and lateral septal nucleus. An additional labeled component is found caudally within the entorhinal and subicular cortices. (From R.E. Watson, R. Troiano, J.J. Poulakos, S. Weiner, C.H. Block, A. Siegel, *Brain Res. Rev.*, 5, 1983b, 1–44. With permission.)

Prefrontal Cortex and Associated Thalamus Nuclei

Principal Findings

In this study, cortical stimulation was applied to the medial prefrontal, infralimbic, and sulcal cortices (Figure 6.23). Thalamic sites selected for stimulation included the mediodorsal thalamic nucleus and midline nuclei, including the nucleus reuniens. These structures were chosen for study because, as indicated in Chapter 5, powerful suppression of predatory attack was obtained following stimulation of the medial or lateral prefrontal cortex of the cat. In addition, it was also shown that fibers arising from sites in the prefrontal cortex from which suppression of

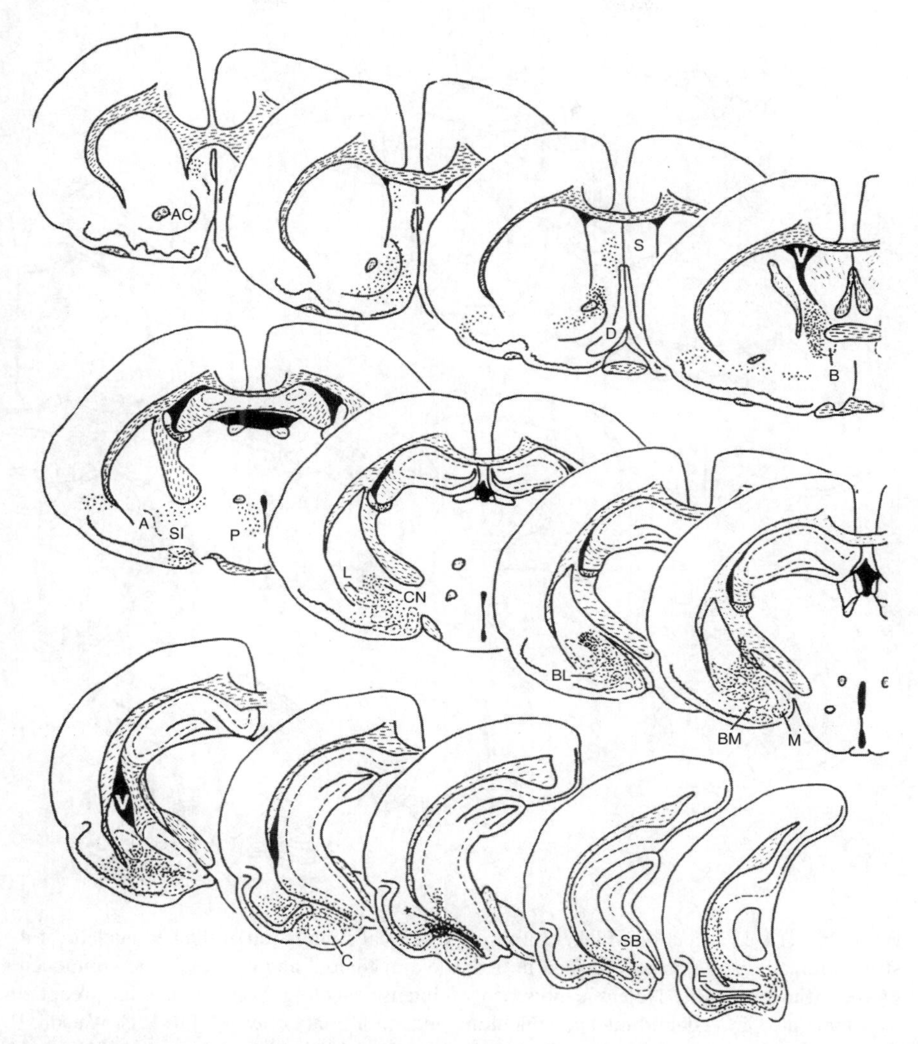

FIGURE 6.13 Distribution of label following electrical stimulation of the amygdalo-hippocampal area. Label is present primarily along the stria terminalis and its primary target nuclei — the bed nucleus of the stria terminalis and anteromedial hypothalamus. Label is also present in the substantia innominata, nucleus accumbens, and lateral septal nucleus. Caudally, label is present over the subiculum and adjacent entorhinal cortex.

attack had been elicited projected to the mediodorsal nucleus, and that fibers from the mediodorsal nucleus pass to the anterior lateral hypothalamus via a chain of neurons located in the midline thalamus. Thus, this study was designed to determine whether similar pathways would be activated following stimulation of the prefrontal cortex and medial thalamic nuclei, or whether different pathways would be involved by stimulation of these sites.

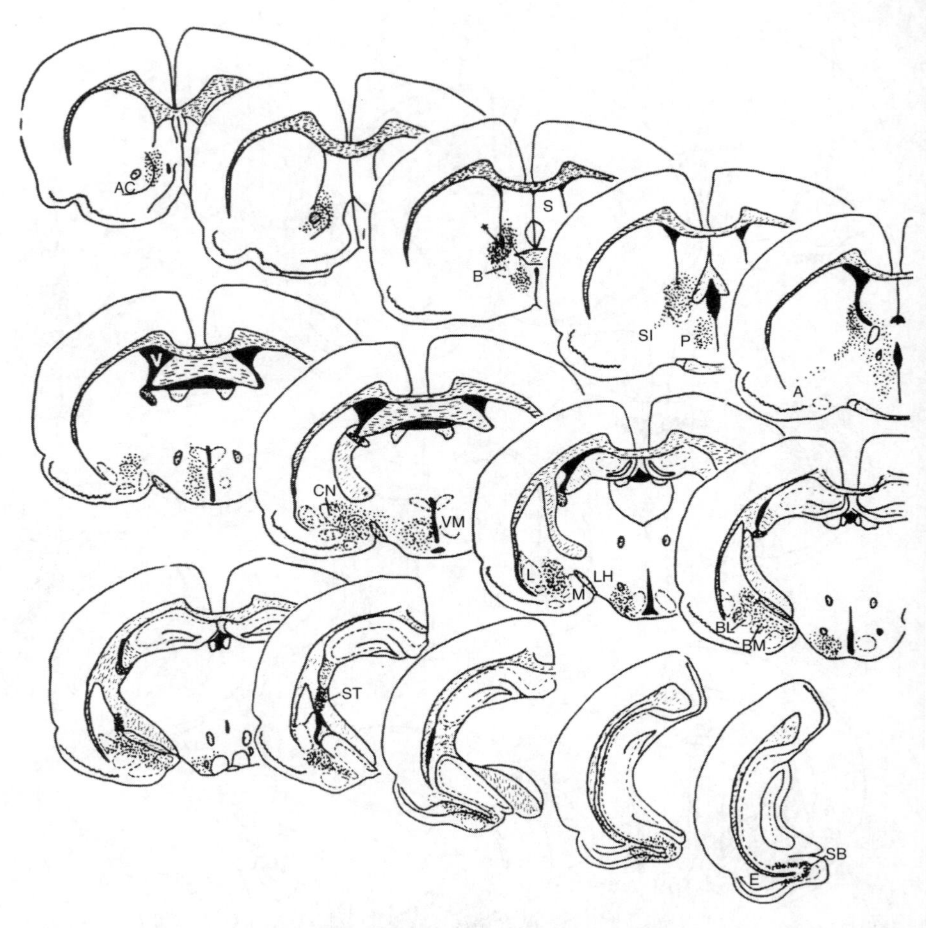

FIGURE 6.14 Distribution of label following electrical stimulation of the bed nucleus of the stria terminalis. Label is pronounced in both the amygdaloid and hypothalamic components of the stria terminalis. Note the presence of intense labeling over the medial preoptico-hypothalamus, amygdaloid nuclei, subiculum, and entorhinal cortex. (From R.E. Watson, R. Troiano, J.J. Poulakos, S. Weiner, C.H. Block, A. Siegel, *Brain Res. Rev.*, 5, 1983b, 1–44. With permission.)

Medial Prefrontal, Sulcal, and Infralimbic Cortices

Stimulation of the ventromedial aspect of the medial prefrontal cortex results in local activation of the sulcal cortex. Concerning other regions of the forebrain, stimulation of this region of prefrontal cortex stimulation produced activation in the medial aspect of the neostriatum, ventromedial and dorsomedial thalamic nuclei, and basolateral nucleus of amygdala (Figure 6.24). These regions correspond to areas that have been shown to receive projections from the prefrontal cortex. Stimulation of the dorsomedial aspect of prefrontal cortex also showed activation in the same regions with the exception that no label could be detected in any region of the

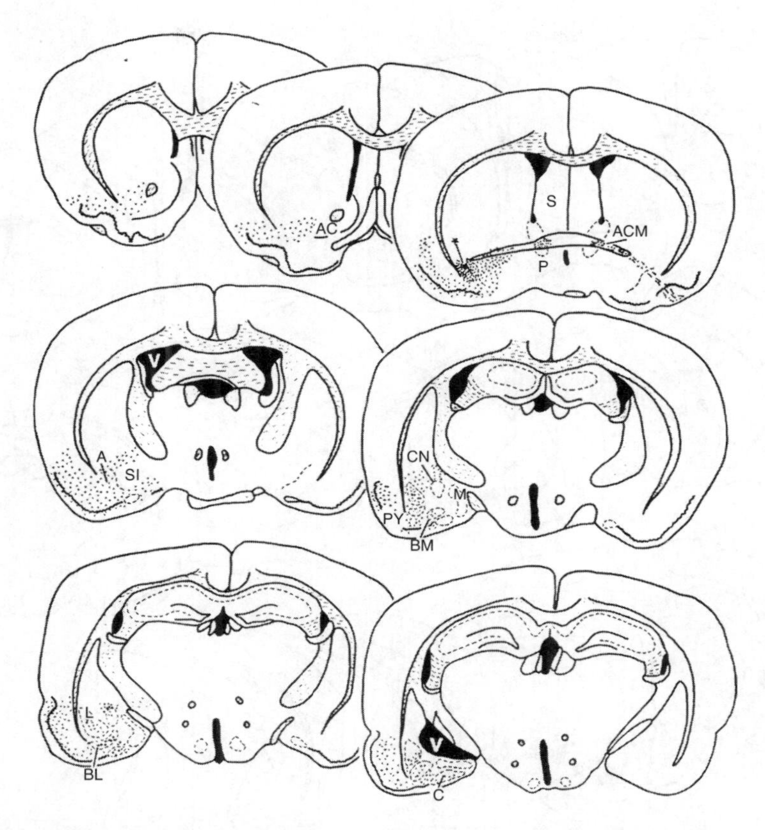

FIGURE 6.15 Distribution of label following electrical stimulation of the anterior amygdaloid area. Labeling throughout the forebrain is sparse, and is restricted to the basolateral amygdala, adjacent prepyriform area, and the temporal cortical component of the anterior commissure. (From R.E. Watson, R. Troiano, J.J. Poulakos, S. Weiner, C.H. Block, A. Siegel, *Brain Res. Rev.*, 5, 1983b, 1–44. With permission.)

amygdala (Figure 6.25). Stimulation of the sulcal cortex, shown in Figure 6.26, resulted in a slightly different pattern of activation. The primary regions activated by sulcal cortical stimulation included the medial prefrontal cortex, medial septal-diagonal band region, ventromedial thalamic nucleus (but not the dorsomedial nucleus), and prepyriform and pyriform cortices. Stimulation of the infralimbic cortex resulted in labeling throughout much of the septal area and mediodorsal thalamic nucleus. Diffuse labeling was noted over much of the caudate nucleus and nucleus accumbens (Figure 6.27). No label could be detected in the amygdala or any part of the hypothalamus. *Thus, from the studies described above, it may be concluded that the primary regions activated by prefrontal cortical stimulation include the neostriatum and the medial thalamus.* In the studies described below, the regions activated following stimulation of the mediodorsal thalamic nucleus and midline nuclei are identified.

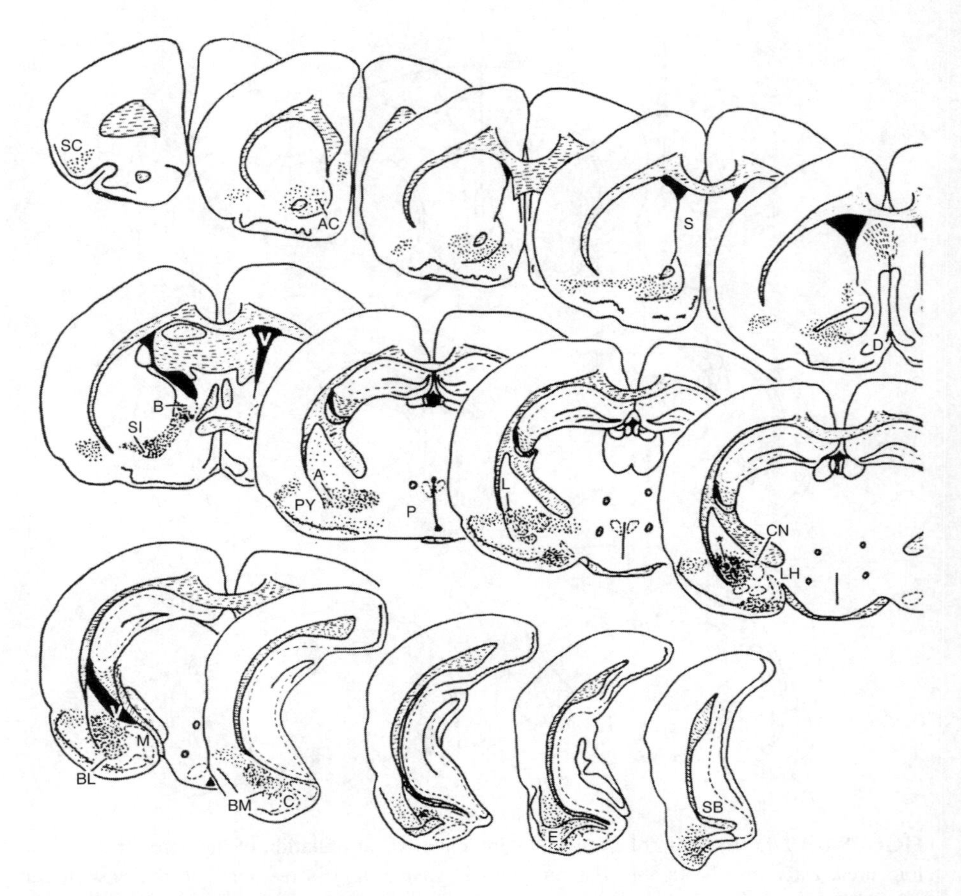

FIGURE 6.16 Distribution of label following electrical stimulation of the lateral amygdaloid nucleus. Label is prominent over the ventral amygdalofugal pathway and several of its target structures — the substantia innominata, nucleus accumbens, and bed nucleus of the stria terminalis. (From R.E. Watson, R. Troiano, J.J. Poulakos, S. Weiner, C.H. Block, A. Siegel, *Brain Res. Rev.*, 5, 1983b, 1–44. With permission.)

Mediodorsal Thalamic Nucleus

Stimulation of the mediodorsal thalamic nucleus produced intense activation of the midline and intralaminar thalamic nuclei, but the labeling did not extend into the hypothalamus. Activation could also be seen at more rostral levels of the forebrain, which included the medial aspect of the neostriatum and sulcal and medial prefrontal cortices (Figure 6.28).

Midline Thalamic Nuclei

It was hypothesized that prefrontal cortical inputs to the hypothalamus are mediated primarily through the mediodorsal thalamic nucleus. As seen in Figure 6.28,

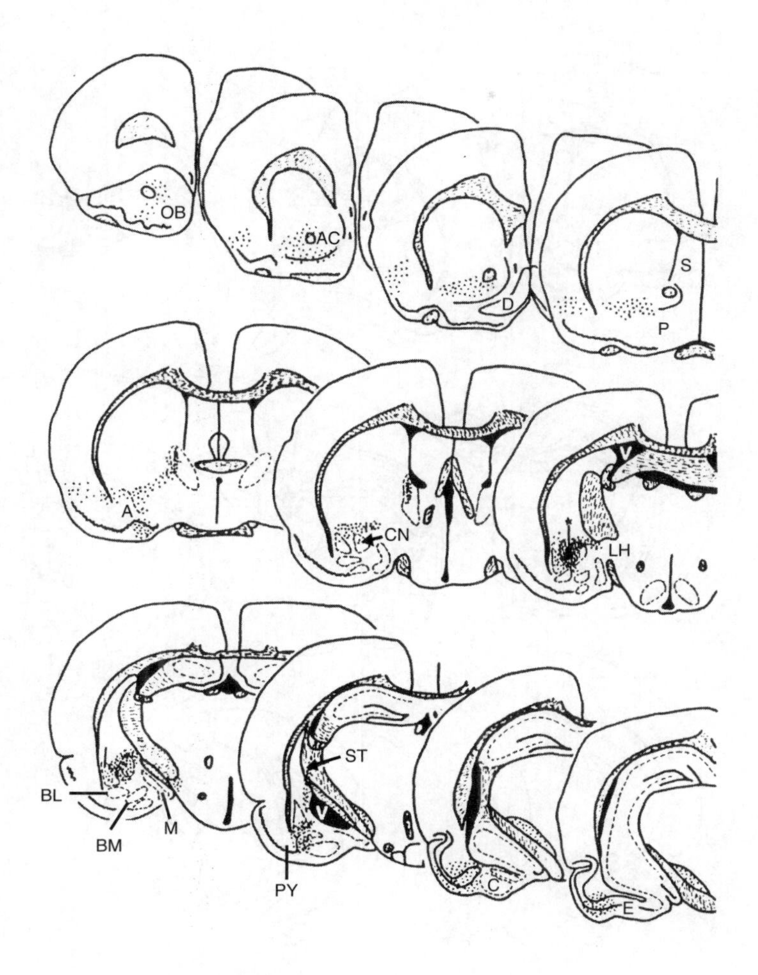

FIGURE 6.17 Distribution of label following electrical stimulation of the central amygdaloid nucleus. Note the presence of label along the ventral amygdalofugal pathway which involves the substantia innominata, nucleus accumbens, the far lateral hypothalamus, and substantia nigra. Label is also present in the bed nucleus of the stria terminalis and olfactory bulb. (From R.E. Watson, R. Troiano, J.J. Poulakos, S. Weiner, C.H. Block, A. Siegel, *Brain Res. Rev.*, 5, 1983b, 1–44. With permission.)

stimulation of the mediodorsal nucleus produced activation in midline thalamic nuclei but not in the hypothalamus. The next stage of this study was to stimulate the midline nuclei to determine whether such stimulation could produce activation in the hypothalamus.

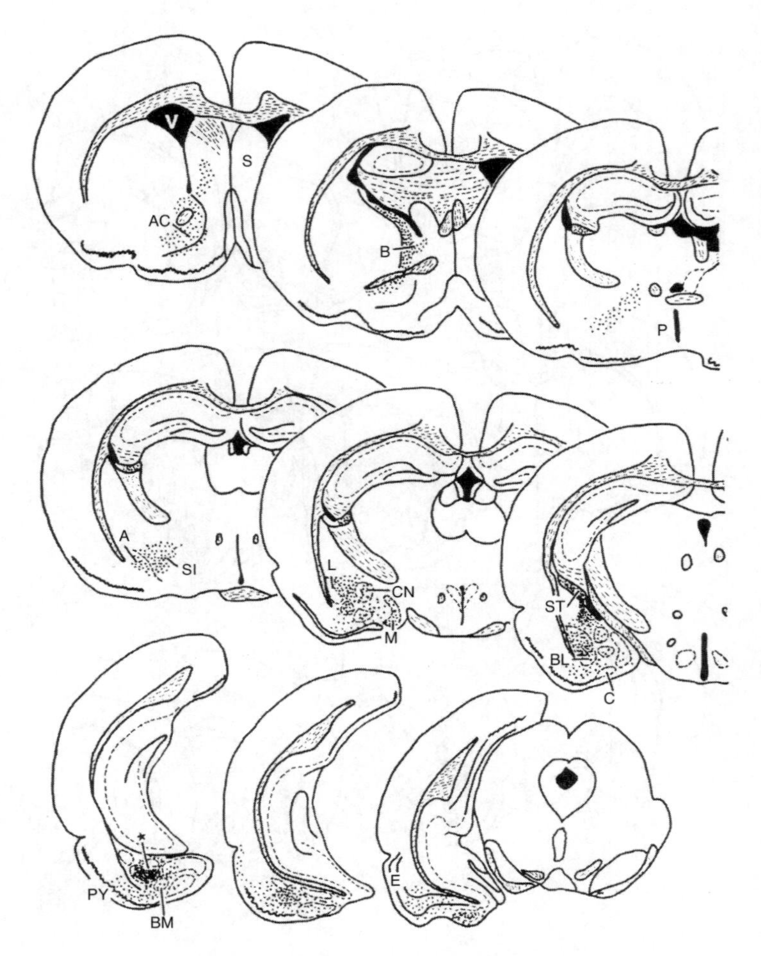

FIGURE 6.18 Distribution of label following electrical stimulation of the basolateral amygdaloid nucleus. Label involves both the stria terminalis and ventral amygdalofugal pathways, as activation is present within the bed nucleus of the stria terminalis, substantia innominata, and nucleus accumbens. (From R.E. Watson, R. Troiano, J.J. Poulakos, S. Weiner, C.H. Block, A. Siegel, *Brain Res. Rev.*, 5, 1983b, 1–44. With permission.)

The results, shown in Figure 6.29, reveal that stimulation of the nucleus reuniens produced activation of the midline thalamus mainly rostral to the site of stimulation. This pattern of activation could be followed from the rostral midline thalamus into the preoptic region. From the preoptic region, activation could be detected through much of the medial and lateral hypothalamus, extending caudally into the ventral tegmental area of the midbrain. Additional label passing from the posterior lateral hypothalamus extended through the central and basal nuclei of amygdala into the

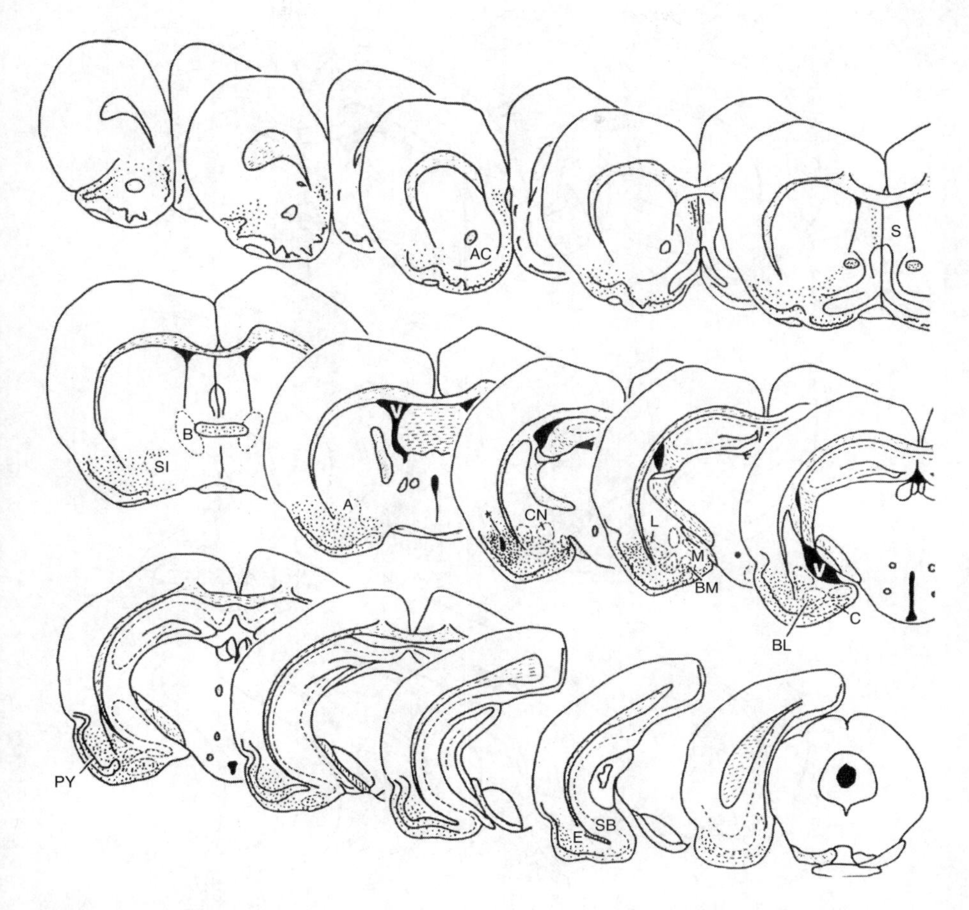

FIGURE 6.19 Distribution of label following electrical stimulation of the prepyriform cortex. Anteriorly, label is present in the olfactory bulb, prelimbic, and infralimbic cortices. Label can be observed over the basolateral amygdala, but not over other regions of the forebrain. (From R.E. Watson, R. Troiano, J.J. Poulakos, S. Weiner, C.H. Block, A. Siegel, *Brain Res. Rev.*, 5, 1983b, 1–44. With permission.)

posterior aspects of the subicular cortex and through it into the molecular layers of the dentate gyrus and hippocampus. A second pattern of activation could also be followed rostrally through the dorsal thalamus into the septal area and diagonal band. Additional labeling could be followed through the septal area into the olfactory bulb and rostrodorsally into the prefrontal cortex.

Conclusions

Thus, these findings indicate that the prefrontal cortex can produce activation of wide regions of the hypothalamus. But it does so mainly indirectly via its synaptic connections with midline thalamus whose axons can ultimately activate much of the hypothalamus.

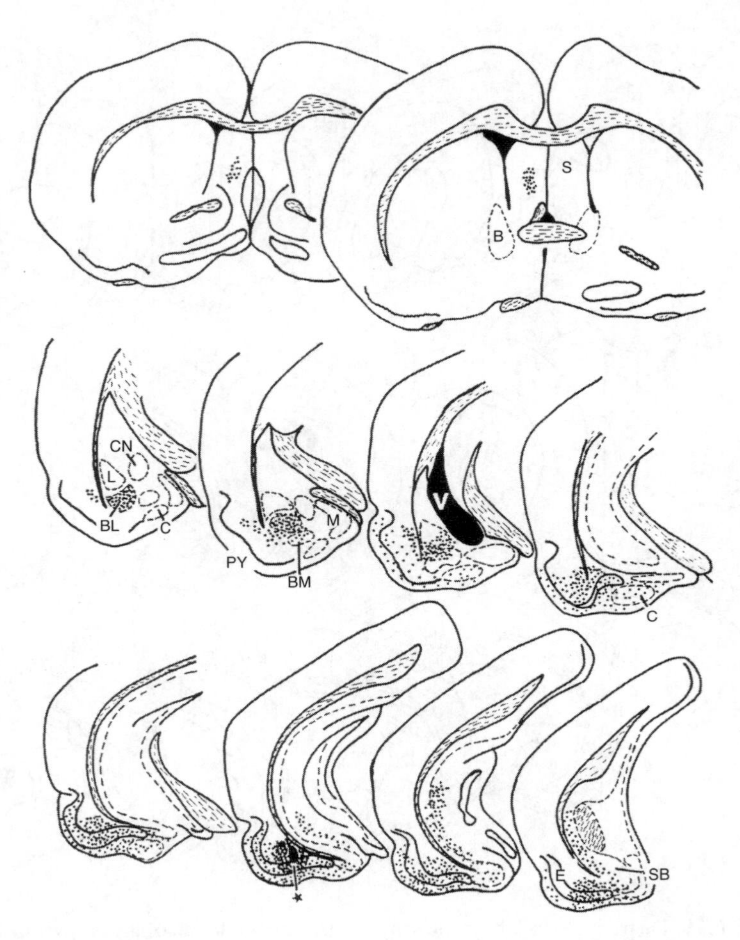

FIGURE 6.20 Distribution of label following electrical stimulation of the entorhinal cortex. Label is prominent over the basolateral nucleus, subiculum, and to a lesser extent, the septal area. (From R.E. Watson, R. Troiano, J.J. Poulakos, S. Weiner, C.H. Block, A. Siegel, *Brain Res. Rev.*, 5, 1983b, 1–44. With permission.)

OVERALL CONCLUSIONS

Results of the 2-DG studies reinforce the conclusions suggested in Chapter 5 concerning a basic functional property of the limbic system — its relationship to the hypothalamus. The feature common to each of the limbic structures is that when stimulation is applied, there is subsequent direct or indirect activation of the hypothalamus. These findings thus provide further evidence that each of these limbic structures can modulate hypothalamic functions (such as aggression and rage) and, in addition, they identify the pathways that likely mediate these effects.

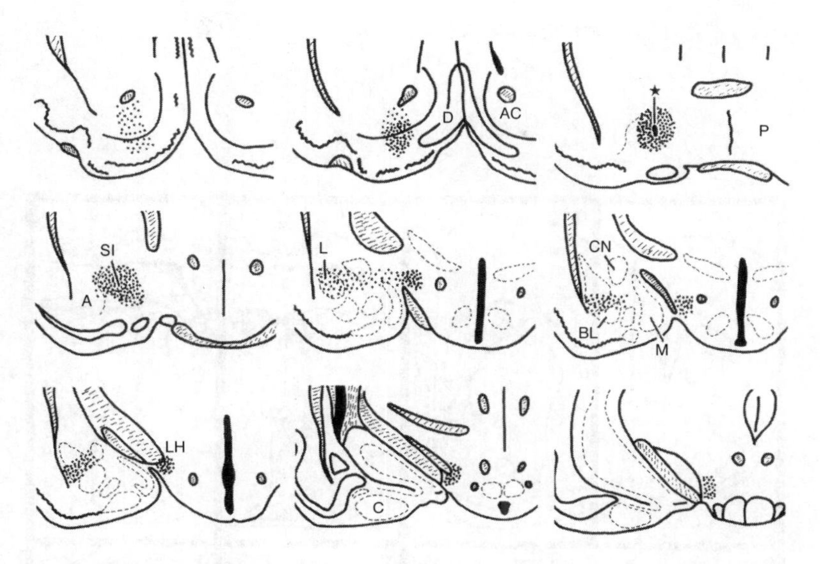

FIGURE 6.21 Distribution of label following electrical stimulation of the substantia innominata. Label is prominent over the medial forebrain bundle within the lateral hypothalamus. Additional label is present within the lateral amygdaloid complex and nucleus accumbens. (From R.E. Watson, R. Troiano, J.J. Poulakos, S. Weiner, C.H. Block, A. Siegel, *Brain Res. Rev.*, 5, 1983b, 1–44. With permission.)

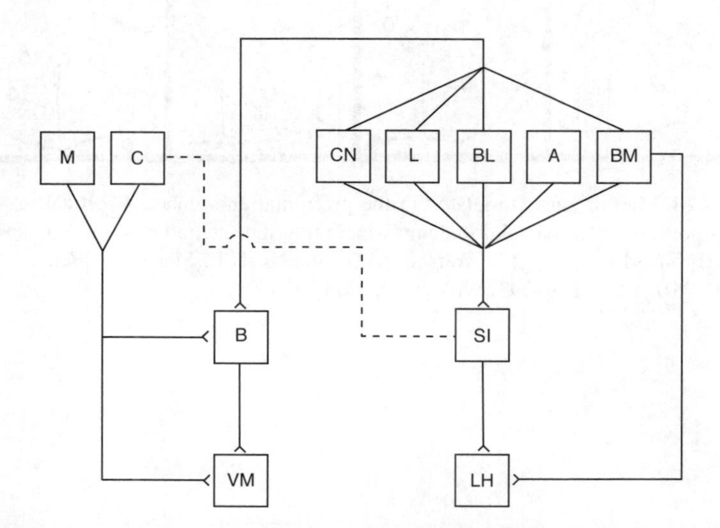

FIGURE 6.22 Summary diagram illustrating the organization of systems activated following electrical stimulation of amygdaloid nuclei and associated structures. Dashed lines represent activation of considerably less magnitude than that resulting from stimulation of other sites. B refers to the BNST. (From R.E. Watson, R. Troiano, J.J. Poulakos, S. Weiner, C.H. Block, A. Siegel, *Brain Res. Rev.*, 5, 1983b, 1–44. With permission.)

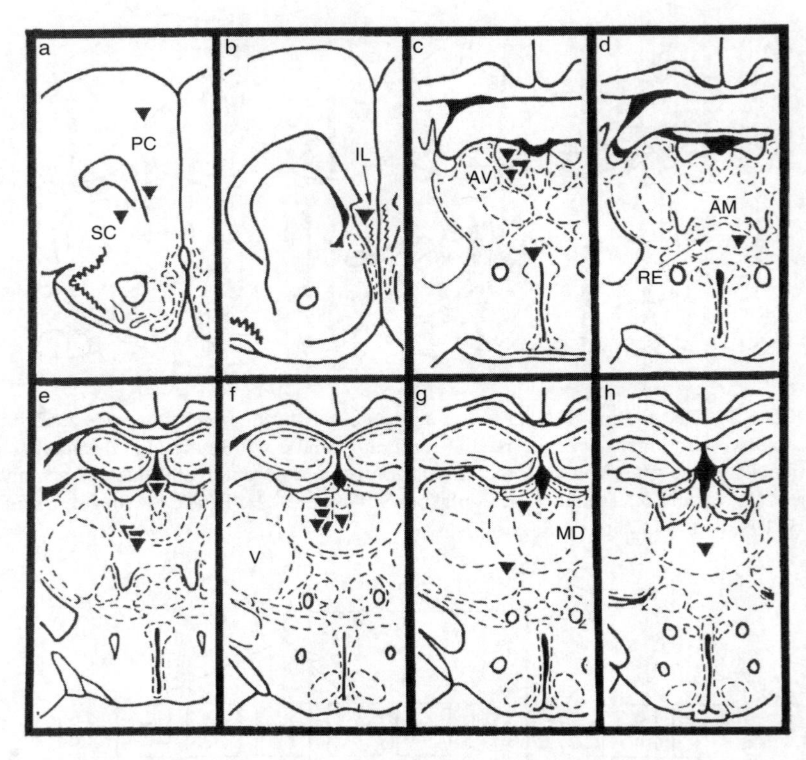

FIGURE 6.23 Map of sites stimulated in the prefrontal and adjacent cortical tissue and in associated portions of the medial thalamus, including the mediodorsal nucleus and nucleus reuniens. (From M. Brutus, R.E. Watson, M.B. Shaikh, H.E. Siegel, S. Weiner, A. Siegel, *Brain Res.*, 310, 1984, 279–293. With permission.)

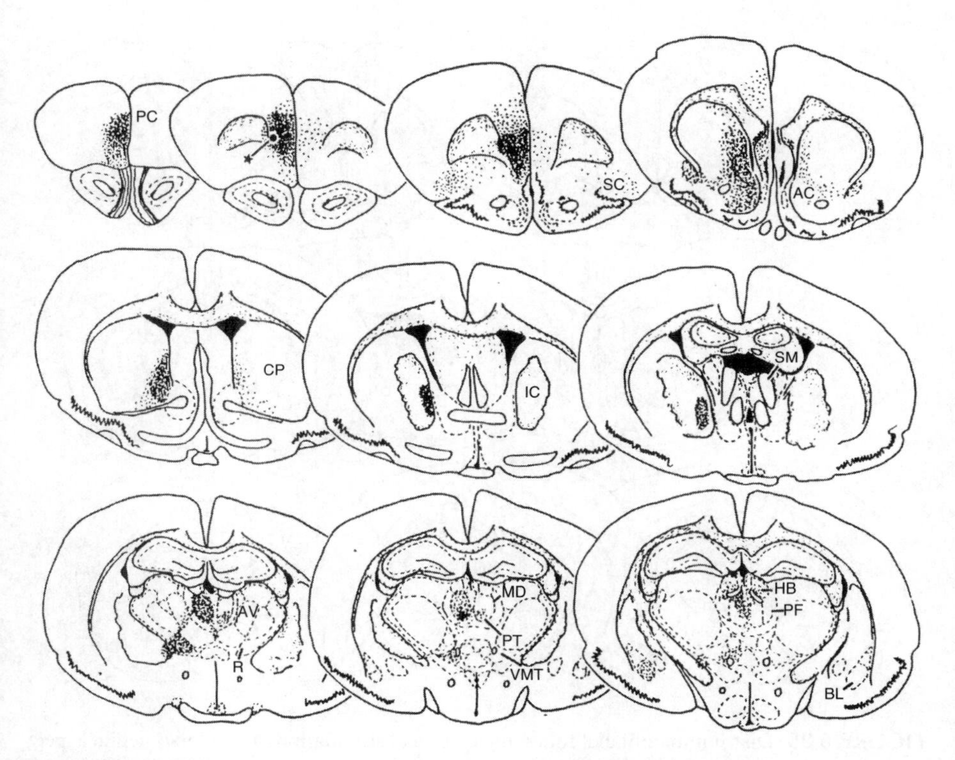

FIGURE 6.24 Distribution of label following electrical stimulation of the ventromedial aspect of the prefrontal cortex. Note the presence of label within adjacent regions of sulcal cortex, dorsomedial aspect of the head of the caudate nucleus, basolateral amygdaloid nucleus, medial segment of the mediodorsal nucleus, and reticular thalamic nucleus. No label was observed in any region of the preoptico-hypothalamus. (From M. Brutus, R.E. Watson, M.B. Shaikh, H.E. Siegel, S. Weiner, A. Siegel, *Brain Res.*, 310, 1984, 279–293. With permission.)

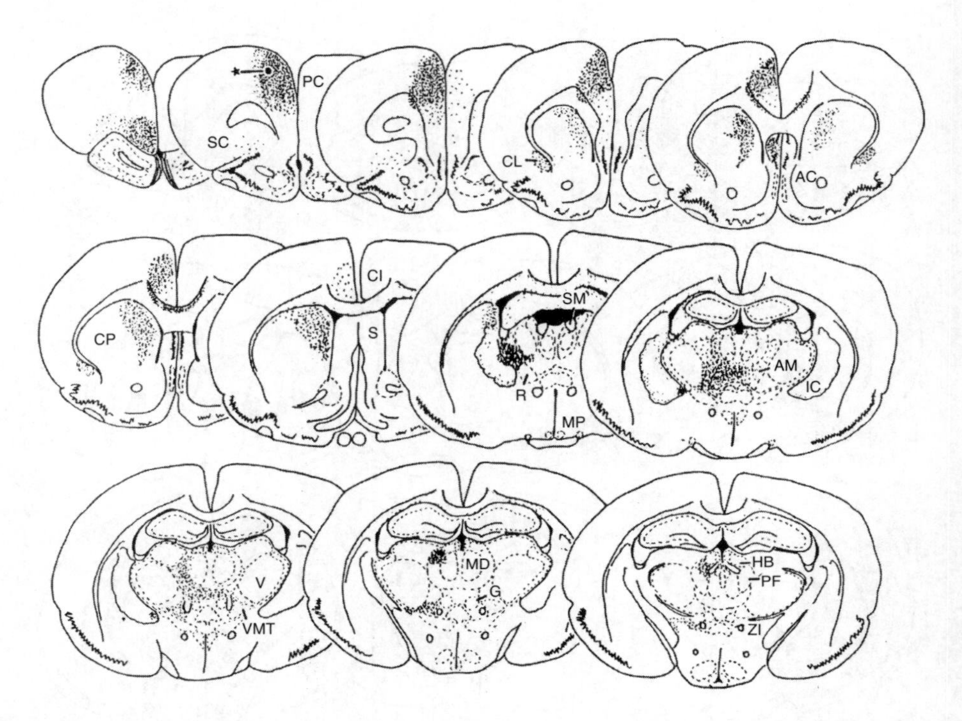

FIGURE 6.25 Distribution of label following electrical stimulation of the dorsomedial aspect of the prefrontal cortex. Intense labeling is noted within the ventromedial prefrontal cortex and dorsomedial aspect of the head of the caudate nucleus. Label is present in the lateral aspect of the mediodorsal thalamic nucleus, and is also diffusely distributed to the midline thalamus dorsal to the nucleus reuniens. Again, no label is present in any region of the preoptico-hypothalamus. (From M. Brutus, R.E. Watson, M.B. Shaikh, H.E. Siegel, S. Weiner, A. Siegel, *Brain Res.*, 310, 1984, 279–293. With permission.)

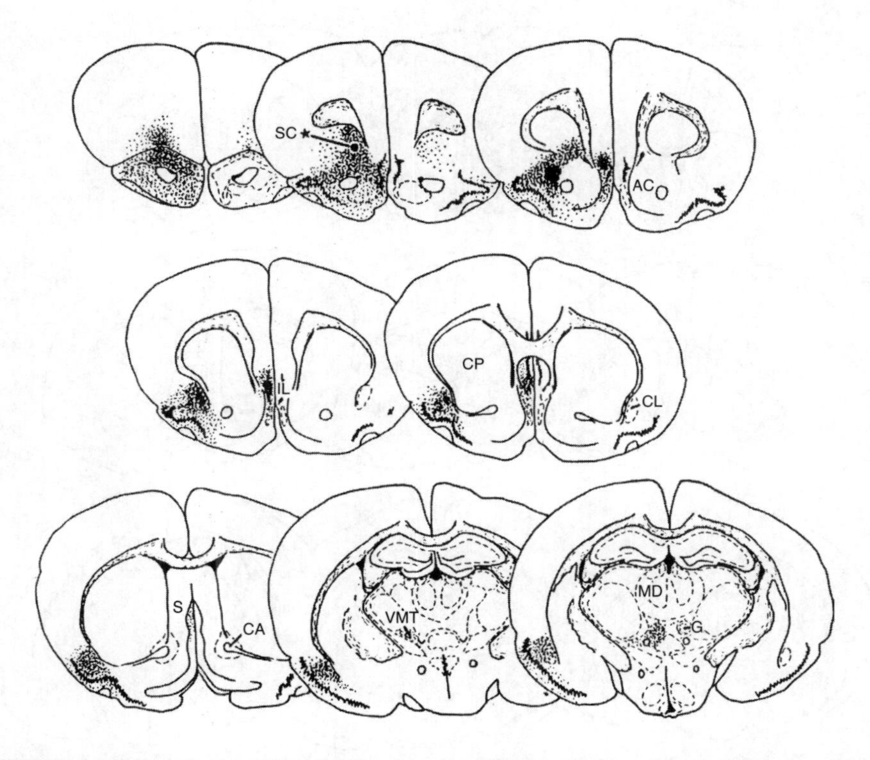

FIGURE 6.26 Distribution of label following electrical stimulation of the sulcal cortex. Label is present within the rostral aspect of the prepyriform area, endopyriform nucleus, infralimbic area, and ventromedial thalamic nucleus. Label is not present elsewhere in the diencephalon. (From M. Brutus, R.E. Watson, M.B. Shaikh, H.E. Siegel, S. Weiner, A. Siegel, *Brain Res.*, 310, 1984, 279–293. With permission.)

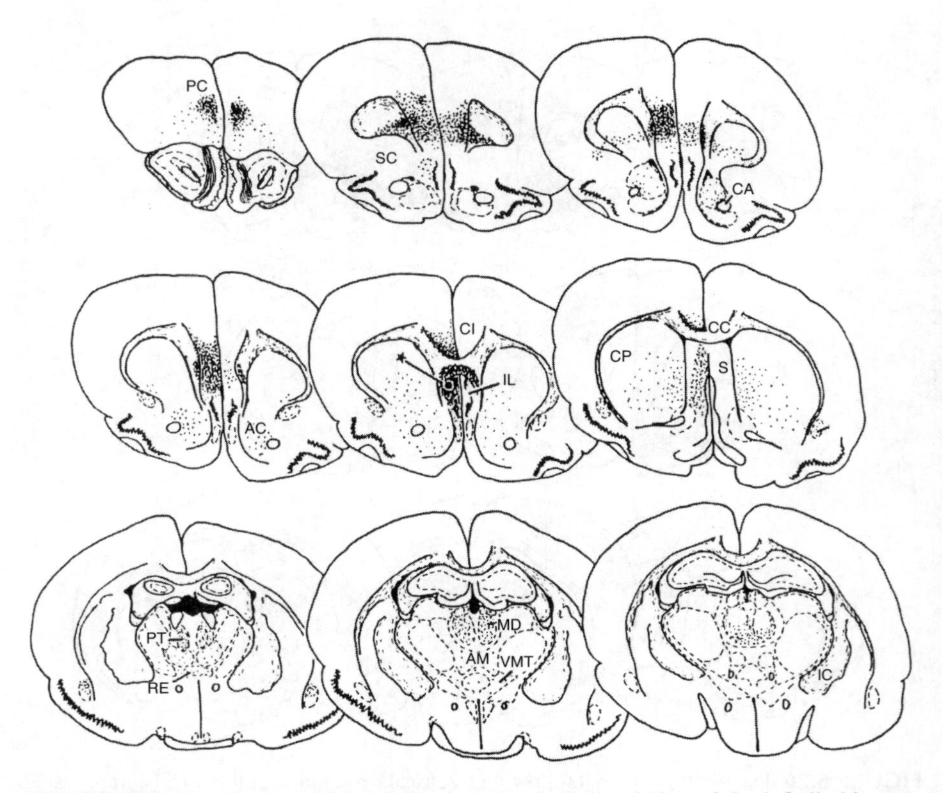

FIGURE 6.27 Distribution of label following electrical stimulation of the infralimbic area. Label is extensive in the medial prefrontal cortex, bilaterally, and lateral septal nucleus, ipsilaterally. Label is diffuse within the caudate nucleus. At the level of the diencephalon, label is limited to the medial segment of the mediodorsal nucleus at rostral levels of this nucleus and is quite diffuse in adjacent thalamic nuclei. (From M. Brutus, R.E. Watson, M.B. Shaikh, H.E. Siegel, S. Weiner, A. Siegel, *Brain Res.*, 310, 1984, 279–293. With permission.)

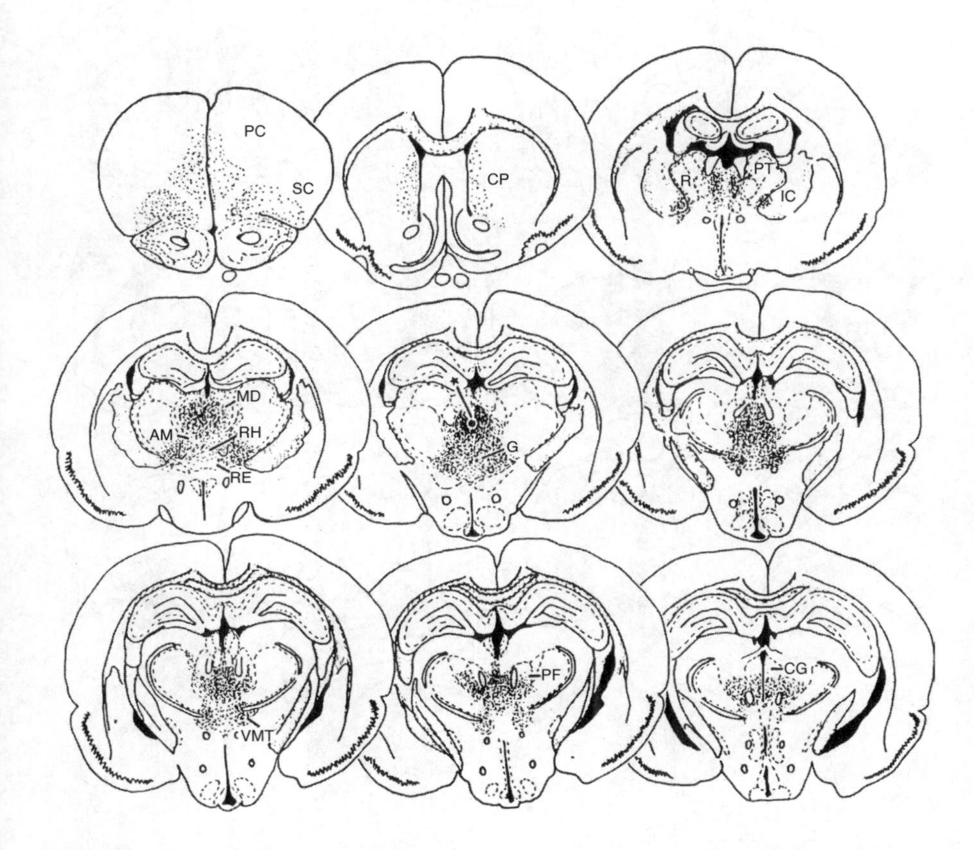

FIGURE 6.28 Distribution of label following electrical stimulation of the posterior aspect of the mediodorsal nucleus. Again, label is intense over both medial and sulcal prefrontal cortices. Label is also concentrated over midline thalamic nuclei, including the nucleus reuniens and the reticular thalamic nucleus. While no label is seen within any region of hypothalamus, it is clearly noted within the midbrain PAG and parafascicular-centromedian complex. (From M. Brutus, R.E. Watson, M.B. Shaikh, H.E. Siegel, S. Weiner, A. Siegel, *Brain Res.*, 310, 1984, 279–293. With permission.)

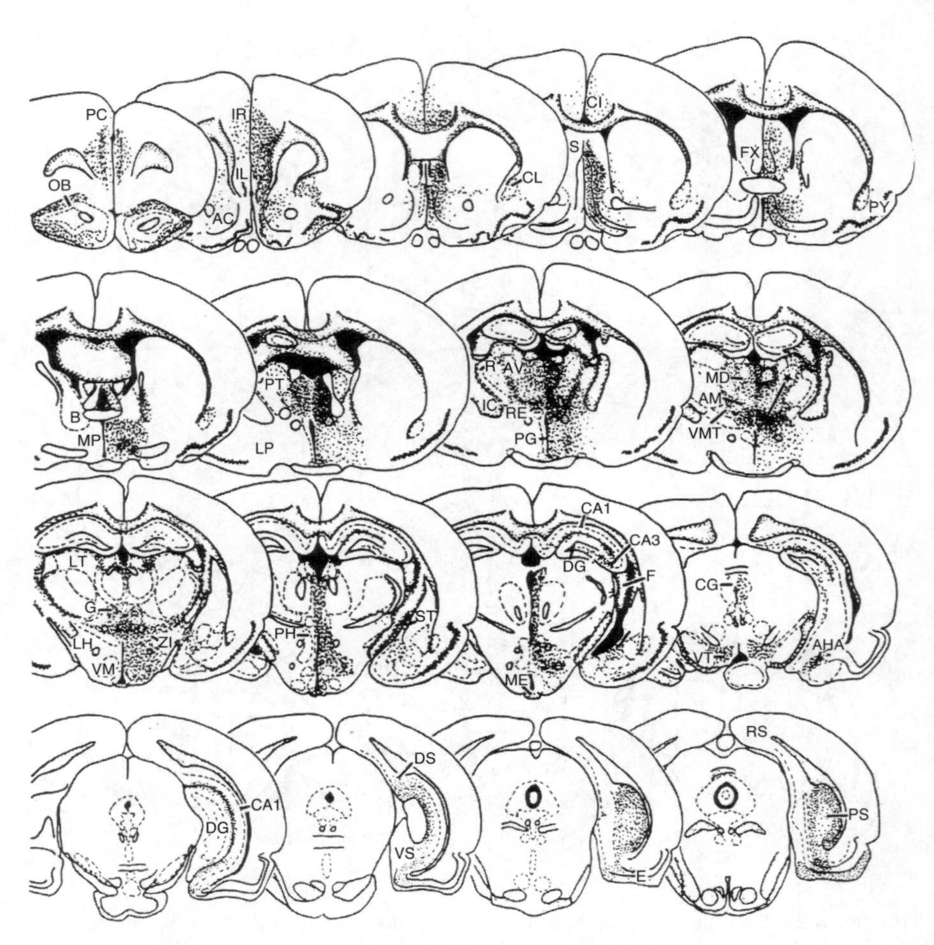

FIGURE 6.29 Distribution of label following electrical stimulation of the nucleus reuniens. Note the presence of intense label over the medial prefrontal cortex, area infraradiata, olfactory bulbs, septal area, diagonal band nuclei, anterior thalamic nuclear complex, amygdalo-hippocampal area, basolateral amygdaloid nucleus, and midbrain midbrain periaqueductal gray. Significantly, label is extensive throughout both lateral and medial preoptico-hypothalamus, indicating the existence of a massive functional projection from the thalamus to the hypothalamus (and indirectly from the prefrontal cortex) that likely modulates functions of the hypothalamus such as aggression and rage. (From M. Brutus, R.E. Watson, M.B. Shaikh, H.E. Siegel, S. Weiner, A. Siegel, *Brain Res.*, 310, 1984, 279–293. With permission.)

ABBREVIATIONS USED IN FIGURES

A, anterior hypothalamic area
AC, or CA, anterior commissure or nucleus accumbens
ACM, anterior commissure
AH, anterior hypothalamus
AHA, amygdalo-hippocampal area
AM, anteromedial thalamic nucleus
AR, arcuate nucleus of hypothalamus
AV, anteroventral thalamic nucleus
B, bed nucleus of stria terminalis
BL, basolateral nucleus of amygdala
BM, basomedial nucleus of amygdala
CA1, CA3, cornu Ammonis fields of hippocampus
C, cortical nucleus of amygdala
CC, corpus callosum
CG, midbrain periaqueductal, central, gray
CI, cingulate cortex
CL, claustrum
CN, central nucleus of amygdala
CP, caudate-putamen, neostriatum
D or DB, diagonal band of Broca
DF, dorsal fornix
DG, dentate gyrus of hippocampus
DH, dorsal hypothalamus
DS, dorsal subiculum
E, entorhinal cortex
F or FX, fornix
FI, fimbria
G, nucleus gelatinosus of thalamus
HB, habenular complex
IC, internal capsule
IL, infralimbic cortex
IP, interpeduncular nucleus
IR, infraradiate cortex
L, lateral nucleus of amygdala
LH, lateral hypothalamus
LP, lateral preoptic area
LS, lateral septal nucleus
LT, lateral thalamic nucleus
M or MM, mammillary bodies
M or ME, medial nucleus of amygdala
MD, mediodorsal thalamic nucleus
ME, median eminence
MP, medial preoptic area
MS, medial septal nucleus

MT, mammillothalamic tract
OB, olfactory bulb
P, preoptic area
PAR, parasubiculum
PC, prefrontal cortex, medial aspect
PF, parafascicular thalamic nucleus
PH, posterior hypothalamus
PL, prelimbic cortex
PRO, prosubiculum
PS, posterior subiculum
PT, parataenial thalamic nucleus
PV, periventricular gray of hypothalamus
PY, pyriform cortex
R, reticular thalamic nucleus
RE, nucleus reuniens
RH, rhomboidal thalamic nucleus
RS, retrosplenial cortex
S, septal area
SB, subiculum
SC, suprachiasmatic nucleus or sulcal cortex
SF, septofimbrial nucleus
SI, substantia innominata
SM, stria medullaris
ST, stria terminalis
V, ventricle or ventral thalamic nucleus
VC, ventral hippocampal commissure
VM, ventromedial nucleus of hypothalamus
VMT, ventromedial thalamic nucleus
VS, ventral subiculum
ZI, zona incerta

REFERENCES

Brutus, M., R.E. Watson, M.B. Shaikh, H.E. Siegel, S. Weiner, A. Siegel, A (14C)2-deoxyglucose analysis of the functional neural pathways of the limbic forebrain in the rat. IV. A pathway from the prefrontal cortical-medial thalamic system to the hypothalamus, *Brain Res.*, 310, 1984, 279–293.

Meibach, R. and A. Siegel, Efferent connections of the hippocampal formation in the rat, *Brain Res.*, 124, 1977, 197–224.

Siegel, A., H. Edinger, The organization of the hippocampal septal axis, in J. De France, Ed., *International Symposium on the Septal Nuclei*, Wayne State University Press, Detroit, Mich., 1976, pp. 241–250.

Siegel, A., H. Edinger, S. Ohgami, The topographical organization of the hippocampal projection to the septal area: a comparative neuroanatomical analysis in the gerbil, rat, rabbit and cat, *J. Comp. Neurol.*, 157, 1974, 359–378.

Siegel, A., S. Ohgami, H. Edinger, The topographical organization of the hippocampal projection to the septal area in the squirrel monkey, *Brain Res.*, 99, 1975, 247–260.

Siegel, A. and J.P. Tassoni, Differential efferent projections from the ventral and dorsal hippocampus of the cat, *Brain Behav. Evol.*, 4, 1971, 185–200.

Watson, R.E., H.E. Edinger, A. Siegel, A 14C-2-deoxyglucose analysis of the functional neural pathways of the limbic forebrain in the rat: III. The hippocampal formation, *Brain Res. Rev.*, 5, 1983a, 133–176.

Watson, R.E., A. Siegel, H.E. Siegel, A (14C)2-deoxyglucose analysis of the functional neural pathways of the limbic forebrain in the rat: V. The septal area, *Brain Res.*, 346, 1985, 89–107.

Watson, R.E., R. Troiano, J.J. Poulakos, S. Weiner, C.H. Block, A. Siegel, A (14C)2-deoxyglucose analysis of the functional neural pathways of the limbic forebrain in the rat: I. The amygdala, *Brain Res. Rev.*, 5, 1983b, 1–44.

Watson, R.E., R. Troiano, J.J. Poulakos, S. Weiner, A. Siegel, A 14 C-2-deoxyglucose analysis of the neural pathways of the limbic forebrain in the rat: II. The hypothalamus, *Brain Res. Bull.*, 8, 1982, 459–476.

7 The Neurochemistry of Rage and Aggression

Of all the areas of investigation concerning the neurobiology of aggression and rage, studies involving neurochemistry and neuropharmacology have attracted the lion's share of attention as seen in the large numbers of research papers published in this field. A goal of this chapter is to provide a representative sampling of the literature, which typically reflects the major conclusions regarding the relationships of the various neurotransmitters to aggression and rage.

Investigators have used diverse approaches in their attempts to relate neurochemical processes and mechanisms to states of rage and aggression. The most common approach has been to examine the effects on aggression of peripheral administration of a drug, which could be a specific or nonspecific receptor agonist or antagonist, or some other class of drug, such as a tranquilizer, anxiolytic, hallucinogen, or drug of abuse. Related approaches that have been used less frequently involved studies testing the effects of similar drugs when applied intracerebroventricularly or intracerebrally into specific brain sites.

Another approach involved the attempt to determine how aggressive reactions may alter brain chemistry. Initially, these studies examined whole brain neurochemical changes following aggressive encounters. Later studies used more refined methods such as the "punch" technique to sample tissue from selected regions of the brain; and, most recently, there has been some attempt to use microdialysis to analyze the effects of aggressive responses on highly localized regions of the brain. An additional approach has been to use newer genetic methods involving transgenic animals such as mice in which the gene expressing a specific neurotransmitter receptor is made defective. Such a preparation is referred to as a "knockout" and the role of the receptor in question has been suggested by the extent to which the experimental preparation displays aggression relative to control animals.

My objective in this chapter is to review and summarize the findings of studies representative of the research strategies described above with respect to the different neurotransmitters and various classes of drugs. Two basic questions addressed include: (1) What are the roles of different neurotransmitters and their receptors? (2) Where in the brain do they act in regulating aggression and rage?

In attempting to answer these questions, the strengths as well as weaknesses of these approaches are briefly noted here. Concerning the use of systemic injections of drugs, one positive feature is that it provides an overall assessment of the role of the neurotransmitter in question in aggressive behavior. A second advantage is that it follows the clinical route of drugs such as carbamazepine, anxiolytic, and antidepressive compounds. However, this approach raises several concerns. One is that it is not readily discernible whether the effects of drug administration by this route are mediated by peripheral or central mechanisms. A second issue is that systemic administration of drugs provides little or no information concerning where in the brain the drug may be acting to produce its effects on aggressive behavior. A third concern is the specificity of the effects resulting from systemic administration of drugs. In particular, if the drug suppresses a given form of aggression, a number of control experiments are required to determine whether the suppressive effects generated are specific to the form of aggressive behavior examined, or whether the effects are nonspecific and common to wide varieties of behaviors.

There is an important advantage associated with the approach using microinjections of drugs into specific regions of the brain. Specifically, it may enable the investigator to identify the receptor–neurotransmitter mechanism operative at a given site in the brain to alter a behavioral process. However, similar to the problems associated with systemic administration of drugs, other concerns relate to intracerebral administration of drugs as well that require the employment of careful controls in the design of the experiment. For example, one aspect of the experimental design should be included to control for difficulties in interpretation of the data due to diffusion of the drug from the site of delivery. This may be achieved by the placement of microinjections into neighboring regions and by comparing the effects obtained from microinjections at this site with those obtained following microinjections into the original site of interest. A second concern, similar to that described above for peripheral administration of drugs, is that the compound may have a nonspecific effect on the behavioral response under examination, such as causing general inhibition of all motor responses. Here, it would be useful for the investigator to compare the effects of intracerebral drug delivery on several different types of behavior in order to rule out this possibility. A third concern is the limited selectivity of various receptor agonists and antagonists, especially when used at higher doses where the selective properties can be somewhat diminished. Investigators can cope with this issue through the generation of dose–response curves and by pretreating a given site with a selective receptor antagonist prior to administration of the selective agonist for that drug.

ACETYLCHOLINE

Synthesis and Removal

Acetylcholine (ACh) is a small-molecule neurotransmitter. It is synthesized from acetylcoenzyme-A and choline in the following way. Initially, pyruvate is derived from glucose (which reaches the nerve terminal by passive transport) by glycolysis and is transported into the mitochondria. Acetylcoenzyme-A is formed from the combination of coenzyme-A found in the mitochondria and an acetyl group derived from pyruvic acid, and is transported back into the cytoplasm. The second component, choline, is present in the plasma and enters the nerve terminal via active transport. The synthesis of choline and acetylcoenzyme-A takes place in the cytoplasm of the nerve terminal by the enzyme, choline acetyltransferase. ACh is then transported into the synaptic vesicles where it is stored.

ACh is removed as a result of the actions of the enzyme, acetylcholinesterase, which is synthesized in the endoplasmic reticulum of cell body, and is found on the outer surface of the nerve terminal. Acetylcholinesterase then hydrolyzes ACh in the extracellular spaces, thus reducing its concentration, and leading to dissociation from its receptor-binding site. Choline is now released, which enables it to re-enter the nerve and be used again for the synthesis of ACh.

Distribution and Functions

Cholinergic neurons are found widely throughout the brain and spinal cord. Examples of specific groups of neurons have been clearly found in such regions as the basal nucleus of Meynert of rostral forebrain, pedunculopontine region of the pontine tegmentum, cerebral cortex, neostriatum, hippocampal formation, and ventral horn cells of the spinal cord. ACh is generally considered to be an excitatory neurotransmitter and plays an important role in the sleep–wakefulness cycle, visceral functions involving the hypothalamus and limbic structures, as well as certain disease states such as in Alzheimer's disease, which is linked to atrophy of basal forebrain cholinergic neurons and ACh levels in the cortex.

Role of ACh in Aggression and Rage

Effects of Systemic Administration of Cholinergic Agents

A number of early studies used the systemic administration of cholinergic drugs in order to determine the role of ACh in aggressive behavior. Most of these studies were conducted in either rodents or cats, and provide evidence that cholinergic agents generally have facilitatory effects on rage and rage-related processes. This is not particularly surprising in view of the fact that, as indicated above, ACh is viewed as an excitatory neurotransmitter.

In studies conducted in mice, muscarinic antagonists atropine or scopolamine reduce or eliminate threat attack (Valzelli et al., 1967; Barnett et al., 1971). A parallel study conducted in rats showed similar results following administration of scopolamine (Van der Poel and Remmelts, 1971). However, not all experiments yielded consistent findings. When rats were tested for defensive behavior after exposure to a predator cat and scopolamine delivery, major changes were not apparent (Rodgers et al., 1990). These authors concluded that there is a lack of significant involvement of muscarinic mechanisms in the regulation of defensive patterns in rats. A similar experiment conducted in the cat revealed that with respect to predatory attack, scopolamine administration causes an elevation of the threshold for this response (Katz and Thomas, 1975).

Consistent with the general pattern of the findings described above, a number of studies examined the effects of systemic administration of cholinergic agonists on aggressive behavior. Gay and Leaf (1976) administered pilocarpine, a potent muscarinic agonist, and observed an induction in mouse killing in different strains of rats. Likewise, systemic injections of muscarinic agonists in facilitated defensive rage in the cat (George et al., 1962; Zablocka and Esplin, 1964; Leslie, 1965).

In addition, Berntson and Leibowitz (1973, 1976) also observed that administration of the muscarinic agonist, arecoline, can induce components of both biting attack and defensive rage in the cat.

Effects of Intracerebral Microinjections of Cholinergic Agents

The second approach used to examine the effects of acetylcholine has been to microinject cholinergic compounds into specific regions of the brain normally associated with aggression and rage. In most cases, the primary target has been the hypothalamus. Here, the basic principle tested was that aggression and rage behavior can be induced or altered following cholinergic receptor activation or blockage in selective regions of the brain.

A number of studies conducted mainly in rats and cats examined the effects of central cholinergic receptor activation on aggression or rage responses. One common approach has been to microinject the muscarinic cholinergic agonist, carbachol, into specific regions of the hypothalamus or neighboring aspects of the forebrain. One problem of interpretation of experiments that examine the effects of administration of receptor agonists on a given behavioral response should be noted here. Receptor-mediated changes in a behavioral response do not demonstrate the presence of a specific neurotransmitter, but only of the presence of the receptor in question and how it may affect the behavioral response under examination. Nevertheless, studies using such an approach provide useful and perhaps important data concerning the functional relationships of the receptor under consideration to specific forms of aggression. The remainder of this subsection summarizes a variety of studies that provide support for a role of ACh in mediating aggression and rage.

Several reports have shown that microinjections of carbachol into the anterior hypothalamus of the rat can induce vocalizations similar to those that follow foot shock (Brudzynski and Bihara, 1990; Brudzynski et al., 1991; Brudzynski, 1994). The authors indicated that the vocalizations represented forms of emotional excitement. Such excitement could be interpreted to reflect anxiety, distress or perhaps a component of defensive behavior. Several investigators have demonstrated that microinjections of carbachol can also induce mouse killing in rats (Smith et al., 1970). Moreover, administration of atropine was shown to block predatory killing in killer rats (Smith et al., 1970).

Other investigators have attempted to link cholinergic receptor activation or blockade in several limbic structures to the regulation of aggression in the rat. For example, cholinergic stimulation of the septal area produces an increase in irritability in rats (Igic et al., 1970). In contrast, administration of scopolamine into the medial amygdala was shown to block mouse killing (Leaf et al., 1969). Likewise, cholinergic blockade (with scopolamine) in the basolateral amygdala and hippocampal formation reduced fighting in rats (Rodgers and Brown, 1976).

Parallel studies have been conducted in the cat involving intracerebral microinjections of cholinergic agents placed mainly into the hypothalamus. One of the first investigators to demonstrate that cholinergic activation of the feline hypothalamus with carbachol can induce defensive rage responses was Myers (1964). Baxter (1968a, 1968b) expanded on these findings by first identifying hypothalamic defensive rage sites in the cat by electrical stimulation and then obtained the same responses after microinjecting carbachol into the same sites. Similar observations were made in subsequent years by other investigators such as Allikmets (1974), Romaniuk et al. (1973, 1974), and later by Brudzynski (1981a, 1981b), Brudzynski and Eckersdorf (1988), and Brudzynski et al. (1993) following cholinergic stimulation of either the hypothalamus or PAG. Several investigators characterized the receptor properties of this response. Romaniuk et al. (1973) showed that pretreatment with the muscarinic antagonist, atropine, into the same medial hypothalamic site where carbachol was injected, blocked the defensive rage response induced by carbachol. However, the carbachol-induced rage response was not affected when the hypothalamic sites was pretreated with a nicotinic receptor antagonist (betamon). That the cholinergic-induced defensive rage response is mediated in part through muscarinic receptors was further supported by a subsequent study by Brudzynski (1981a) who showed that the rage response elicited by carbachol could not be induced by nicotine.

The studies described above have examined the role of cholinergic receptors in defensive rage behavior induced directly by chemical stimulation of the medial hypothalamus. A somewhat different approach was used by Brudzynski et al. (1990) in order to demonstrate the importance of muscarinic receptors in defensive rage. In this study, defensive responses were induced by exposing a cat to a dog, and atropine sulfate was either administered systemically or injected bilaterally into the ventromedial aspect of the borderline region of the anterior hypothalamus and preoptic area. Pretreatment by the cholinergic antagonist was shown to block or suppress the naturally occurring defensive rage response, thus providing additional support for the role of muscarinic receptors in the expression of defensive rage behavior.

Genetic Approach

A different approach to the study of the role of cholinergic mechanisms in aggression was used by Pucilowski et al. (1991). These authors used two rodent lines. The first, called the Flinders Sensitive Line, were selectively bred for increased sensitivity to cholinergic agonists presumably due to increased muscarinic binding and ACh synthesis rates. The second, called the Flinders Resistant Line, appear to lack these properties. When these two groups of rats were compared with respect to shock-induced and apomorphine-induced fighting tests, the Flinders Sensitive Line were considerably more aggressive. Interestingly, the authors also reported that the Flinders Sensitive Line rats displayed a higher pain threshold, indicating that one could not attribute the increase in aggression to a greater sensitivity to pain. Thus, this finding provides additional evidence of cholinergic involvement in aggressive behavior, which is consistent with the results of studies using classical pharmacological approaches presented above.

Summary

The studies described above offer data supporting a cholinergic (excitatory) mechanism involved in the overall expression of defensive rage behavior. In particular, there is good evidence confirming the presence of muscarinic receptors in this behavior. That defensive rage behavior can be suppressed or eliminated by cholinergic receptor blockade provides support for the presence of cholinergic neurotransmitter function as well in this process. A central question that remains to be answered is how cholinergic receptors and neurons are organized within the hypothalamus and elsewhere in the limbic-hypothalamic-midbrain axis in mediating the expression of defensive rage behavior. In particular, it would be important to know whether cholinergic neurons in the medial hypothalamus serve as interneurons or whether they contain long-descending axons that directly mediate the output of defensive rage behavior.

NOREPINEPHRINE

SYNTHESIS AND REMOVAL

The stages of synthesis are briefly described here. Tyrosine (hydroxy-phenylalanine), present in food, is synthesized from phenylalanine. After tyrosine enters the neuron by active transport, it is first converted into dihydroxyphenalanine (DOPA) by the enzyme, tyrosine hydroxylase, and then converted into dopamine by the enzyme, L-amino acid decarboxylase (DOPA-decarboxylase). Dopamine is then actively transported into storage vesicles where it is converted into norepinephrine by the enzyme, dopamine-β-hydroxylase (DBH). At this point, norepinephrine is released by exocytosis. After norepinephrine is released into the synaptic cleft, it is removed *primarily* by re-uptake into the synaptic terminal. Other mechanisms that somewhat contribute include inactivation by catechol-O-methyltransferase (COMT) in the effector cell or by destruction in the liver by COMT and monoamine oxidase (MAO).

DISTRIBUTION AND FUNCTIONS

The greatest concentration of noradrenergic neurons is located within the pons in the locus ceruleus. Other groups of noradrenergic neurons are located in different regions of the medulla. These neurons project to the forebrain, including the limbic system, diencephalon, and cerebral cortex. A well-known function of norepinephrine is that it is released from postganglionic sympathetic nerve terminals. Norepinephrine plays important roles in many behavioral processes, such as in states of wakefulness, cortical activation, and the treatment for depression.

ROLE OF NOREPINEPHRINE IN AGGRESSION AND RAGE

Effects of Systemic and Intracerebroventricular Administration of Noradrenergic Agents

A number of studies have been conducted in several species that have addressed the question of the relationship of norepinephrine to aggressive behavior. These studies have addressed this problem through the use of peripheral or intracerebroventricular (icv) administration of drugs, which have either noradrenergic agonistic or antagonistic properties. However, findings do not appear to be entirely consistent, especially when compared with other experiments using different methodological approaches (described below). A number of investigators employing systemic administration of noradrenergic agents have suggested that norepinephrine has an inhibitory effect on aggressive processes. For example, icv administration of norepinephrine (Geyer and Segal, 1974) seemed to suppress shock-induced fighting in rodents, and icv delivery of 6-hydroxydopamine (6-OHDA), which depletes catecholamines, causes a permanent increase in irritability in rats. Likewise, chronic systemic treatment with methyldopa, which depletes norepinephrine (and presumably serotonin and dopamine), can induce aggressive behavior in mice (Dominic and Moore, 1971). In contrast, Kemble et al. (1991) showed that administration of the α_2-adrenergic antagonist, yohimbine, inhibited isolation-induced aggression in mice. Likewise, several studies where amphetamine was administered facilitated brain stimulation–induced aggression in cats (Sheard, 1967; Marini et al., 1979). Further confusing the picture here was the observation that following catecholamine depletion by alpha methyl paratyrosine (AMPT), reflexive biting remains intact (Katz and Thomas, 1975).

Neurochemical Measurements

One of the earlier approaches to the study of the relationship of neurotransmitters to aggressive behavior was to determine whether the expression of aggressive behavior could affect changes in regional neurotransmitter levels in the brain. In a series of experiments, Reis et al. (Reis and Gunne, 1965; Reis and Fuxe, 1968, 1969; Reis, 1974) observed depletions of norepinephrine in the brainstem following the expression of rage behavior in the cat, suggesting an inverse relationship between rage and norepinephrine levels. In contrast, Lamprecht et al. (1972) reported that rats subjected to extended periods of immobilization stress were more aggressive as determined from shock-induced fighting. Here, such animals revealed higher serum levels of both dopamine beta hydroxylase as well as tyrosine hydroxylase, suggesting a possible relationship between catecholamine synthesis and aggressive behavior.

Other studies also examined the relationship of norepinephrine turnover and aggression. Welch and Welch (1965, 1968) showed that brain norepinephrine levels of isolated mice (which tends to make them more aggressive) were higher than in aggregated ones. In addition, Eichelmann (1973) provided data consistent with those of Welch and Welch, suggesting that an increase in norepinephrine turnover rates is associated with an increase in aggression as determined by the shock-induced aggression paradigm. When aggression in rats is induced by olfactory bulbectomy (Ueki and Yoshimura, 1981), there was an increase in norepinephrine content in the hypothalamus relative to intact or sham operated rats. However, changes in regional brain norepinephrine levels as a result of fighting were not observed, leading these authors to suggest that norepinephrine may not participate in mediating mouse-killing behavior.

Thus, as seen in the results described above, it is difficult to draw conclusions concerning the unity of the effects of norepinephrine on aggression and rage. Several factors may account for these divergent conclusions. One is possibility that the relationships of norepinephrine to different forms of aggression may differ and that clear-cut distinctions between affective (defensive), predatory, and perhaps other forms of aggression may not have been clearly indicated. A second possibility involves the roles played by different subtypes of noradrenergic receptors. It is possible that when different drugs are employed in various studies, different receptor subtypes may have become activated, thus causing different actions on the attack mechanism.

A third approach involves the examination of the effects of direct infusion of noradrenergic compounds into specific regions of the brain on aggression and rage. Studies using this approach are described here.

Intracerebral Injections

In general, intracerebral injections of drugs have been directed mainly into regions of the brain normally associated with aggression and rage in rats and cats. These include the hypothalamus and occasionally, into limbic structures such as the amygdala and septal area. A representative sampling of the studies conducted in this manner follows.

The literature contains very few studies in which a transmitter was directly injected into a specific region of the brain. However, in an early study, Myers (1964) microinjected epinephrine directly into different regions of the hypothalamus of the cat, and observed drowsiness at low doses and sleep-like behavior at higher doses. In another study, Torda (1976) microinjected catecholamines, which included norepinephrine, epinephrine, and dopamine, into the ventromedial hypothalamus of the rat. This author, likewise, reported inhibition of shock-induced aggression in rats after administration of epinephrine but, when a cocktail of dopamine and norepinephrine was injected into this region, facilitation of this form of aggression was noted. In this study, facilitatory effects were also reduced or eliminated following microinjections of α and β adrenergic blocking agents or after catecholamine depletion by AMPT. On the other hand, Romaniuk and Golebiewski (1979) failed to observe any changes in behavior following microinjections of norepinephrine into sites within the medial hypothalamus from which carbachol injections had elicited defensive rage in the cat.

The general underlying assumption on which these studies were based is that noradrenergic fibers project from the pontine tegmentum to the hypothalamus and limbic structures and interact with noradrenergic receptors in these regions to modulate aggression and rage behavior. In one study, the hypothesis that noradrenergic fibers facilitate defensive rage behavior in the cat at the level of the anterior hypothalamus was tested directly (Barrett et al., 1987, 1990). In order to test this hypothesis, stimulating electrodes were implanted into the region of the ventromedial hypothalamic nucleus and cannula-electrodes were implanted into the anterior medial hypothalamus preoptic area, which receives major inputs from the ventromedial nucleus (Figure 7.1). In this study, noradrenergic compounds were microinjected into the anterior hypothalamic sites from which defensive rage could be elicited, and defensive rage was elicited by electrical stimulation at sites in the ventromedial nucleus. Administration of either norepinephrine or the α-2 agonist, clonidine, into the anterior hypothalamic defensive rage site facilitated the occurrence of this response when elicited from the ventromedial nucleus (Figure 7.2). These investigators further showed that pretreatment with the α-2 antagonist, yohimbine, blocked the effects of both norepinephrine and clonidine. In contrast,

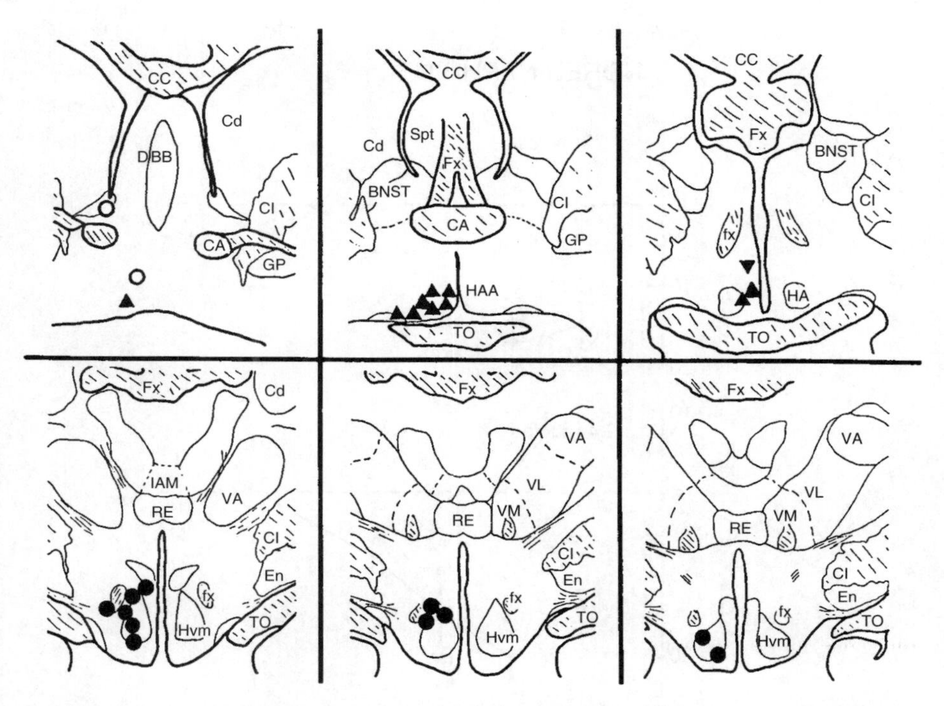

FIGURE 7.1 Loci of infusion cannulae and stimulating electrodes. Cannula electrode sites within the anterior hypothalamus where infusion of norepinephrine reduced hissing threshold (closed triangles facing up). Site where norepinephrine elevated hissing threshold (closed triangle facing down). Site at which electrical stimulation failed to evoke hissing, and norepinephrine failed to alter hissing threshold (open circles). Sites within ventromedial hypothalamus from which defensive rage was elicited (closed circle). BNST, bed nucleus of stria terminalis; CA, anterior commissure; cc, corpus callosum; CI, internal capsule; DBB, diagonal band of Broca; Fx, fornix; Hvm, ventromedial nucleus of hypothalamus; RE, nucleus reuniens; TO, optic tract. (From J. Barrett, H. Edinger, A. Siegel, *Brain Res.*, 525, 1990, 285–293. With permission.)

β-adrenergic compounds had little or no effect on the defensive rage response. It was concluded from this study that activation of α-2 adrenergic receptors in the anterior medial hypothalamus facilitates defensive rage in the cat.

While the majority of the studies considered the effects of the infusion of catecholaminergic compounds into the hypothalamus, several studies examined the effects of microinjections of these drugs into limbic regions. When epinephrine was injected into the basolateral amygdala of the cat made aggressive by ventromedial hypothalamic lesions, a taming effect was observed (Mark et al., 1975). These observations would appear to suggest an inhibitory role for this transmitter in aggression. Consistent with this finding was the observation that microinjections of α-adrenergic antagonists into the ventral septal area of the rat were effective in increasing muricide and intermale aggression (Albert and Richmond, 1977; Albert et al., 1979).

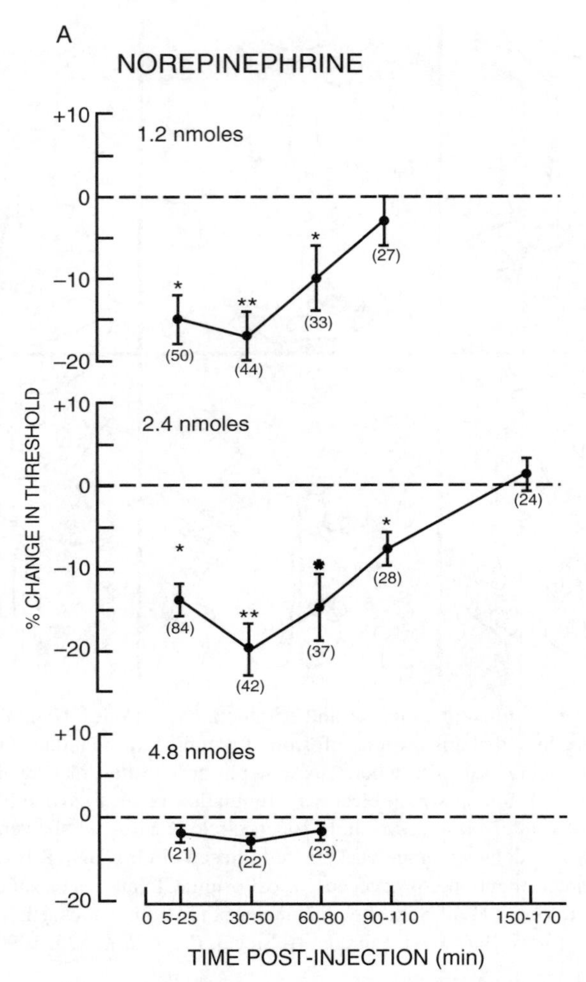

FIGURE 7.2 Effects of (A) norepinephrine and (B) clonidine microinjected in the anterior hypothalamus on hissing threshold. Responses are expressed as percentage change from baseline threshold. The dashed line represents baseline threshold. * Significant difference from threshold ($p < 0.05$); ** $p < 0.01$. Numbers in parentheses indicate number of threshold determinations. (From J. Barrett, H. Edinger, A. Siegel, *Brain Res.*, 525, 1990, 285–293. With permission.)

Summary

In contrast to the highly consistent findings based on studies of cholinergic compounds on aggression and rage, the overall direction of the effects of norepinephrine on aggression and rage were more variable. The reasons for such variability were suggested above. The most likely possibility is that various noradrenergic receptor subtypes mediate different effects on aggressive processes. If this is, indeed, the case, then it would be important for future experiments concerning the role of noradrenergic mechanisms on aggression to attempt to clarify how various

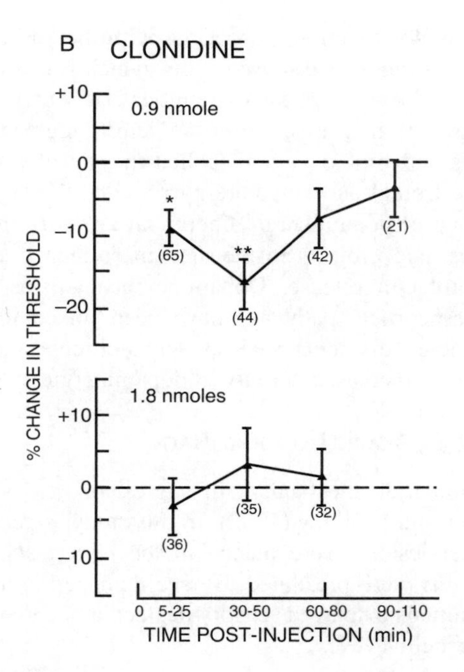

FIGURE 7.2 (Continued).

noradrenergic receptors can alter the different forms of aggression and rage. One aspect of this research does seem to be reasonably clear — namely, that activation of noradrenergic α-2 adrenergic receptors in the anterior medial hypothalamus facilitates defensive rage in the cat. What is not known about this effect is whether activation of these receptors in the hypothalamus selectively affects aggressive behavior, or whether they serve to facilitate other ongoing motivational processes as well. Again, it would be useful if future experiments could address this issue.

DOPAMINE

Synthesis and Removal

The biosynthesis of dopamine is described above as the initial steps in the biosynthesis of norepinephrine (up to the point where dopamine is stored in vesicles). Like norepinephrine, dopamine is removed mainly by re-uptake back into the terminals of the presynaptic neuron. Similar to norepinephrine described above, dopamine may also be removed by the action of COMT.

Distribution and Functions

There are several major groups of dopaminergic neurons. One group arises from the substantia nigra and projects to the neostriatum. These fibers play an important role in the regulation of motor functions associated with the basal ganglia. When such inputs are damaged, *Parkinson's disease* results, which is characterized by a tremor at rest, hypokinesia, and rigidity of the extremities. The other major group of fibers has two components. The first arises from the ventral tegmental area and projects to the hypothalamus and limbic system (called the *mesolimbic pathway*) and the second passes to the frontal and cingulate gyri (called the *mesocortical pathway*). There is a third group of dopaminergic fibers that arises from the arcuate nucleus of the hypothalamus, and projects to the median eminence and anterior pituitary where it inhibits prolactin release. Dopamine mechanisms associated with the mesolimbic and mesocortical pathways have been linked to various psychiatric conditions. In particular, disorders such as schizophrenia and anxiety have been thought to be related to increased activity at dopaminergic synapses.

Role of Dopamine in Aggression and Rage

A role for a dopaminergic mechanism in aggression was suggested in a study published by Bandler and Halliday (1982). In this study, predatory attack in the cat was diminished after lesions were placed in the region surrounding the medial forebrain. Such lesions were presumed to have damaged ascending dopaminergic fibers because the animals displayed sensory neglect, a phenomenon associated with dopamine loss (Marshall, 1979).

However, for the most part, more direct approaches were taken by most investigators in examining the role of dopamine in aggression and rage, and they are basically identical to those described above for the analysis of norepinephrine. Similar to the studies conducted with norepinephrine, the overwhelming numbers of studies conducted with dopaminergic compounds involved the systemic administration of these drugs.

Effects of Systemic and Intracerebroventricular Administration of Dopaminergic Agents

A number of studies have attempted to assess the role of dopamine using indirect approaches. For example, amphetamine, which releases biogenic amines from the nerve terminals, was employed in a number of studies. Since it is difficult to determine whether amphetamine's effects on aggression are due to release of dopamine or norepinephrine, the use of this drug to evaluate transmitter function is imperfect at best. Not surprisingly, the results of these studies have been inconsistent. Several studies suggested facilitatory effects following amphetamine delivery. These included experiments by Hasselager et al. (1972) who reported that administration of amphetamine to mice induced aggressive responses, and those of Lal et al. (1971) who showed that aggressive responses in rats induced by morphine withdrawal can be potentiated by amphetamine. In the cat, methamphetamine, which is similar to amphetamine but which is a more potent stimulant, was shown to facilitate defensive rage elicited from the medial hypothalamus of the cat (Maeda et al., 1985). However, other investigators provided evidence that amphetamines suppress aggressive responses. Examples of such studies include the findings of Malick (1975) and Gay et al. (1975, 1977) who found that muricidal behavior in rats can be antagonized by amphetamines, and Barr et al. (1979) who reported similar results with the use of phenethylamine, which acts somewhat like amphetamine.

A considerable numbers of studies have reported the effects of using of other dopaminergic agents in the study of aggression and rage. The large majority of these studies provided evidence in support of the view that dopamine facilitates aggression and rage. A number of these studies are indicated at this time. Messeri et al. (1975) were able to induce conspecific fighting in rats following peripheral administration of the dopamine agonist, apomorphine. Direct intraventricular injections of dopamine can increase fighting in rats (Geyer and Segal, 1974). Consistent with these findings were the observations of Garmendia et al. (1992) who showed that administration of the neuroleptic, clozapine, which acts as a dopamine D_1 antagonist, caused a significant decrease in fighting in mice. Several other studies conducted in mice also indicated that dopamine receptors facilitate aggressive processes. Here, administration of the dopamine D_2 agonist, LY 171555, increased defensive behavior (Cabib and Puglisi-Allegra, 1989), while delivery of the dopamine D_2 antagonist, sulpiride, reduced this response (Puglisi-Allegra and Cabib, 1988).

When dopamine receptors were blocked by peripheral administration of chlorpromazine or haloperidol, hypothalamically elicited defensive rage behavior in the cat was suppressed (Maeda, 1976). Likewise, Maeda and Maki (1986) observed that when electrolytic lesions of the basolateral amygdala of cats attenuated the facilitatory effects of visual provocation on attack elicited from the hypothalamus, subsequent administration of apomorphine abolished the suppressive effects of the lesions. Two other studies conducted in the cat further elucidated the role of dopamine D_2 receptors in facilitating aggressive processes. In one study, it was shown that peripheral administration of haloperidol alone elevated thresholds for defensive rage behavior, or if injected prior to apomorphine delivery, blocked the facilitatory effects of the dopamine agonist (Sweidan et al., 1990). These authors further showed that administration of the dopamine D_2 agonist, Quinpirol, facilitated defensive rage, while the dopamine D_1 agonist, SKF 38393, was largely ineffective. They also showed that the facilitatory effects of apomorphine were selectively blocked by the dopamine D_2 antagonist, spiperone, but not with the selective dopamine D_1 antagonist, SCH 23390. In a parallel study, these procedures were employed to examine how these dopamine receptors affect predatory attack in the cat (Shaikh et al., 1991b). The interesting finding associated with this study was that dopamine D_2 receptors had similar effects on predatory attack. In brief, dopamine D_2 but not D_1 receptor activation facilitated predatory attack and that spiperone administration resulted in an elevation of predatory attack response thresholds. Thus, it appears that dopamine receptors modulate both defensive rage and predatory attack in similar ways.

Not all findings supported a facilitatory role for dopamine on aggressive processes. Abe et al. (1987) reported that peripheral administration apomorphine inhibited muricide in rats. Thoa et al. (1972) and Nakamura and Thoenen (1972) administered 6-OHDA by an intraventricular route in rats and observed an increase in footshock-induced fighting and irritability, respectively. However, as noted above, it is difficult to determine from this study whether the effects could be attributed to changes in dopamine or norepinephrine levels in the brain. Consistent with these findings, Beleslin et al. (1985) observed that intracerebroventricular administration of compounds, which have dopaminergic receptor antagonist properties, induced aggressive displays in cats.

Neurochemical Measurements

As mentioned previously, another approach to the study of neurotransmitters and aggression involves examination of the tissue content of the neurotransmitter in question in specific regions of the brain of an animal in response to its elicitation of an aggressive response. A number of studies have been conducted using this approach, and a few of them with respect to dopamine are summarized here.

In one study, Tizabi et al. (1979) reported regional increases in dopamine levels in the striatum of mice after isolation-induced aggressive behavior. Hadfield (1981, 1983) conducted a series of studies in which he examined the regional neurochemical kinetics as a result of acute fighting in isolated male mice. In contrast, his major finding was that K_m and V_{max} for dopamine were significantly increased in the prefrontal cortex but not in the neostriatum, pointing to the importance of the mesocortical dopaminergic pathway in aggressive behavior.

Further support for a linkage between dopamine and prefrontal cortical mechanisms in aggression and rage come from several more recent studies. Cabib et al. (2000) observed that mice which were defeated tend to become more defensive. Subsequent interactions involving these mice were associated with increased dopamine metabolism in the prefrontal cortex. Using microdialysis measurements, van Erp and Miczek (2000) reported increases in dopamine levels in the prefrontal cortex (and nucleus accumbens) during aggressive encounters.

Genetic Approaches

Several reports used a genetic approach to test the role of dopamine in aggressive behavior. Pradhan (1973) compared two strains of mice, one of which was bred for high levels of aggressiveness. Here, it was observed that the more aggressive group had higher levels of dopamine than the group that was not as aggressive. In a more recent study, Lewis et al. (1994) also bred mice for high and low levels of aggression. These authors reported that treatment with the selective dopamine D_1 receptor agonist, dihydrexidine, reduced isolation-induced aggression in the aggressive strain of rats, and that pretreatment with the selective dopamine D_1 antagonist, SCH 23390, blocked the effects of dihydrexidine. While the findings of this study support the role of dopamine in aggressive behavior, it is not consistent with several others (described earlier as well as in the next section), which points to the importance of dopamine D_2 receptors in aggression and rage.

Intracerebral Administration of Dopaminergic Compounds

Since it is known that dopaminergic fibers project to wide areas of the forebrain, including the hypothalamus, a study was performed by Sweidan et al. (1991) to determine what role dopaminergic innervation of the medial hypothalamus has in regulating defensive rage behavior in the cat. The rationale for this study was basically the same as that described above for the analysis of noradrenergic mechanisms in the hypothalamus (Barrett et al., 1990). In brief, stimulating cannula-electrodes were initially implanted into sites within the ventromedial and anterior hypothalamus, respectively, from which defensive rage could be elicited. Then, dopaminergic compounds were microinjected into defensive rage sites in the anterior hypothalamus, and the effects of such drug administration on defensive rage elicited from the ventromedial hypothalamus were assessed. Microinjections of either apomorphine or of the selective dopamine D_2 agonist, LY 171555, facilitated defensive rage behavior (Figure 7.3). However, microinjections of the selective dopamine D_1 agonist, SKF 38393, had little effect on the attack response. Of further interest were the observations that administration of either the nonselective dopamine antagonist, haloperidol, or the selective dopamine D_2 antagonist, sulpiride, alone (but not the selective dopamine D_1 antagonist, SCH 23390) into the anterior hypothalamus, elevated response thresholds, and blocked the effects of apomorphine or LY 171555 when delivered prior to delivery of the agonist. These findings demonstrate that dopaminergic stimulation in the anterior hypothalamus, acting via dopamine D_2 receptors, facilitate defensive rage in the cat.

Summary

Overall findings from the studies described above support the view that dopamine facilitates various forms of aggression and rage behavior. Moreover, evidence was presented which indicates that the facilitatory effects of dopamine are mediated through dopamine D_2 receptors in the anterior medial hypothalamus (and perhaps elsewhere). As indicated in the discussion of norepinephrine, it is indeed possible that the dopaminergic effects on aggression and rage may not be selective for these processes, but may serve to enhance a variety of motivational processes associated with hypothalamic processes. Answers to this and related questions can only be achieved through a systematic analysis of the manner in which dopaminergic mechanisms may influence the varieties of functions of the hypothalamus and related regions of the brain.

SEROTONIN

Synthesis and Removal

Since serotonin does not cross the blood–brain barrier, serotonin in the brain results from the synthesis in selective cell bodies in the brainstem. The substrate for serotonin synthesis is dietary tryptophan. Tryptophan in plasma passes through the blood–brain barrier. It is hydroxylated by tryptophan hydroxylase at the 5 position, then decarboxylated by L-amino acid decarboxylase, forming serotonin, and stored in vesicles for future release. Two mechanisms are involved in the removal of serotonin. One involves re-uptake, which is facilitated by high-affinity re-uptake sites within the serotonin nerve terminals. The other is by metabolism. Serotonin is deaminated to 5-hydroxyindoleacetaldehyde by monoamine oxidase, and is then oxidized to form 5-hydroxyindoleacetic acid by aldehyde dehydrogenase. It is ultimately excreted in the urine.

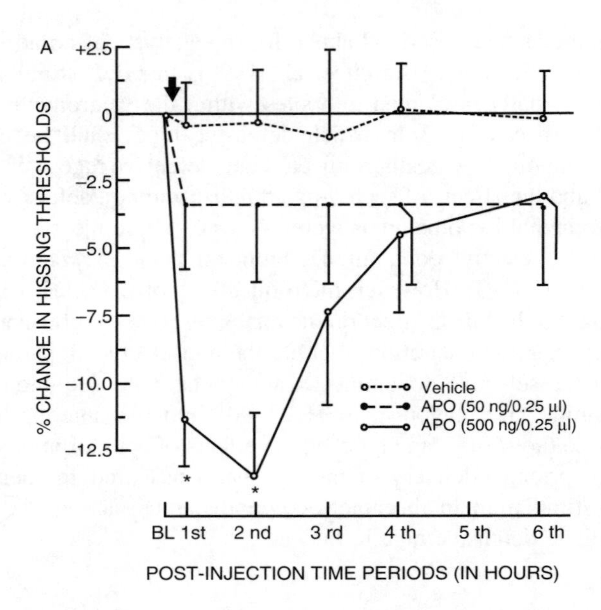

FIGURE 7.3 Time course of the effects of microinjections of (A) apomorphine or vehicle, and (B) the selective dopamine D_2 agonist, quinpirole (LY 171555), into the anterior hypothalamus on ventromedial hypothalamically elicited hissing. Each point represents the average change in hissing threshold current expressed as a percentage relative to preinjection baseline threshold (BL). * $p < 0.05$ compared to vehicle. (From S. Sweidan, H. Edinger, A. Siegel, *Brain Res.*, 549, 1991, 127–137. With permission.)

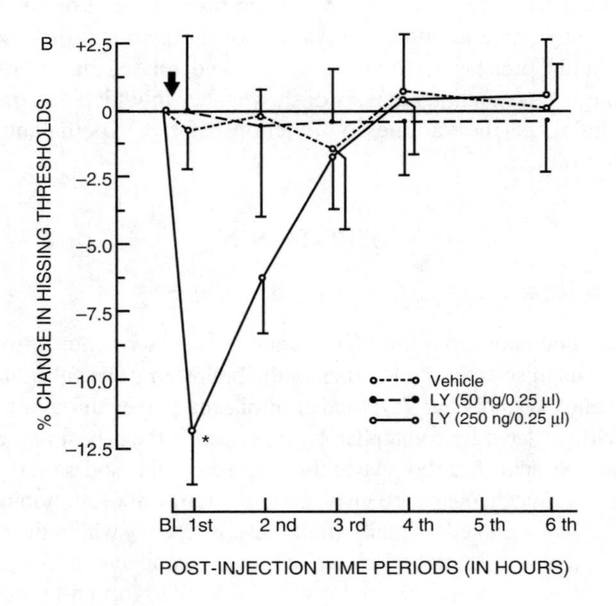

FIGURE 7.3 (Continued).

Distribution and Functions

Various groups of serotonin containing neurons are located mainly along the midline of the medulla, pons, and midbrain. Neurons from the dorsal and median pontine and midbrain raphe nuclei project through the medial forebrain bundle to wide areas of the forebrain, including the diencephalon, limbic system, and cerebral cortex. Serotonin neurons arising from the medulla project to the dorsal horn of the spinal cord. The projections to the diencephalon and limbic system most likely modulate functions associated with these regions. These include visceral processes such as feeding, drinking, endocrine, affective, thermoregulatory and autonomic functions, as well as aggression and rage, which are considered in depth below. Of considerable importance has been the linkage between serotonin and emotional disorders, including depression, anxiety, and obsessive-compulsive disorders. In fact, one of the most popular pharmacological treatments for these disorders involve the use of serotonin re-uptake inhibitors to increase serotonin levels in the brain. Projections to the cerebral cortex have been shown to be associated with sleep and wakefulness and descending projections to the spinal cord are related to the descending brainstem pathway that inhibits pain impulses.

Role of Serotonin in Aggression and Rage

Of all the neurotransmitters that have been investigated with respect to aggressive behavior, none has received more attention than serotonin. Some studies considered how stimulation or electrolytic or neurochemical lesions of serotonergic neurons affect aggression and rage. However, the overwhelming numbers of studies have used peripheral administration of serotonergic compounds, including drugs that deplete serotonin, and conducted mainly in rodents, to identify how serotonin receptor activation or blockade affects these processes. In contrast, few studies have attempted to determine the effects of intracerebral microinjections of serotonergic compounds. Other approaches have also been applied. One such approach involved the measurement of central or peripheral levels of serotonin and correlating those levels with changes in the propensity for the expression of the responses under consideration. A second approach has examined the role of dietary manipulation of tryptophan levels and a third has considered how genetic factors affect aggression and rage. The discussion below reviews a representative number of these studies in an attempt to gain an understanding of the overall role of serotonin receptors in the regulation of aggression and rage.

Lesions or Stimulation of Serotonergic Nuclei

As indicated by the large number of studies summarized below, most researchers studying the serotonin system have manipulated this neurotransmitter by pharmacological means. In contrast, several earlier investigators attempted to do the same by examining the effects of either destruction (by the placement of lesions) or activation (by electrical stimulation) of serotonin neurons and then observing the effects of such procedures on aggression and rage.

Several studies examined the effects of raphe lesions on aggression. One study, conducted by Sheard (1973), compared the effects of electrolytic lesions of the raphe nuclei with those seen with *p*-chlorophenylalanine (PCPA), which depletes brain serotonin, on aggressive responsive in rats. Sheard reported that the increase in aggressive behavior seen following PCPA lesions was not seen after electrolytic lesions. In this study, the difference in effects following each of these procedures could possibly have been attributed to the fact that a more complete depletion of serotonin occurred after administration of PCPA. However, several other studies, positive effects of electrolytic lesions were reported. Vergnes et al. (1973, 1974) showed that destruction of the dorsal and medial raphe nuclei (but not lesions limited to one of these two nuclei) induced mouse-killing behavior in rats. Jacobs and Cohen (1976) examined the effects of electrolytic lesions in the dorsal or median raphe nuclei of rats, which resulted in significant reductions in forebrain serotonin levels, on a different form of aggression — that is, pain-elicited aggression. With respect to behavioral effects, dorsal raphe lesions produced long-lasting increases in this form of aggression, while median raphe lesions were ineffective.

Concerning the effects of electrical stimulation of the raphe complex, Kostowski et al. (1980) reported that such stimulation of the dorsal raphe nuclei markedly reduced muricide in rats. This author went on to provide additional evidence of serotonergic involvement in the modulation of aggression by demonstrating that activation of 5-HT receptors by quipazine, which is generally characterized as a nonselective agonist, strongly decreased mouse-killing behavior.

Thus, the findings from lesion and stimulation studies suggest that serotonin suppresses several different forms of aggressive behavior in rodents. The studies described below overall provide additional support for these initial findings.

Neurochemical Depletion of Serotonin

While depletion of serotonin can be achieved by the placement of lesions in the brainstem that destroy large numbers of serotonin neurons, a more sophisticated approach applies a chemical agent that selectively depletes serotonin neurons by blocking synthesis or by acting as a neurotoxin. As indicated above, PCPA is such a drug (as it prevents serotonin synthesis by inhibiting tryptophan hydroxylase), and has been used widely to examine the effects of reductions in serotonin on various forms of aggression as well as other behaviors. Another drug that has been used in a parallel manner is 5,7 dihydroxytryptamine (5,7 DHT), which acts as a selective serotonin neurotoxin. The findings using these drugs are summarized below in this section of the chapter but reflect some inconsistencies among investigators. The basis for such inconsistencies is considered at the end of this section.

Based on the data derived from lesions and stimulation described above, it would be predicted at a simplistic level that depletion of serotonin should increase the propensity for aggression. In fact, the majority of studies using neurochemical methods for depleting brain serotonin have yielded results consistent with this prediction. Several independent investigations reported that serotonin depletion caused by PCPA administration induced mouse killing in rats (Miczek et al., 1975; Vergnes, 1980; Valzelli et al., 1981). Likewise, increases in aggressive behavior in rats following PCPA treatment were also described by Sheard (Sheard, 1969, 1970) and later by Knutson et al. (1979). In contrast to these findings, Malick and Barnett (1976) induced aggression in mice by isolating them, and then observed that the aggressiveness was reduced by administration of PCPA. Similar approaches have also been used by Katz and Thomas (1976) and by Ferguson et al. (1970) in several studies conducted in cats. Here, the results were parallel to the majority of studies described above in the rat; namely, that peripheral administration of PCPA resulted in potentiation of defensive rage and predatory attack, respectively.

With respect to 5,7-DHT, File and Deakin (1980) placed microinjections of this neurotoxin into the dorsal and median raphe nuclei of rats. Animals with lesions of the median (but not the dorsal) raphe nucleus displayed increases in dominance behaviors to intruders placed in the home cages. Similar facilitation of mouse-killing behavior in rats was reported by Vergnes and Kempf (1982) following administration of 5,7-DHT into the lateral hypothalamus. Several studies conducted in cats also produced similar results. When 5,7-DHT was administered peripherally, cats showed an enhancement of growling induced by carbachol infusion into the hypothalamus (Koprowska and Romaniuk, 1997). These authors further showed that cats that were previously submissive displayed more dominance and fighting behavior after the placement of the chemical lesions. Likewise, when 5,7-DHT lesions were placed in the dorsal raphe complex, there was a consequent increase in growling induced by carbachol injections into the hypothalamus. When viewed collectively, the evidence indicates that reductions in serotonin levels by either PCPA or 5,7-DHT administrations are associated with an increase in different forms of aggressive behavior. As previously noted, in this context, these data are clearly consistent with other approaches described above suggesting an inhibitory role of serotonin in the regulation of aggressive processes.

Depletion of Serotonin Levels by Diet

We have seen from the studies described above that a most common way of causing serotonin depletion involves the application of a neurochemical toxin or synthesis inhibitor. Since tryptophan is the primary amino acid essential for the synthesis of serotonin and it is also commonly present in the diet, several investigators have attempted to reduce serotonin levels by the use of a low-tryptophan diet given to rats and mice over a period of weeks (Kantak et al., 1979a, 1979b; Vergnes and Kempf, 1981).

In several studies, Kantak et al. (1979a, 1979b) examined the effects of administration of a tryptophan-free diet on cricket killing in mice. Interestingly, these authors reported differential effects of reduced brain serotonin levels on isolation-induced killing and predatory responses. Specifically, there was a reduction in the duration of isolation-induced fighting, but an increase in the number of mice expressing cricket killing coupled with decreased latencies to attack and to kill the mice. In another study, Vergnes and Kempf (1981) placed rats on a low-tryptophan diet for several weeks, causing a 75% reduction in brain serotonin levels while catecholamine levels remained relatively the same. When rats were placed in an environment that allowed for interactions between rats and mice prior to initiation of dietary constraints, there was a subsequent suppression of mouse killing in these rats. However, after the rats were placed on the low-tryptophan diet for several weeks, there was a subsequent increase in mouse killing that persisted even to the period after they were returned to a normal diet. Consistent with these findings is the observation that peripheral administration of L-tryptophan results in inhibition of muricide in rats (Broderick and Bridger, 1984). Thus, these observations provide further support to the view presented thus far that serotonin has a suppressive effect on aggressive behavior.

Genetic Manipulation of Brain Serotonin Levels

There are several approaches to the study of genetic factors involved in aggressive behavior. One source of support for a genetic basis comes from studies in which animals are specifically bred for high levels of aggression. In one study, Korte et al. (1996) showed a correlation between mice expressing high levels of aggression, as demonstrated by short latency attacks, and elevated levels of 5-HT_{1A} mRNA in the dorsal hippocampus relative to levels observed in low-aggressive mice that required longer latencies to express attack. In essence, this study indicated that genetic selection for high offensive aggression also selects at the same time for increased postsynaptic 5-HT_{1A} receptors in limbic and related regions of the forebrain. In another study, Lyons et al. (1999) explored the role of brain-derived neurotrophic factor (BDNF), which has a trophic effect on 5-HT neurons in the central nervous system (CNS), on aggressiveness in mice. Here, animals that were bred for deficiencies in BDNF displayed losses in 5-HT axons and lower 5-HIAA/5-HT ratios in the hypothalamus. These same animals also showed increases in intermale aggressiveness as applied in the resident–intruder model.

An alternate strategy in recent years has been to employ knockout mice — that is, mice created to be deficient in selected genes — to study the role of various neurochemical systems on behavioral processes. As this procedure relates to the present discussion, one group of investigators, using homologous recombination, were able to create mutant knockout mice devoid of the 5-HT_{1B} receptor (Saudou et al., 1994). These animals did not display any behavioral or development defects, but did exhibit heightened aggressiveness as determined from a resident–intruder paradigm. While certain pitfalls in the use of this methodology (similar to most other experimental methods) need to be addressed, this approach certainly represents a new and possibly heuristic direction in the study and analysis of the role of specific neurotransmitter receptors in aggression and rage. The use of knockout mice has also been applied in a recent study by Chiavegatto et al. (2001). Here, mice were genetically engineered in order to disrupt the neuronal nitric acid sythase gene, which the authors believed inhibited aggression, by acting through 5-HT_{1A} and 5-HT_{1B} receptors. These authors showed that such genetic manipulation resulted in decrements in 5-HT turnover and reduced 5-HT_{1A} and 5-HT_{1B} receptor function. To further test this possibility, agonists for 5-HT_{1A} and 5-HT_{1B} receptors were administered systemically and caused a dramatic reduction in aggressive behavior when tested in a resident–intruder paradigm.

Peripheral Administration of Serotonergic Compounds

As noted earlier, the most common approach to the study of serotonin receptors in aggression and rage has involved the examination of the effects of peripheral administration of serotonin compounds. Because this literature is so extensive, and since it is not practical to review all of these studies here, a representative number, which reflect the overall findings, are presented in this section.

Attempts to identify specific serotonin receptors that play important roles in regulating aggression and rage have yielded consistent findings in pointing to the functional importance of the 5-HT_{1A} receptor. White et al. (1991) examined the effects of administration of a variety of serotonergic compounds to mice in an isolation-induced aggression paradigm. They observed that 5-HT_{1A} receptor agonists such as buspirone and gepirone reduced aggression, as did the 5-HT_2 receptor antagonist, ritanserin, while the 5-HT_3 receptor antagonist, zacopride (which also has some affinity for 5-HT_4 receptors) had little effect on the attack response. Interestingly, administration of gepirone also attenuates low frequency vocalization, which likely communicates "affective states" in rodents (Vivian and Miczek, 1993), as well as defensive postures in rats (Blanchard, 1989). Further discussion of the role of 5-HT_2 receptors in rage will be considered later in this section.

Other studies have attempted to test the role of 5-HT receptors more directly by using more specific 5-HT receptor compounds. In a series of studies carried out by Kruk and colleagues, and summarized in Olivier et al. (1990a), Kruk (1991), and Siegel et al. (1999), administration of "serenic" compounds such as fluprazine caused an elevation of threshold responses for hypothalamic stimulation-induced attack by one rat on another. Fluprazine delivery also was shown to inhibit intermale attack in mice (but not prey killing) (Parmigiani and Palanza, 1991). The key point to note here is that these serenic drugs have a high affinity for 5-HT_1 receptors, therefore providing additional support that this receptor plays an important role in serotonin suppression of aggressive responses. Moreover, a number of other investigators have obtained results that indicate that related behaviors such as isolation-induced aggression (Sánchez et al., 1993), intermale aggression (Cologer-Clifford et al., 1996), offensive aggression (De Boer et al., 2000; Korzan et al., 2000), and defensive behaviors associated with anxiety (Blanchard et al., 1993b) are similarly affected 5-HT_1 receptor agonists. These findings thus suggest that with respect to different components of aggression and defense, the inhibitory effects of 5-HT_1 receptor activation are rather extensive.

Studies concerning the roles of other serotonin subtypes have been very few in number, perhaps because of the absence of the lack of receptor specificity for receptors other than 5-HT_1 and possibly 5-HT_2 receptors. In one study, m-chlorophenylpiperazine (m-CPP), a 5-HT receptor agonist, which has a high affinity for 5-HT_2 receptors (and some affinity for 5-HT_{1C} and 5-HT_{2C} receptors) was injected peripherally into male mice (Navarro et al., 1999). It was observed that isolation-induced aggression was decreased following administration of this drug. In another study, described below, opposing effects of delivery of this 5-HT_2 agonist on defensive rage were noted. Thus, delineation of the functions of the 5-HT_2 and other receptors on aggression and rage will require considerable more research coupled with the development of more highly specific compounds.

Central Administration of Serotonergic Compounds

As noted above, peripheral administration of serotonergic drugs represents the primary route of administration for the study of this neurotransmitter. Only a handful of studies have considered the effects of delivery of serotonergic compounds in different regions of the brain, and they are reviewed in this section.

Numerous studies have been conducted in the rat. In one study, 5-HT_{1A} agonists including eltoprazine and 8-OH-DPAT were microinjected into the dorsal raphe nucleus of rats and reduced aggression (Mos et al., 1993). These authors concluded that these drugs activated serotonin 5-HT_1 autoreceptors in the dorsal raphe, thus reducing serotonergic transmission and a nonspecific reduction in aggression. However, the one difficulty with this interpretation of the data is that the larger numbers of findings described earlier indicate that increases (and not deceases) in serotonin transmission are linked to suppression of aggressive behaviors. It is difficult to provide an adequate explanation of these data at the present time. Several studies also conducted in the rat have examined how serotonergic compounds microinjected into the amygdala may affect aggressive responses. In one study, Rodgers (1977) reported that bilateral administration of 5-HT into the cortico-medial amygdala (whose activation excites neurons in the medial hypothalamus) depressed fighting induced by electric shock. Similar findings of suppression of aggressive postures in shock-induced fighting were observed by Pucilowski et al. (1985) following bilateral injections of 5-HT or the nonselective agonist, quipazine, into the cortico-medial amygdala. This would suggest that the release of serotonin acts mainly on 5-HT_{1A} receptors in the medial amygdala to suppress neuronal activity and thus reduce its excitatory inputs into the medial hypothalamus which is associated with this form of aggression.

Microinjections of serotonergic compounds have also been placed in sites in the preoptico-hypothalamus of the rodent and cat and in the PAG of the cat. In one study, the 5-HT_{1A} agonist, 8-OH-DPAT, was infused into the preoptic region of the mouse and decreased aggression (Cologer-Clifford et al., 1997). In a study conducted in the cat, Golebiewski and Romaniuk (1985) induced components of defensive rage by injecting carbachol into the medial hypothalamus. They then injected either 5-HT or the 5-HT antagonist, methysergide, into these sites and observed that 5-HT administration suppressed the level of hissing and growling, while methysergide enhanced vocalization induced by carbachol delivery.

In a more recent study, Shaikh et al. (1997) sought to determine the effects of serotonergic receptor activation in the PAG on hypothalamically elicited defensive rage in the cat. In this experiment, the 5-HT_{1A} agonist, 8-OH-DPAT, was microinjected into PAG sites from which defensive rage could also be elicited and was found to powerfully suppress defensive rage elicited from the medial hypothalamus. The specificity of this effect was further demonstrated when pretreatment with a 5-HT_{1A} antagonist blocked the suppressive effects of the agonist. In contrast, administration of the 5-HT_2 agonist, m-CPP, facilitated the occurrence of the defensive rage response. These authors further showed by immunocytochemical methods that serotonin-positive fibers were present in the region of the PAG where the compounds were infused.

Results of these studies generally support the view that 5-HT_{1A} receptors in regions of the brain that are normally associated with the expression of rage behavior as well as regions linked to its potentiation serve to suppress this response. However, the finding that 5-HT_2 receptor activation facilitates defensive rage indicates the presence of a differentiation of function by these receptors on the rage mechanism. It would be of considerable interest to determine whether activation of these serotonin receptors in other regions such as the hypothalamus can also differentially modulate defensive rage.

Neurochemical Measurements and Approaches

Several other approaches have been used to gain a better understanding of the role of serotonin in aggression and rage. Studies summarized briefly here involved indirect measurements of serotonin levels and were carried out in rodents, monkeys, and humans.

With respect to rodents, an early study demonstrated that decreased levels of 5-HT turnover and 5-HIAA in mice, which were subject to isolation, were also correlated with heightened aggressiveness (Kostowski and Valzelli, 1974). In a more recent study, mice were exposed to low levels of x-rays, and then, changes in brain turnover of 5-HT were correlated with their tendencies to display aggression. They observed that rats which were irradiated had faster turnover rates of 5-HT and showed a calming effect in comparison to sham-irradiated animals which displayed lower turnover rates and were more aggressive (Miyachi et al., 1994).

A somewhat different approach was taken by Higley et al. (1996a, 1996c) in which they studied monkeys in natural group settings while measuring 5-hydroxyindole-acetic acid (5-HIAA), the major metabolite of serotonin. In several related studies, female macaques with low levels of 5-HIAA in the cerebrospinal fluid (CSF) exhibited higher rates of spontaneous aggression, and were more likely to be removed from the group because of their wounding of other monkeys or for treatment of injuries. Moreover, animals that had a high ranking in the social group tended to have lower levels of 5-HIAA in the CSF (Higley et al., 1996a, 1996c). In another study, it was shown that low concentrations of 5-HIAA in the CSF of monkeys, when quantified early in life, served as an effective biological predictor of future excessive levels of aggression (Higley et al., 1996b).

There is rather extensive literature concerning similar studies conducted in humans in which 5-HIAA in blood was used as an index of serotonin levels and compared to levels of aggressiveness. Because of the size of this literature, the general findings will be briefly summarized at this time. It is interesting to note that studies by Coccaro et al. have shown a similar relationship between 5-HIAA levels and aggression to what was described above in monkeys (Coccaro, 1989; Coccaro et al., 1990, 1997). The general observation from these and related studies indicate that there is a reduction in 5-HIAA levels in individuals who are classified as antisocial. Related characteristics such as impulsive behavior and dyscontrol were also linked to low 5-HIAA levels. In their recent review of this literature, Moore et al. (2002) reached a similar conclusion and further suggested that reduced serotonin levels may be characteristic of a wide variety of individuals who express antisocial behavior, even though they may commit different types of crimes or may be associated with different psychiatric diagnoses.

SUMMARY

Studies summarized in this section involve various species, including humans, and a variety of methodologies. In spite of the diverse approaches used, it is of interest to note that the findings from these studies seem to point to the same conclusion, namely, that serotonin has a suppressive effect on aggression, rage, and related processes. It also appears that this phenomenon extends over different species, ranging from rodent to humans. The studies further suggest that different serotonin receptors do not function in the same way. The literature clearly indicates that 5-HT_{1A} receptor activation is associated with suppression of aggression and rage, while the more limited data regarding 5-HT_2 receptors suggest that activation of this receptor potentiates these same processes. Further studies are definitely required to confirm the role of 5-HT_2 receptors, as well as to identify the roles of other serotonin receptors.

PEPTIDES

Many peptides have been identified. However, only a few have been implicated in the regulation and control of aggression and rage behavior. These include opioid peptides, substance P (SP), and cholecystokinin (CCK). With respect to aggression and rage, few studies have been conducted concerning SP and CCK, while greater attention has been given to the role of opioid peptides. The discussion below provides an overview of the studies conducted on opioid peptides and summarizes the few studies concerning other peptides.

OPIOID PEPTIDES

CHARACTERIZATION

The three classes of opioid peptides include β-endorphins, enkephalins, and dynorphins.

β-endorphins are derived from pre-proopiomelanocortin, which is present in the rough endoplasmic reticulum of cells in the anterior and intermediate lobes of the pituitary as well as in the arcuate nucleus of the hypothalamus. The propeptide, proopiomelancortin, is generated after removal of the signal peptide from the prepeptide within the endoplasmic reticulum. It is then packaged into vesicles in the Golgi apparatus of the cell and transported down the axon. In the axon terminal, further proteolytic processing occurs, resulting in the formation of the active peptides, β-lipotropin and ACTH. Beta-lipotropin is cleaved into β-endorphin and β-lipotropin These endorphins are packaged in synaptic vesicles and released by exocytosis as neurotransmitters.

Enkephalins are formed in the following way: the pre-propeptide, pre-proenkephalin A, is present in the rough endoplasmic reticulum of neurons in different regions of the brain. When the signal peptide from the propeptide, proenkephalin A, is removed, the propeptide, proenkephalin A, is generated and packaged in vesicles in the Golgi apparatus. The vesicles are then transported down the axon to the synaptic terminal and further proteolytic processing at this site results in the formation of the active peptides, methionine and leucine enkephalin. Similar to β-endorphin, these enkephalins are packaged in vesicles in the synapse and released as neurotransmitters by exocytosis.

Dynorphins are derived from prodynorphin peptides, which are C-terminal extended forms of Leu^5-enkephalin. In general, the actions of the opioid peptides end by the presence of peptidases, which lead to their degradation.

Distribution and Functions

Enkephalinergic neurons are found in different areas of the CNS. These regions include the spinal cord and PAG, where they contribute to the modulation of pain impulses; hypothalamus and limbic structures, where they are involved in processes such as temperature regulation, blood pressure, feeding, sexual behavior, and aggression and rage (discussed below); and basal ganglia, where they contribute to the control of motor functions.

Role of Enkephalins in Aggression and Rage

Studies Conducted in Rodents

Most studies concerning opioid peptides have been conducted in rodents and involved systemic administration of drugs. The general pattern of findings suggests that they have antiaggressive properties (Gianutsos and Lal, 1978). One line of supporting evidence comes from experiments showing that morphine withdrawal induces aggression (Thor, 1971; Davis and Khalsa, 1971; Moyer and Crabtree, 1973; Gianutsos and Lal, 1976; Gianutsos et al., 1976; Gianutsos and Lal, 1978). In contrast, findings from a number of investigators who have used the shock-induced and predatory aggression models have not been as clear-cut as the data obtained from morphine administration. For example, several authors studying the effects of opioid peptides on shock-induced aggression have reported potentiation of aggression following naloxone administration (Fanselow et al., 1980; Puglisi-Allegra et al., 1982; Tazi et al., 1983), while no direct effects were observed in a third study (Rodgers and Hendrie, 1982). With respect to predatory attack in rodents, two papers suggested that there was little relationship between opioid blockade (by naloxone) and prey killing (Kromer and Dum, 1980; Walsh et al., 1984).

It is difficult to account for the variations in effects of opioid peptides on a given model of aggression. However, one may conjecture that the directionality of the effects of opioids may be related to a specific form of aggression. A possible reason for thinking along these lines is that predatory attack and defensive rage are associated with positive and negative reinforcing properties, respectively, and that both forms of aggression are also associated with different anatomical pathways. In this manner, differential effects of opioid peptides may be linked to the different physiological and anatomical properties of each of these forms of aggression.

STUDIES CONDUCTED IN CATS

Systemic Administration of Opioid Compounds

Three studies addressed the question of how nonspecific opioid receptor blockade would affect defensive rage and predatory attack (Shaikh and Siegel, 1989; Brutus and Siegel, 1989). In one earlier study, fighting was induced in cats by delivery of carbachol into the lateral ventricles (Kromer and Dum, 1980). These authors then tested the role of opioid receptor activation by injecting morphine (which activates primarily μ receptors) into the lateral ventricles and observed an attenuation of the fighting. This study, similar to others described above, provided additional support to the view that opioid receptors mediate suppression of aggressive responses.

In two other studies, systemic administration of naloxone resulted in a dose-dependent decrease in the threshold for defensive rage elicited from either the hypothalamus (Brutus and Siegel, 1989) or PAG (Shaikh and Siegel, 1989). In contrast, when tested against predatory attack elicited from the lateral hypothalamus, naloxone suppressed this response while having little or no effect on contralateral circling elicited from other regions of the hypothalamus (Brutus and Siegel, 1989). Thus, on the basis of these findings, there is consistent evidence that opioid receptors mediate suppression of defensive rage behavior, but it also suggested that these receptors may in fact facilitate predatory attack behavior. In view of the relative paucity of data obtained with respect to opioid peptides and predatory attack, it would seem quite useful for future studies to replicate and extend these findings. The importance of such a study would be that it could help to establish whether opioid receptors have uniform or differential effects on different forms of aggressive behavior.

Intracerebral Administration of Opioid Compounds

Several basal forebrain structures, which include the bed nucleus of the stria terminalis and nucleus accumbens, have been implicated in the regulation of rage and aggression. Evidence for this conclusion was presented in Chapters 5 and 6. In several experiments, the question was asked whether opioid receptor activation in these regions could affect these forms of aggression (Brutus et al., 1988, 1989). The rationale for these studies was based on several lines of evidence. The first is that both regions have been shown to modulate aggression and rage (Block et al., 1980b; Shaikh et al., 1986); and the second is that both regions receive opioidergic fibers that appear to arise from the central nucleus of amygdala (Uhl et al., 1978). The paradigm for the studies was to initially establish stable baseline response thresholds elicited by stimulation of the medial hypothalamus. Then, the nonspecific enkephalin agonist, D-ala^2-met^5-enkephalinamide (DAME) was microinjected into these sites in separate experiments, and the effects of administration of this drug on defensive rage were assessed. The results revealed a dose–response suppression of defensive rage following delivery of this agonist into each of these regions. Moreover, the suppressive effects of DAME were blocked by pretreatment with naloxone into these sites, indicating that these effects were mediated through opioid receptors.

From these studies, one may conclude that enkephalinergic neurons arising from the central nucleus of amygdala project to the bed nucleus of the stria terminalis and nucleus accumbens. When opioid receptors in these regions are activated, there is suppression of the defensive rage response elicited from the medial hypothalamus. It is of interest to recall that stimulation of the bed nucleus facilitates defensive rage, presumably by fibers that project directly to either the medial hypothalamus or PAG. Thus, it may be concluded that the mechanism for modulation here involves central nucleus–induced opioid suppression of neurons in the bed nucleus, which normally facilitates defensive rage behavior. As discussed below, the inhibitory actions of the opioid projections from the central nucleus on defensive rage are quite powerful and involve yet another, and perhaps more critical input to the PAG.

Thus far, the studies described above focused on opioid receptor activation of limbic forebrain structures as they impacted on defensive rage elicited from the medial hypothalamus. In the remainder of this section, results of a number of studies conducted with respect to the PAG are reviewed as they provide evidence that opioid receptors in this structure play an important role in modulating defensive rage behavior.

One such study involved the implantation of a stimulating electrode in the medial hypothalamus from which defensive rage could be elicited and a cannula-electrode in the PAG that would either suppress or facilitate defensive rage (Pott et al., 1986). When naloxone was microinjected into these inhibitory (but not facilitatory) sites, the inhibitory effects of PAG stimulation on hypothalamically elicited defensive rage was blocked. In addition, infusion of DAME into the inhibitory sites in the PAG suppressed hypothalamically elicited defensive rage. These data can be interpreted to mean that electrical stimulation of an inhibitory site in the PAG-activated enkephalinergic neurons arose from either the central nucleus of amygdala (see below) or from interneurons within the PAG itself. Simply put, when naloxone was infused into the inhibitory sites, it served to block enkephalins from interacting with opioid receptors in the PAG to suppress defensive rage behavior.

It should be pointed out that there are also sites in the PAG that, when stimulated, can facilitate or suppress predatory attack elicited from the lateral hypothalamus (Weiner et al., 1991). The key point to note here is that these modulatory sites are affected by naloxone, indicating that an enkephalinergic mechanism contributes to the regulation of predatory attack within the PAG as well. From the studies described above, two important questions required attention. The first concerned the specific receptor subtypes that mediate opioid functions within the PAG. The second concerned the origin of the opioidergic pathway that projects to the PAG and which mediates suppression of defensive rage behavior. Answers to the first question were addressed in an experiment that involved examination of the effects of infusion of opioid compounds through a cannula-electrode into a site within the PAG from which defensive rage had been elicited (Shaikh et al., 1988, 1991a). In an initial study, the nonspecific opioid agonist, DAME, was microinjected into the PAG, and clearly suppressed defensive rage behavior elicited from that site, while having no effect on circling behavior that was also elicited from the PAG. In addition, the suppressive effects of DAME were blocked following pretreatment with naloxone at the same PAG site. In a subsequent study, specific opioid receptor agonists were microinjected into defensive rage sites in the PAG and their effects on defensive rage were identified. The results indicated that administration of the selective μ receptor agonist, morphiceptin, into the PAG powerfully suppressed defensive rage behavior. Lesser magnitudes of suppression were noted following administration of the selective δ agonist, DPDPE. However, there were no changes in response thresholds following administration of the selective κ agonist, U-50488H. In this study, pretreatment with the selective μ and δ antagonists, β-funaltrexamine (β-FNA) and ICI 174864, blocked the suppressive effects of morphiceptin and DPDPE, respectively. The significance of this study is that it indicated the potency of these two receptors in mediating suppression of defensive rage within the PAG.

From the studies that have just been described, one may conclude that the opioidergic fibers mediating suppression of defensive rage constitute interneurons within the PAG that are activated from an external source. An alternative interpretation of the data is that the opioidergic mechanism within the PAG involves opioid peptides that are synthesized in cell bodies in a distal structure that projects to the PAG. When activated, these neurons would then suppress defensive rage. This latter possibility was tested in the following experiment (Shaikh et al., 1993). In this study, it was hypothesized that the central and lateral nuclei of the amygdala served just such a function, since it is known that stimulation of this region suppresses defensive rage behavior (Brutus et al., 1986), and that these nuclei have been identified as enkephalinergic neurons in rodents (Uhl et al., 1978; Rao et al., 1987). Initially, electrical stimulation applied to the central or lateral nucleus caused powerful suppression of defensive rage elicited from the PAG. This suppression was so potent that it, in fact, lasted for over 30 minutes. Then, dual stimulation including the central or lateral nucleus was again applied following administration of nonselective (naloxone) and specific μ (β-FNA) and δ (ICI 174864) opioid antagonists. The suppressive effects of central amygdaloid stimulation were completely blocked by naloxone or β-FNA but not by ICI 174864. That the suppressive effects of the central and lateral nuclei on defensive rage are mediated through a descending pathway to the PAG and that act on opioid μ receptors constitutes the key conclusion suggested from the data in this study (see Figure 7.4) Alternatively, it might be concluded from these data that central amygdaloid neurons are not opioidergic, but instead, project to interneurons within the PAG that are opioidergic and then project to PAG neurons associated with the expression of defensive rage. In order to determine which of these possibilities is correct, the following additional experiment was conducted. The retrograde tracer, Fluoro-Gold, was microinjected into defensive rage sites within the PAG, and after a suitable survival period, the animals were sacrificed and the brain tissue was processed for both Fluoro-Gold and met-enkephalin immunoreactivity. The findings revealed the presence of considerable numbers of enkephalin-positive neurons situated within the central and lateral nuclei of amygdala, which were also labeled with Fluoro-Gold. These findings thus provided further support for the view that enkephalinergic neurons in these nuclei of amygdala project directly to the PAG and suppress defensive rage behavior via an enkephalinergic mechanism (Figure 7.5).

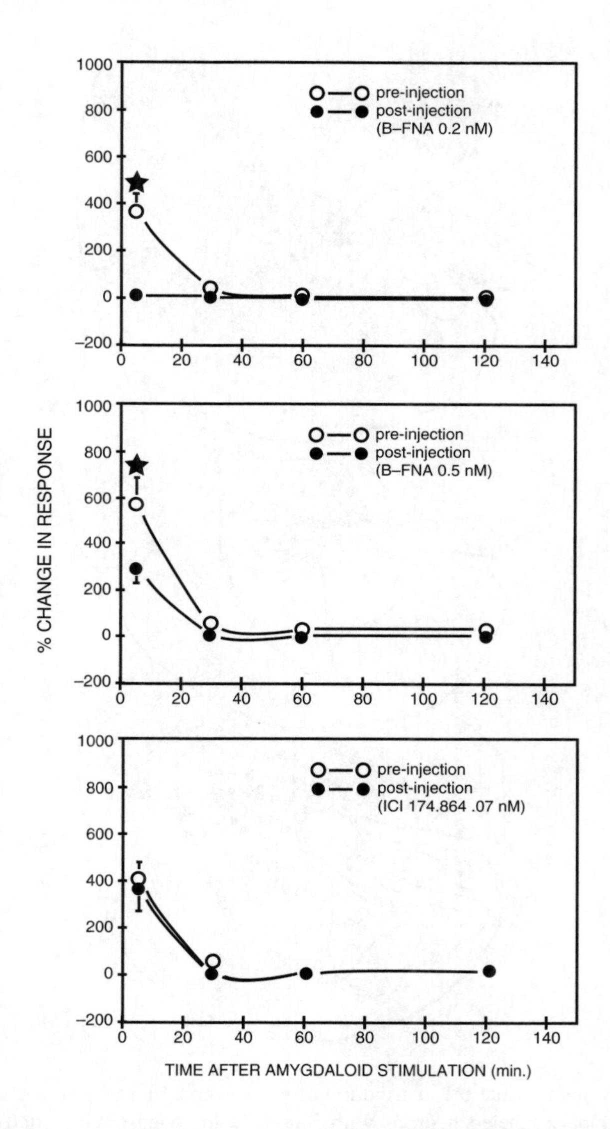

FIGURE 7.4 Open circles indicate the time course for amygdaloid induced suppression of defensive rage as determined by the percent change in response latencies from baseline values obtained prior to amygdaloid stimulation. Closed circles indicate the effects of drug delivery. Note that when the selective μ receptor antagonist, β-FNA, was administered at a dose of 0.2 nM, it completely blocked the suppressive effects of amygdaloid stimulation (top panel). A partial blockade was noted at a lower dose of this drug (middle panel), and the selective δ antagonist, ICI 174,864, failed to block the suppressive effects of amygdaloid stimulation when microinjected at a dose of 0.7 nM into the PAG (bottom panel). * $p < 0.05$. (From Shaikh et al., 1991c. With permission.)

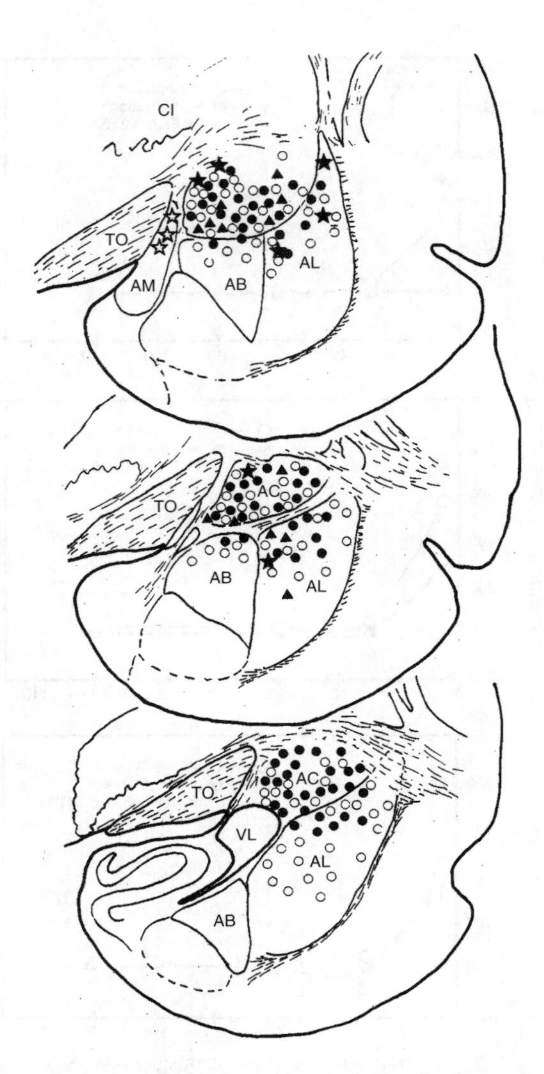

FIGURE 7.5 Maps indicate the distribution of met-enkephalin immunoreactive cells (closed circles), retrogradely labeled neurons with Fluoro-Gold (open circles), and double-labeled neurons (triangles). Immunoreactive cells are principally located with the central nucleus (CE) and retrogradely labeled cells are present in both the central and adjoining parts of the lateral (AL) and basal (AB) nuclei. Neurons that are double labeled with Fluoro-Gold and met-enkephalin are present in the rostral half of CE and adjoining portions of AL and AB. Closed stars depict the electrode sites from which stimulation suppressed defensive rage and open stars depict sites from which stimulation facilitated defensive rage. Note that sites associated with suppression of defensive rage were located mainly in the central nucleus in juxtaposition to the cell groups which display double labeling. In contrast, the sites associated with facilitation of defensive rage are located in the medial amygdala, (AM) a region linked to the origin of stria terminalis fibers which project to the medial hypothalamus and whose potentiating effects are mediated by substance P acting through NK_1 receptors. CI, internal capsule; TO, optic tract; VL, lateral ventricle. (From Shaikh et al. 1991. With permission.)

Summary

The preponderance of evidence concerning the role of opioid peptides in aggressive behavior indicates that activation of opioid receptors suppresses aggressive reactions in general, and those associated with affective aggression (or defensive rage) in particular. Of considerable importance is the role of opioid μ receptors, which appear to be of considerable significance within the PAG in mediating response suppression. In addition, it should be noted that specific groups of enkephalinergic neurons arising from the central and lateral nuclei of amygdala project directly to the PAG and mediate their suppressive actions via μ opioid receptors in the PAG. Other opioid receptors, which include δ and κ receptors, seem to be of lesser importance in the regulation of rage and aggression.

SUBSTANCE P

Characterization, Distribution, and Functions

Substance P (SP) belongs to a family of tachykinins that also includes neurokinin A and neurokinin B. Substance P is an undecapeptide and is found in different regions of the brain such as the hypothalamus (see below), amygdala, basal ganglia, cerebral cortex, and dorsal horn of the spinal cord. Three kinds of neurokinin receptors have been identified: neurokinin-1 (NK_1), neurokinin-2 (NK_2), and neurokinin-3 (NK_3). It was previously believed that SP interacts with NK_1 receptors, neurokinin A with NK_2 receptors, and neurokinin B with NK_3 receptors. Although this view is somewhat oversimplified and, in fact, several of the neurokinins may interact with several neurokinin receptors, SP is generally believed to interact mainly with NK_1 receptors.

SP is an excitatory neurotransmitter, producing depolarization of neurons for an extended period of time. The most well-known function of SP in the CNS is that it is released from the endings of first-order sensory axons mediating pain inputs into the spinal cord. More recently, it has been implicated in affective processes and recent evidence is presented below that links SP as an important neurotransmitter in the regulation of rage and aggression.

Role of SP in Rage and Aggression

While considerable effort has been given to the study of the role of SP in the mediation of pain impulses at the level of the dorsal horn of the spinal cord, little attention has been directed at other possible functions of this neurotransmitter associated with affective processes. In recent years, however, several studies have been directed at examining this possible relationship. Results of these experiments are described in this section.

There are several reasons why attention was directed at SP's role in rage and aggression. The first is that SP and NK_1 receptors are present in regions of the brain associated with rage and aggression, such as the amygdala, hypothalamus and PAG (Langevin and Emson, 1982; Warden and Young, 1988; Gallagher et al., 1992a, 1992b). One of these regions, the medial amygdala, powerfully facilitates defensive rage and suppresses predatory attack (Stoddard-Apter and MacDonnell, 1980; Block et al., 1980a). It is reasonable to assume that the mechanism underlying such facilitation of defensive rage would require that the neurotransmitter in question be excitatory. Thus, the second reason is that SP does, in fact, have an excitatory action on neurons (Mayer and MacLeod, 1979; Otsuka and Yoshioka, 1993). A third reason is that axon terminals are present in the medial hypothalamus that displays immunopositive staining for SP.

In order to test directly the hypothesis that medial amygdaloid facilitation of defensive rage behavior is mediated through SP-NK_1 receptors in the medial hypothalamus, the following series of experiments were conducted (Shaikh et al., 1993). The strategy for conducting this study was basically identical to that used for the analysis of opioidergic mechanisms from the central and lateral nuclei of amygdala described above. In the first stage of this study, sites in the medial amygdala were identified which, on dual stimulation (of the medial amygdala and medial hypothalamus), facilitated defensive rage. In the second stage, the dual stimulation procedure was repeated following administration of the SP NK_1 receptor antagonist, CP 96,345, either peripherally or directly through the cannula-electrode situated in the medial hypothalamus from which defensive rage was elicited. The results, shown in Figure 7.6, indicate that peripheral or central administration of the CP 96,345 blocked the facilitating effects of medial amygdaloid stimulation on defensive rage. In the third stage of the study, the retrograde tracer, Fluoro-Gold, was microinjected into the medial hypothalamus and, when the animal was sacrificed, the brain was processed for immunocytochemical labeling of SP-positive neurons in the amygdala. Results revealed the presence of considerable quantities of SP-positive neurons situated in the medial amygdaloid nucleus and basal complex. However, Fluoro-Gold-labeled neurons were located in the medial amygdala and adjoining parts of the medial aspect of the basal complex. In particular, neurons labeled for both Fluoro-Gold and SP were situated in the medial amygdaloid nucleus. Further, axons located in the stria terminalis, also stained positively for SP, and these neurons could be followed directly into the medial hypothalamus. These findings thus provided anatomical evidence that was consistent with the pharmacological data presented above. Results support the view that SP neurons located in the medial amygdala project directly to the medial hypothalamus and act on NK_1 receptors in the medial hypothalamus to facilitate the defensive rage mechanism in the medial hypothalamus.

In separate studies, Han et al. (1996a, 1996b) addressed the question of how the medial amygdala may modulate predatory attack and whether such modulation was mediated through SP-NK_1 receptors in the medial hypothalamus. The paradigm for this experiment was similar to that conducted above for the study of NK and opioid receptors on defensive rage behavior. In Han's study, a stimulating electrode was implanted in the medial amygdala of the cat and cannula-electrodes were implanted in the medial and lateral hypothalamus from which defensive rage and predatory attack could be elicited, respectively, and where specific NK and other receptors agonists and antagonists could be microinjected into these sites. In the first part of the study, it was observed that in contrast to the findings with defensive rage, stimulation of the medial amygdala suppressed predatory attack elicited from the lateral hypothalamus as determined by the dual stimulation procedure including the medial amygdala.and lateral hypothalamus. Then, the SP-NK_1 receptor antagonist, CP 96,345, was microinjected into the medial hypothalamus. The results, again

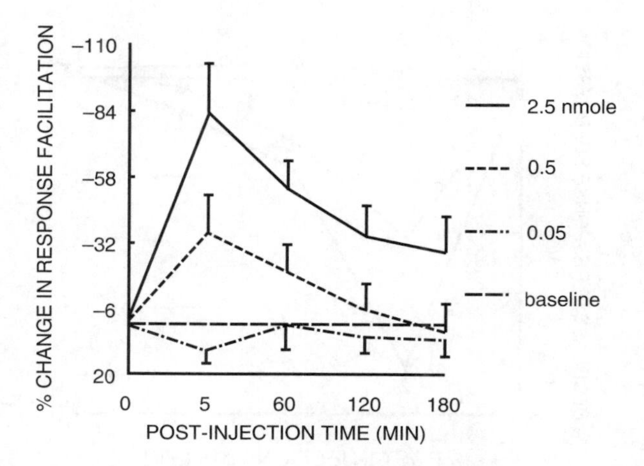

FIGURE 7.6 Graphs indicate that microinjections of the selective NK_1 antagonist, CP 96,345, into the medial hypothalamic sites from which defensive rage can be elicited block the facilitatory effects of medial amygdaloid stimulation. (From M.B. Shaikh, A. Steinberg, A. Siegel, *Brain Res.*, 625, 1993, 283–294. With permission.)

determined from dual stimulation, indicated that administration of the NK_1 receptor antagonist blocked the suppressive effects of medial amygdaloid stimulation dual stimulation (Figure 7.7). In a separate but related experiment, SP was microinjected directly into the medial hypothalamus and was shown to suppress predatory attack elicited from the lateral hypothalamus (Figure 7.8). The suppressive effects of either SP administration into the medial hypothalamus or medial amygdaloid stimulation were then blocked by infusion of the $GABA_A$ receptor antagonist, bicuculline, into the lateral hypothalamic predatory attack site.

Results of these experiments can be accounted for in the following way. The medial amygdala contains, at least in part, SP neurons that project through the stria terminalis to the anterior half of the medial hypothalamus, and their excitatory effects on the medial hypothalamus are mediated through NK_1 receptors. The medial hypothalamus contains at least two classes of neurons that are excited by these NK_1 receptors. The first class of receptors includes neurons that mediate defensive rage and which project to the dorsolateral PAG as part of the descending circuit for the expression of defensive rage. The second class of neurons includes smaller GABAergic neurons, which project locally to the lateral hypothalamus and serve to inhibit lateral hypothalamic functions such as predatory attack when excited by afferent sources such as the medial amygdala. Consequently, as a result of NK_1 receptor activation of these classes of medial hypothalamic neurons from a source such as the medial amygdala, there is a consequent potentiation of the defensive rage response because of the excitation involving the descending pathway to the PAG, and, concurrently, a suppression of predatory attack because of the activation of GABAergic interneurons whose cell bodies lie in the medial hypothalamus and which project to the lateral hypothalamus.

Further confirmation of this circuitry was provided in several additional studies. The first was conducted by Han et al. (1996b). In this study, Fluoro-Gold was

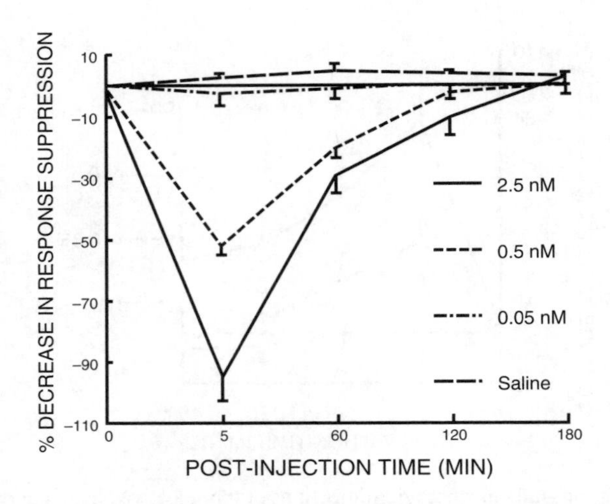

FIGURE 7.7 Effects of microinjections of CP 96,345 into the medial hypothalamus on medial amygdaloid suppression of predatory attack behavior elicited from the lateral hypothalamus. CP 96,345 blocked the suppressive effects of the medial amygdala on predatory attack in a dose-dependent manner. The pre-injection baseline represents the percent increase in response latency following dual stimulation whose value was then converted to 0%. Following drug or vehicle administration, decreases in the magnitude of medial amygdaloid modulation are expressed as a negative percentage change and reflect an enhancement of the suppressive effects of medial amygdaloid stimulation. (From Y. Han, M.B. Shaikh, A. Siegel, *Brain Res.*, 716, 1996a, 59–71. With permission.)

microinjected into lateral hypoothalamic predatory attack sites, and the brain tissue was processed by immunocytochemical staining for GABA-positive neurons as well as for Fluoro-Gold-positive neurons. Results showed the presence of neurons labeled for both Fluoro-Gold and GABA situated in the medial hypothalamus, indicating that these GABAergic neurons project directly to the lateral hypothalamus. In several further studies, Yao et al. (1999, 2001) used *in situ* hybridization to label NK_1 receptors as well as GABA and glutamate-positive neurons in the hypothalamus. The findings of these studies demonstrated that indeed, NK_1 receptors are situated on both GABA and glutamate-positive neurons in the medial hypothalamus. Evidence presented in studies described below revealed that the glutamate neurons in the medial hypothalamus project to the PAG, and as just noted, GABAergic neurons project to the lateral hypothalamus. The overall findings of these studies are summarized in Figure 7.9.

CHOLECYSTOKININ (CCK)

Characterization, Distribution, and Functions

CCK, a gut hormone, is converted from procholecystokinin to peptides where it acts on the gallbladder and liver. Smaller CCK-related peptides are found in the brain. An interesting feature of CCK is that it is frequently found co-localized with other

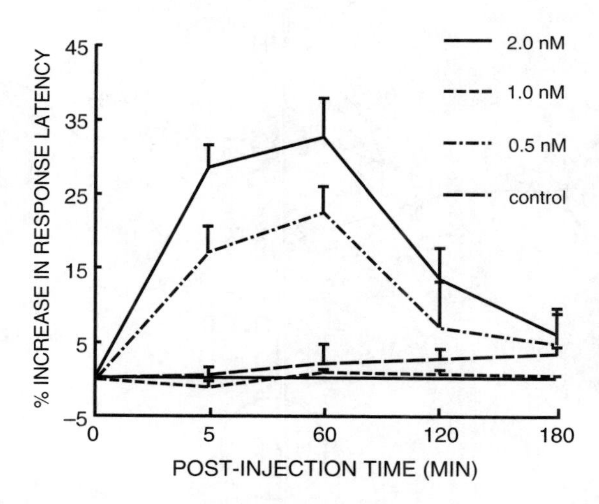

FIGURE 7.8 Effects of microinjections of (Sar^9,$Met(O_2)^{11}$)-substance P into the medial hypothalamus on predatory attack elicited by single stimulation of the lateral hypothalamus. Preinjection response latencies were converted to 0%. Following drug delivery, response latencies were determined again over a 180-minute period. Increases in response latencies indicated the suppressive effects of drug infusion and are expressed as a positive percentage change on the graph. Administration of the substance P agonist resulted in suppression of predatory attack in a dose-dependent manner. (From Y. Han, M.B. Shaikh, A. Siegel, *Brain Res.*, 716, 1996a, 59–71. With permission.)

neurotransmitters. These include pro-enkephalin-derived peptides, SP, 5-HT, dopamine, neuropeptide Y, and GABA. CCK is found in wide regions of the brain that include the cerebral cortex, diencephalon, brainstem, and dorsal root ganglion. There are two types of CCK receptors: CCK_A and CCK_B. It is generally accepted that satiety mechanisms are mediated via CCK_A receptors and that reactions such as anxiety and panic attacks are mediated via CCK_B receptors. It is also believed that CCK_B receptor activation serves to have excitatory effects on neuronal activity. In this section, evidence is presented supporting the view that defensive rage is also potentiated via CCK_B receptors.

Role of CCK in Defensive Rage Behavior

Behavioral experiments have focused mainly on the roles of CCK on feeding behavior and anxiety reactions. In contrast, very little research has been conducted relating to its possible role in rage and aggression. The results of one such study are described in this section.

Evidence had indicated that CCK fibers and terminals (Lanaud et al., 1989; Liu et al., 1994) as well as CCK_B receptors (Mercer et al., 1996) are present in the PAG. Inasmuch as activation of CCK_B receptors have been shown to induce anxiety reactions in rodents (Adamec et al., 1997), the hypothesis was proposed that CCK_B receptors in the PAG would potentiate defensive rage behavior elicited from the medial hypothalamus of the cat. In this study, stimulating electrodes were placed in

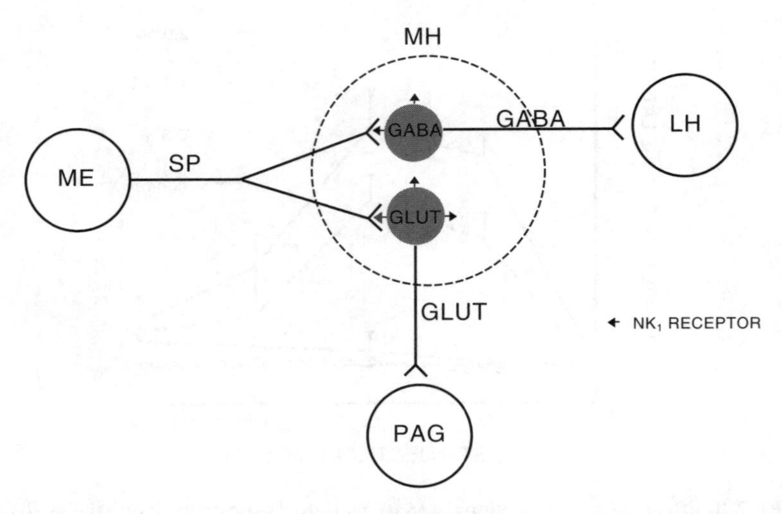

FIGURE 7.9 Schematic diagram depicting the significance of substance P (SP)-NK_1 receptors present on both GABA and glutamate (GLUT) neurons. GABA neurons from the medial hypothalamus (MH) project to the lateral hypothalamus (LH) and GLUT neurons in the MH project to the midbrain PAG. Thus, NK_1 activation of MH-GABA neurons causes suppression of predatory attack elicited from the LH, while NK_1 activation of MH-GLUT neurons causes facilitation of defensive rage which is mediated through the MH-PAG circuit. (From R.H. Yao, P. Rameshwar, R.J. Donnelly, A. Siegel, *Brain Res. Mol. Brain Res.*, 71, 1999, 149–158. With permission.)

sites in the medial hypothalamus from which defensive rage could be elicited, and a cannula-electrode was placed in the midbrain PAG from which stimulation could also elicit defensive rage and where specific CCK compounds could be microinjected into the defensive rage site (Luo et al., 1998). Microinjections of the selective CCK_B receptor antagonist, LY 288513, directly into the PAG defensive rage site suppressed defensive rage elicited from the medial hypothalamus in a dose- and time-dependent manner. When a specific CCK_B receptor agonist, pentagastrin, was microinjected into the PAG in a separate experiment, there was significant facilitation of defensive rage elicited from the medial hypothalamus (Figure 7.10). The specificity of these effects was further demonstrated when the PAG sites was pretreated with CCK_B receptor antagonist prior to microinjection of pentagastrin. Pretreatment with the CCK_B receptor antagonist blocked the facilitating effects of pentagastrin. In contrast, pretreatment with a CCK_A receptor antagonist, PD 140548, had no effect on the facilitating effects of pentagastrin on defensive rage elicited from the medial hypothalamus.

Summary

While our overall understanding of the role of CCK receptors on aggression and rage remains incomplete because of the relative absence of research directed at this problem, the limited data provide data supporting the notion that CCK_B receptor activation, in particular within the PAG, potentiates defensive rage in the cat. It

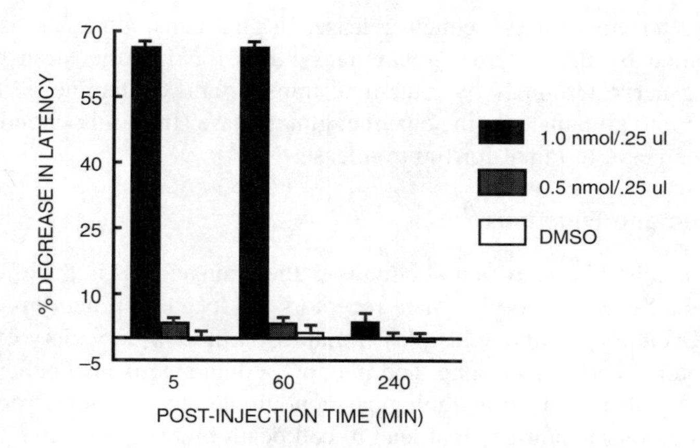

FIGURE 7.10 Graph indicates that microinjections of the CCK_B receptor agonist, pentagastrin, into the dorsal aspect of the midbrain periaqueductal gray from which defensive rage could be elicited facilitated this response when elicited from the medial hypothalamus. (From B. Luo, J.W. Cheu, A. Siegel, *Brain Res.*, 796, 1998, 27–37. With permission.)

appears that CCK_A receptors play little or no role in the process of defensive rage. There are several basic questions that remain unanswered. One is whether CCK receptors in other regions of the forebrain such as the hypothalamus or elsewhere contribute to the expression of the defensive rage response. A second is whether CCK receptors play any role in the regulation of predatory attack. A third question is that we really have little idea about the origins of CCK neurons that project to the PAG and interact with CCK receptors. One possibility is that CCK is co-localized with other neurotransmitters, such as glutamate, which is present in descending neurons from the medial hypothalamus or basal amygdala that project to the PAG and play a significant role in the expression of defensive rage (see discussion below). However, such a possibility remains speculative at the present time.

AMINO ACIDS

Excitatory Amino Acids (Glutamate)

Synthesis and Removal

Of the various excitatory amino acids, glutamate is the most common and will be discussed here as an example of this type of amino acid transmitter. It is synthesized in the nerve terminal in the following manner. After glucose enters the cytoplasm of the neuron, it undergoes glycolysis to generate pyruvic acid which enters the mitochondria. Within the mitochondria, pyruvic acid produces an acetyl group that combines with coenzyme-A to form acetylcoenzyme-A. The acetyl group is reformed from acetylcoenzyme-A and enters the Krebs cycle in mitochondria. From the Krebs cycle, ketoglutaric acid is formed and transaminated into glutamate. Glutamate is released into the synaptic cleft and brought back into the terminal by high-affinity neuronal or glial glutamate transporters. Glutamate is then repackaged in the nerve terminal for subsequent release. In glial cells, glutamate is converted into glutamine by the enzyme glutamine synthase, and is then transported into neighboring nerve terminals by glutamine transporters. Glutamine is then reconverted back into glutamate by the enzyme, glutaminase (from mitochondria), and is stored in the nerve terminal for future release.

Distribution and Functions

Glutamate is found in many neural circuits in the brain such as in the efferent fibers of the cerebral cortex, and glutamate receptors are located on neurons throughout the CNS. Excitatory amino acids play a major role in a wide variety of processes such as motor functions, learning, and memory. Glutamate is also believed to play roles in a number of neuropathological conditions such as amyotrophic lateral sclerosis and other conditions that lead to cell death by excitotoxicity.

Role of Excitatory Amino Acids in Defensive Rage Behavior

Perhaps the most important pathway governing defensive rage behavior in the cat is that which arises from the medial hypothalamus and projects to the dorsolateral aspect of the PAG. Thus, knowledge of the neurotransmitter(s) associated with this pathway would be extremely important in our quest to gain an understanding of the neurochemistry of defensive rage behavior. The rationale for believing that excitatory amino acids could be a good candidate here was based on previous anatomical data suggesting that this pathway may contain glutamate (Beart et al., 1988, 1990; Beitz, 1989).

On the basis of this evidence, several studies were conducted aimed at testing the hypothesis that the function of defensive rage behavior was mediated through excitatory amino acid receptors in association with the pathway from the medial hypothalamus to the PAG in the cat (Lu et al., 1992; Schubert et al., 1996b). In an initial study, the nonspecific excitatory amino acid antagonist, kyneurenic acid, was microinjected into sites within the PAG from which defensive rage could be elicited. Administration of this agonist blocked medial hypothalamic facilitation of PAG-elicited defensive rage (Lu et al., 1992). When the specific NMDA antagonist, AP-7, was microinjected into the PAG, it also blocked medial hypothalamic facilitation of defensive rage. However, when the kainate and quisqualate receptor blocker, CNQX, was microinjected into the PAG in a similar to that of AP-7, it had little or no effect on this response. Further, infusion of NMDA directly into the PAG facilitated defensive rage elicited from that site (Figure 7.11). A replication and extension of this study was carried out by Schubert et al. (1996b). Here, it was observed that microinjections of the NMDA antagonist, AP-7, into PAG defensive-rage sites, blocked medial hypothalamically elicited defensive rage. Another aspect of both of these studies combined retrograde labeling following microinjections of Fluoro-Gold into the PAG and immunocytochemical labeling of glutamatergic neurons. Results indicated that double-labeled cells (for Fluoro-Gold and glutamate) were observed within the medial hypothalamus, mainly dorsal and rostral to the ventromedial nucleus. This finding provided direct support to the pharmacological data and to the

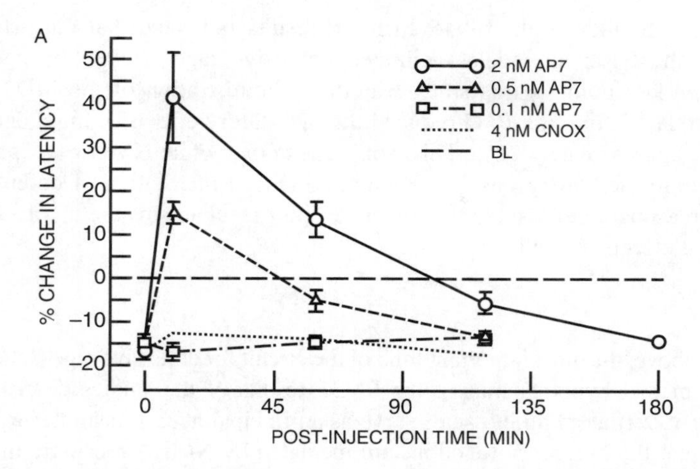

FIGURE 7.11 (A) Microinjections of the selective NMDA receptor antagonist, AP7, into periaqueductal gray (PAG) sites from which defensive rage could be elicited suppressed the facilitatory effects of medial hypothalamic stimulation upon this response in a dose-dependent manner. Note that a higher dose of the CNQX, a non-NMDA receptor antagonist, had little effect on medial hypothalamic facilitation of this response. (B) Infusion of NMDA into defensive rage sites in the PAG resulted in a dose-dependent facilitation of this response, mimicking the effects obtained with dual stimulation of the medial hypothalamus and PAG. (From C.-L. Lu, M.B. Shaikh, A. Siegel, *Brain Res.*, 581, 1992, 123–132. With permission.)

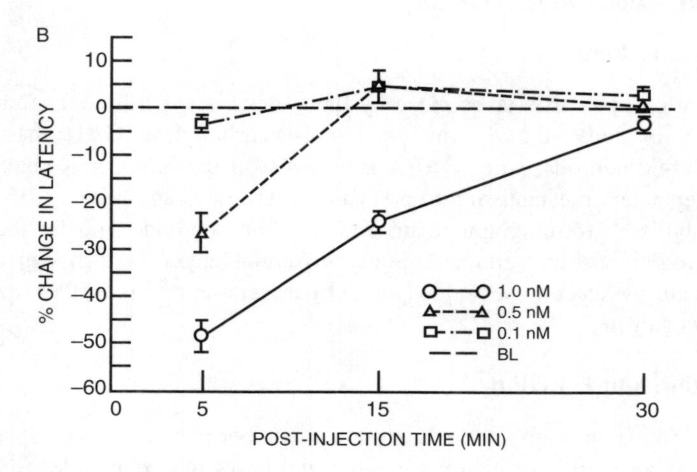

FIGURE 7.11 (Continued).

hypothesis that glutamatergic neurons project from the medial hypothalamus to the PAG, a pathway that subserves the expression of defensive rage behavior and whose functions are mediated by NMDA receptors in the PAG.

In another study, evidence was provided indicating that basal amygdaloid facilitation of defensive rage behavior is also mediated, in part, through NMDA receptors in the PAG (Shaikh et al., 1994). In brief, results indicated that dual stimulation involving the basal amygdala facilitated defensive rage elicited from the PAG. However, a key point in this study was that administration of the MDA receptor antagonist, AP-7, into the PAG reduced the facilitatory effects of amygdaloid stimulation by approximately 22%. This would mean that while NMDA receptors likely play a role in facilitating basal amygdaloid–induced facilitation of defensive rage, it is also reasonable to assume that other receptor mechanisms are involved in these facilitating effects as well.

Summary

As stated above, the most important limb of the circuit for defensive rage is the pathway from the medial hypothalamus to the PAG. Results of the studies described above indicate that excitatory amino acids serve as a principal neurotransmitter in this pathway and that the excitatory functions are mediated by NMDA receptors in the PAG. These medial hypothalamic neurons contain NK_1 receptors, which are excitatory and are typically driven by inputs such as the medial amygdala. In addition, one other pathway associated with excitatory amino acids and defensive rage has been identified. This involves a projection from the basal complex of amygdala to the PAG whose functions are also partially mediated by NMDA receptors. However, another neurotransmitter or neurotransmitters are likely involved as well in mediating the facilitating effects of this region of the amygdala on defensive rage.

Inhibitory Amino Acids (GABA)

Synthesis and Removal

The formation of GABA takes place by the α-decarboxylation of L-glutamate, as the reaction is catalyzed by L-glutamic acid-1-decarboxylase (GAD) that is present in GABAergic neurons. After GABA is released in the brain, it is taken up by a sodium-dependent mechanism into presynaptic terminals and glia. GABA is generally metabolized, forming glutamate and succinic semialdehyde by the enzyme GABA-oxoglutarate transaminase. Succinic semialdehyde is then converted into succinic acid by succinic semialdehyde dehydrogenase, and is further metabolized in the Krebs cycle.

Distribution and Functions

GABA is found throughout the central but not peripheral nervous system. It is classified as an inhibitory neurotransmitter and plays important roles in a number of circuits mediating motor functions. Examples include the inhibitory functions of the Purkinje cell in the cerebellum and nuclei of the basal ganglia (caudate nucleus, putamen, and globus pallidus). Since drugs that inhibit GABA synthesis or release frequently have epileptogenic properties, and those such as valproic acid that inhibit the enzyme, GABA-transaminase which metabolizes GABA, act as anticonvulsants. Reductions in GABA levels are believed to be an important factor in the genesis of seizure activity.

Role of Inhibitory Amino Acids in Defensive Rage and Predatory Attack Behavior

In studying the effects of GABA on aggressive processes, the general procedure has been in most cases to use intracerebral injections of GABAergic compounds. Several studies testing the role of GABA in aggression were conducted in rats by microinjecting GABA antagonists into the hypothalamus. Both studies produced the same general results. When the GABA antagonist, picrotoxin, which acts mainly on $GABA_A$ receptors, was microinjected into the anterior hypothalamus, it was shown to induce offensive patterns of aggression in rats (Adams et al., 1993). Likewise, microinjections of a different $GABA_A$ receptor antagonist, bicuculline, into the ventral hypothalamus of the rat also induced attack (Roeling et al., 1993). These data would appear to suggest that GABA is tonically released onto neurons in the medial hypothalamus, and that aggression in these models reflect release of such an inhibitory mechanism. However, a somewhat different conclusion is drawn from the studies described below that were conducted in the cat.

A number of studies have been conducted using defensive rage and predatory attack models of aggression. In an early study, GABA was injected directly into defensive rage sites in the medial hypothalamus and surprisingly, it was reported that response thresholds were actually lowered (Nakao et al., 1979). In a later study, a slightly different design was employed to determine the role of GABA receptors in defensive rage (Shaikh and Siegel, 1990). Cannula-electrodes were implanted into sites in the PAG from which defensive rage could be elicited and where drugs could be microinjected. Administration of the GABA agonist, muscimol, into these sites in very low doses (i.e., 12 pmol) suppressed defensive rage behavior elicited from that site and pretreatment with bicuculline at that site blocked the suppressive effects of muscimol. Moreover, when tested against predatory attack elicited from the ventral aspect of the PAG, muscimol was reported to have little or no effect on this form of aggression, indicating the specificity of the effects of GABA on defensive rage at the level of the PAG. Although these findings indicate the importance of GABA receptors in the PAG in suppressing defensive rage, they do not tell us anything about the organization of GABA neurons in this region. For example, are the GABA neurons that modulate defensive rage behavior in the PAG interneurons, or do such neurons arise from other regions of the brainstem or forebrain? However, evidence from other studies would appear to suggest that there are many GABAergic neurons situated throughout the PAG in an uneven manner and likely serve mainly as interneurons in this neuropil (Reichling and Basbaum, 1990, 1991).

Several studies carried out with respect to the role of GABA in the hypothalamus have begun to clarify the organization of the GABAergic system in modulating defensive rage and predatory attack. These studies were presented earlier in this chapter in another context and in Chapter 3, and will be reviewed at this time. One study attempted to account for the fact that medial amygdaloid stimulation suppresses predatory attack while facilitating defensive rage (Han et al., 1996a, 1996b). As I have already described, medial amygdaloid facilitation of defensive rage is mediated over a pathway contained in the stria terminalis, which arises in the medial amygdala and terminates throughout wide regions of the medial hypothalamus. The principal neurotransmitter mediating such facilitation is SP acting through NK_1 receptors. These fibers then make contact with glutamate neurons in the medial hypothalamus, which project to the PAG. When these neurons are activated by inputs such as from the medial amygdala, potentiation of the defensive rage response ensues. However, in order to account for the suppression of predatory attack by the same region of amygdala, it was proposed that the same fibers of the stria terminalis that make contact with descending glutamatergic fibers in the medial hypothalamus that project to the PAG also make contact with GABA neurons in the medial hypothalamus. It was further proposed that these neurons act as inhibitory interneurons within the hypothalamus and project to the lateral hypothalamus. In this manner, stimulation of the medial amygdala would excite these GABAergic neurons in the medial hypothalamus, causing them to inhibit neurons in the lateral hypothalamus associated with predatory attack.

This hypothesis was tested in several ways. As noted earlier in this chapter, administration of SP into medial hypothalamic defensive rage sites suppressed predatory attack elicited from the lateral hypothalamus (Han et al., 1996a). In another experiment, it was initially shown that medial amygdaloid stimulation suppressed predatory attack elicited from the lateral hypothalamus. Then, infusion of the $GABA_A$ antagonist, bicuculline, into the lateral hypothalamic attack site blocked the suppressive effects of medial amygdaloid stimulation (Figure 7.12). Moreover, microinjections of the GABA agonist, muscimol, into lateral hypothalamic attack sites suppressed predatory attack and this suppression was, in turn, blocked with pretreatment with bicuculline (Han et al., 1996b). At the completion of these experiments, an anatomical study was conducted in which Fluoro-Gold was microinjected into the lateral hypothalamic predatory attack site and, following sacrifice of the animal, the brain tissue was processed for immunocytochemical labeling of GABA neurons. The results identified neurons in the medial hypothalamus that were labeled for both Fluoro-Gold and GABA, indicating the presence of GABA neurons in the medial hypothalamus that project to lateral hypothalamic predatory attack sites. Recall that previously anatomical evidence was provided that NK_1 are present on both GABA- and glutamate-positive neurons in the medial hypothalamus (Yao et al., 1999, 2001). This anatomical evidence provides support for both the conclusions reached from both the double-labeling and pharmacological experiments described above — namely, that NK_1 receptors situated on GABA neurons in the medial hypothalamus cause activation of these neurons, which, in turn, suppress their target neurons in the lateral hypothalamus, resulting in suppression of predatory attack.

The results of the experiments described above demonstrate the presence of GABAergic interneurons that project from medial hypothalamic defensive rage sites to the lateral hypothalamic predatory attack sites, and which function to suppress predatory attack. A related question may be raised here: Are these GABAergic projections reciprocal with respect to these regions and do they mediate similar functions? In order to test this hypothesis, a cannula-electrode was placed in a defensive rage site in the medial hypothalamus and a stimulating electrode was placed in a lateral hypothalamic predatory attack site (Cheu and Siegel, 1998). In

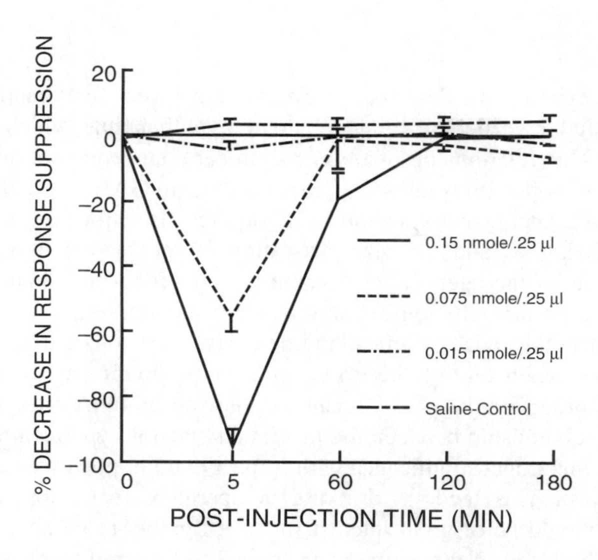

FIGURE 7.12 Microinjections of the $GABA_A$ receptor antagonist, bicuculline, into the lateral hypothalamus (from which predatory attack was elicited) blocked the suppressive effects of medial amygdaloid stimulation upon predatory attack. Zero percent (0%) represents the pre-injection baseline response latencies, and reductions in response latencies following drug administration are expressed as a negative percentage, reflecting blockade of the suppressive effects of medial amygdaloid stimulation. (From Y. Han, M.B. Shaikh, A. Siegel, *Brain Res.*, 716, 1996b, 72–83. With permission.)

this study, stimulation of the lateral hypothalamus suppressed defensive rage was elicited from the medial hypothalamus. Muscimol was microinjected into the medial hypothalamic defensive rage site and suppressed that response and muscimol-induced suppression was then blocked by pretreatment with bicuculline. This experiment established the presence of GABA receptors in the medial hypothalamus which, when activated, suppress defensive rage. In a separate procedure, it was shown that the suppressive effects of lateral hypothalamic stimulation on medial hypothalamically elicited defensive rage were blocked following infusion of bicuculline into the medial hypothalamus. The authors noted that administration of bicuculline into medial hypothalamic defensive rage sites by itself did not alter the latency or threshold for this response in the absence of dual stimulation involving the lateral hypothalamus.

In a manner similar to that described above for the analysis of medial hypothalamic regulation of predatory attack, an immunocytochemical-anatomical analysis revealed that when Fluoro-Gold was microinjected into the medial hypothalamus, double-labeled neurons were observed in the lateral hypothalamus that were labeled for both Fluoro-Gold and GABA. The conclusion from this study was that GABAergic neurons present in the lateral hypothalamus in proximity to predatory attack sites project to medial hypothalamic neurons associated with defensive rage. Such neurons inhibitory to defensive rage are GABAergic, and act through $GABA_A$ receptors.

Summary

Results of the experiments described above indicate a very fundamental and important relationship between the medial and lateral hypothalamus, which was discussed in Chapter 4. This relationship involves reciprocal connections between these two regions of the hypothalamus involving neurons that are GABAergic. In this manner, activation of the lateral hypothalamus will suppress functions associated with the medial hypothalamus, and likewise, activation of the medial hypothalamus will inhibit functions of the lateral hypothalamus. As described in Chapter 4, this relationship has important ethological significance; the importance is that it is not compatible for both predatory attack and defensive rage to occur at the same time. Said otherwise, when one of these responses is predominant, the other must be suppressed in order for the predominant response to be most effective. This basic physiological relationship between the medial and lateral hypothalamus most likely has important functional significance with respect to other processes associated with these regions, such as feeding, drinking, temperature regulation, and endocrine regulation. It would be of great interest for investigators to be able to apply these principles to the study of these functions as well. An overall summary of the basic relationships of the amygdala, the hypothalamus and its intrinsic relationships, and PAG with respect to a number of the neurotransmitters described above are presented in Figures 7.13 and 7.14.

CAN SUBSTANCES OF ABUSE, PSYCHOTROPIC DRUGS, AND ANTIDEPRESSANT DRUGS INFLUENCE AGGRESSION?

SUBSTANCES OF ABUSE

Opioid Peptides

In our discussion of the role of opioid peptides in aggression and rage behavior presented earlier in this chapter, the general consensus reached was that opioid peptides, which act mainly on μ and perhaps δ receptors, generally suppress rage and related responses. However, some investigators have reported a linkage between withdrawal from morphine and aggression (Davis and Khalsa, 1971; Gianutsos and Lal, 1978; Kantak and Miczek, 1986). Although the mechanism underlying morphine withdrawal–induced aggression is not fully understood, it is possible that at least one element of this process involves the absence of critical levels of enkephalins in the CNS. Support for this notion comes, not only from the literature described earlier in this chapter, but also from a recent study (Konig et al., 1996). In this study, the pro-enkephalin gene was disrupted by the use of a homologous recombination of embryonic stem cells, resulting in the generation of enkephalin-deficient mice. Such mice displayed significant behavioral abnormalities, including elevated levels of both anxiety and offensive components of aggression.

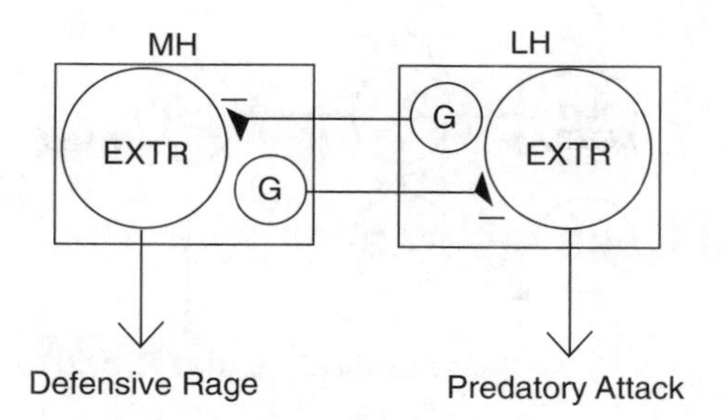

FIGURE 7.13 Schematic diagram illustrating the relationships between the medial and lateral hypothalamus with respect to defensive rage behavior and predatory attack. The medial hypothalamus (MH) contains extrinsic (EXTR) neurons that project caudally to the periaqueductal gray (PAG), and constitutes the primary descending pathway in the brain for the expression of defensive rage behavior. The lateral hypothalamus (LH) also contains extrinsic neurons (EXTR) which project to different brainstem nuclei for the expression of predatory attack behavior. In addition, local GABAergic (G) inhibitory interneurons are present, which project from MH to LH and from LH to MH. Thus, when the medial hypothalamus receives an excitatory input from a region such as the medial amygdala, the descending pathway to the PAG is also activated, resulting in excitation of the circuit underlying the expression of defensive rage. At the same time, excitatory inputs to MH also activate GABA neurons, which project to LH. GABA activation causes these LH neurons to be inhibited, which effectively inhibits predatory attack. A similar mechanism applies to the lateral hypothalamus. Namely, when lateral hypothalamic neurons are excited, the descending neural substrate for predatory attack is activated, resulting in the expression of this response. At the same time, activation of GABA neurons in LH causes inhibition of neuronal activity in MH, thus effectively suppressing defensive rage behavior. This functional anatomical arrangement is likely to be of great importance from an ethological perspective, because the effectiveness of one response is dependent on the suppression of the other.

Alcohol: Linkage between Alcohol Consumption and Violent Behavior

In Humans

The association between alcohol consumption and violent and assaultive behavior in humans is well established (Bushman and Cooper, 1990; Murdoch et al., 1990; Pernanen, 1991; Miczek et al., 1994a). Based on correlational studies, which used case reports and interviews as sources of data, alcohol consumption has been linked to homicide, child abuse, marital assaults, rape, and verbal abuse (Krsiak and Borgesova, 1973; Lindqvist, 1986; Babor et al., 1989). Parallel findings were obtained in experimental studies as well. The results consistently revealed that individuals who received alcohol responded more aggressively than those who received placebos or nonalcoholic beverages (Babor et al., 1989). These authors performed a meta-analysis of 30 studies, and found clear evidence of a causative relationship between alcohol and aggressive behavior (Babor et al., 1989). Other evidence (summarized in Miczek et al., 1994c) suggests that brain mechanisms play a role in this relationship. Such evidence includes:(1) altered serotonin levels in aggressive alcohol abusers, (2) functional alterations in evoked potential responses to sensory stimuli, and

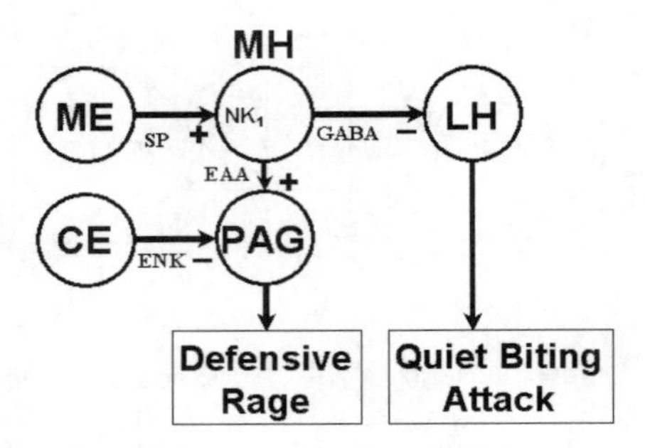

FIGURE 7.14 Summary of functional relationship of several key amygdaloid nuclei with respect to how they regulate defensive rage and predatory attack. As summarized in Figure 7.9, activation of the medial amygdala (ME) excites medial hypothalamic neurons (MH) via a substance P (SP)-NK_1 receptor mechanism. This drives the defensive rage mechanism by virtue of its descending pathway to the periaqueductal gray (PAG), whose effects are mediated largely through excitatory amino acids (EAA) acting on NMDA receptors in the PAG. At the same time, activation of MH causes suppression of LH and the predatory attack mechanism, as described in Figure 7.13. In contrast, activation of neurons in the central amygdala (CE) causes marked suppression of defensive rage via a direct pathway from CE to the PAG. These inhibitory effects are mediated through an enkephalinergic mechanism acting on μ-opioid receptors in the PAG.

(3) a possible common genetic basis for alcoholism and antisocial personality disorder.

In Rodents, Monkeys, and Cats

Various animal studies have, in general, also provided evidence for potentiation of aggressive behavior following alcohol consumption. Such effects were most clearly observed with respect to defensive behavior in rodents (Chance et al., 1973; Krsiak, 1976; Miczek and O'Donnell, 1980; Blanchard et al., 1986, 1987, 1993a; Miczek et al., 1994b), and agonistic responses in a social setting in squirrel monkeys (Winslow et al., 1988; Weerts et al., 1993). However, not all studies have provided uniform results. Other investigators have shown that alcohol may, in fact, suppress aggressive behavior in rodents and primates (Krsiak and Borgesova, 1973; Crowley et al., 1974; Lagerspetz and Ekqvist, 1978; Smoothy et al., 1982; Benton and Smoothy, 1984; Blanchard et al., 1986; Mos and Olivier, 1988). Moreover, several of these studies showed that the effects in rodents appear to be dose dependent: suppression typically follows administration of relatively higher doses of alcohol (e.g., 2 g/kg). Dose-related effects were also observed in cats following administration of ethanol: a low dose (0.37 g/kg) facilitated and a higher dose (1.50 g/kg) suppressed defensive rage behavior elicited by electrical stimulation of the medial hypothalamus (MacDonnell et al., 1971a, 1971b). In light of these observations, I plan to use a wider dose range of ethanol than was used in our pilot experiments in order to more fully examine its modulating effects. It has been further shown that such variations in the effects of alcohol on aggressive behavior may be due to individual differences in the actions of alcohol on the brain (Miczek et al., 1994b,

1994c). These authors further speculated that several of the brain mechanisms mediating aggression and which are affected by ethanol include GABA, 5-HT, and dopaminergic systems.

In a more recent study, it was discovered that ethanol (administered peripherally) oppositely affects two forms of aggressive behavior by potentiating defensive rage and suppressing predatory attack (Schubert et al., 1996a). In this paper, it was suggested that one mechanism underlying alcohol's potentiation of defensive rage behavior is that it acts on the descending pathway from the medial hypothalamus to the PAG. Recall that this pathway uses excitatory amino acids as a neurotransmitter. Confirmation of this hypothesis was obtained from data showing that the potentiating effects of alcohol on defensive rage were blocked following administration of the NMDA antagonist, AP-7, into the PAG. The significance of this finding was that it identified a specific component of the neural circuit for defensive rage behavior over which the potentiating effects of alcohol are mediated.

Technical Concerns

As noted above, progress has been made in demonstrating alcohol's role in potentiating aggressive behavior and some data have further indicated a component of the neural circuit mediating these effects. In spite of the efforts made by various investigators in this field, we still know very little about how alcohol affects brain cells in modulating aggression as well as the overall circuitry and receptors implicated in this process. There are several reasons that may account for the slow progress here. One involves the route of administration. A common approach has been to employ an intraperitoneal approach. Although successful experiments have been generated using this route, it does raise several concerns. In particular, it leaves open the possibility that a component of the observed effects may be due to peripheral rather than central effects of alcohol delivery, raising an issue that is not easily solved. The second is that peripheral administration may result in the absorption of alcohol directly into fat pads, reducing the amount actually taken into the circulatory system. However, this latter concern can be partially handled by directly measuring levels of blood alcohol. A second reason is that it is not easy to administer alcohol by mouth. It is somewhat unpleasant to the animal for it to be force-fed because of the restraint required as well as the fact that animals generally dislike the taste. If alcohol is included in a more pleasant liquid solution, the amount ingested remains dependent on the animal's decision. In this instance, there may be considerable differences in blood alcohol levels among different animals in the same as well as different experimental and control conditions. A third reason is that, in contrast to the many successful studies employing direct intracerebral administration of selective receptor–neurotransmitter compounds into discrete regions of the brain (as described in this chapter), such has not been the case with alcohol. The difficulty stems in part from the fact that alcohol diffuses quite extensively when microinjected into the brain. Therefore, if low concentrations are injected, it is difficult to determine how far it may spread, and it is also difficult if not impossible to tell whether the amount approaches physiological levels. The application of higher levels into the brain raises the specter of introducing a toxic substance in brain cells. Hopefully, new approaches will be developed that will overcome these pitfalls.

Psychotropic and Antidepressant Drugs

Over the years, various investigators have attempted to determine how antidepressants, tranquilizers, and psychotropic drugs affect rage and aggression. Typically, several different strategies have been employed. In one, several forms of aggressive behavior were recorded as part of a battery of psychological and behavioral tests against different drugs. Here, the purpose of using a given drug was to determine its efficacy in treating a specific disorder. In a second strategy, the approach has been to use a specific drug as a tool in order to learn how specific neurotransmitter—receptors may affect aggression and rage. Various studies using such drugs were described earlier in this chapter. One caveat related to our interpretation of these studies is that many of them describe inhibitory effects of these drugs on aggressive processes. The problem in interpreting such data is that it is difficult to discern whether the inhibitory effects were due to specific actions of the drug on aggressive behavior, or include a wider body of motor and behavioral processes. Unfortunately, few of these studies have addressed this issue.

Several early studies considered the role of several antidepressants, including imipramine, desimipramine, and amytryptyline (Baxter, 1968b; Funderburk et al., 1970; Dubinsky and Goldberg, 1971; Dubinsky et al., 1971, 1973). The data from these studies suggested that imipramine and desimipramine suppress predatory attack, but have little effect on defensive rage. However, in another study, it was observed that a lower dose of imipramine facilitated defensive rage, while a higher dose suppressed this response (Penaloza-Rojas et al., 1961). The antidepressant fluvoxamine was tested at a clinically relevant dose in rats, and was found to suppress biting attack elicited by hypothalamic stimulation (Kruk et al., 1987; Olivier et al., 1990b; Kruk, 1991).

Various investigators have explored the role of benzodiazepine drugs acting on the GABA receptor complex on aggression and rage in several species. In general, the effects of administration of this class of drugs, including diazepam, oxazepam, and etizolam, have been to suppress defensive rage and fighting in rats (Panksepp, 1971); cats (Baxter, 1964; Malick, 1970; Funderburk et al., 1970; Fukuda and Tsumagari, 1983); and monkeys (Delgado, 1973). In the rat, it may be that the effects of this drug act mainly on the lateral threat component rather than on biting and fighting, *per se*. While suppression has been the main effect following administration of this drug, in one study using rats, chlordiazepoxide was reported to facilitate predatory attack (Kruk et al., 1979). However, the overall effects of benzodiazepines on aggression appear to mirror the inhibitory effects of GABA agonists on aggression and defense.

Other drugs have also been tested for their possible effects on aggression and rage. In one study, carbamazepine, which resembles tricyclic antidepressants in structure, and has been used for the treatment of both epilepsy and mood disorders, was reported to powerfully suppress defensive rage in the cat while having little or no effect on predatory attack (Shaikh et al., 1987). In contrast, when tested in the rat, carbamazepine appeared to have little effect on biting attack (Kruk et al., 1987).

The general approach to the study of drugs and aggression has been to administer such drugs and observe their effects on aggression and defense. In a recent series of studies, a somewhat different strategy has been adopted in order to identify the relationship between benzodiazepines and aggression (Adamec, 1993a; 1993b, 1997). In these studies, systemic administration of the β-carboline and inverse benzodiazepine agonist, FG 7142 (N-methyl-β-carboline 3 carboxamide), generally potentiated defensive behavior but suppressed predatory behavior in cats. With respect to defensive behavior, the facilitatory effects of FG 7142 were blocked following treatment with the benzodiazepine antagonist, flumazenil.

It is of interest to point out that Adamec showed that the evoked potential recorded in the medial hypothalamus following stimulation of the basal or medial amygdala and administration of FG 7142 resulted in an increase in the size of the evoked potential. Since I indicated previously in this chapter and elsewhere that the strial pathway from the medial amygdala to the medial hypothalamus mediates the facilitatory effects of the medial amygdala, it would appear that the potentiating effects of FG 7142 can be explained in terms of the presumably excitatory actions of this drug on neurons in the medial and basal amygdala that affect the medial hypothalamus via this pathway. One other feature of these findings, namely the differential effects of FG 7142 on naturally occurring defensive rage and predatory attack, may also be accounted for in terms of our understanding of the anatomical relationships of the hypothalamus. The likely explanation for these effects is that if FG 7142 has excitatory effects on medial and basal amygdaloid neurons, which give rise to the stria terminalis, the net effect will be excitation of the medial hypothalamus and potentiation of the defensive rage mechanism. Since I have described earlier in this chapter and elsewhere that the medial hypothalamus contains GABAergic (inhibitory) neurons that supply the lateral hypothalamus, it is appropriate to conclude that the suppressive effects of this drug on predatory attack is the result of indirect activation of these GABAergic neurons by axons of the stria terminalis.

Summary

In this section, I examined results of studies on several classes of drugs of abuse as well as psychotropic and antidepressant drugs. The general findings suggest that opioids tend to suppress aggressive reactions, while alcohol potentiates rage responses. However, individual differences appear to play a significant role in determining whether alcohol will have a potentiating or inhibiting effect on aggressive behavior. Moreover, the effects may be further related to the type of aggression considered, as alcohol appears to potentiate defensive rage but inhibit predatory attack. With respect to psychotropic and antidepressant drugs, the overall picture is that these drugs suppress aggressive reactions. However, a basic difficulty with many of these studies in concluding that such drugs specifically affect aggression and rage is that one may not be able to factor out nonspecific suppressive effects of these drugs on many motor and other functions. Accordingly, at the present time, there is no known drug available that acts specifically against one or more forms of aggressive behavior.

REFERENCES

Abe, Y., T. Tadano, A. Yonezawa, K. Kisara, Suppressive effects of intraventricular injected dopamine and nomifensine on muricide induced by thiamine deficiency, *Pharmacol. Biochem. Behav.*, 26, 1987, 77–81.

Adamec, R.E., Transmitter systems involved in neural plasticity underlying increased anxiety and defense — implications for understanding anxiety following traumatic stress, *Neurosci. Biobehav. Rev.*, 21, 1997, 755–765.

Adamec, R.E., Lasting effects of FG-7142 on anxiety, aggression and limbic physiology in the cat, *J. Psychopharmacol.*, 7, 1993a, 232–248.

Adamec, R.E., Partial limbic kindling — brain, behavior, and the benzodiazepine receptor, *Physiol. Behav.*, 54, 1993b, 531–545.

Adamec, R.E., T. Shallow, J. Budgell, Blockade of CCK_B but not CCK_A receptors before and after the stress of predator exposure prevents lasting increases in anxiety-like behavior: implications for anxiety associated with posttraumatic stress disorder, *Behav. Neurosci.*, 111, 1997, 435–449.

Adams, D.B., W. Boudreau, C.W. Cowan, C. Kokonowski, K. Oberteuffer, K. Yohay, Offense produced by chemical stimulation of the anterior hypothalamus of the rat, *Physiol. Behav.*, 53, 1993, 1127–1132.

Albert, D.J. and S.E. Richmond, Reactivity and aggression in the rat: induction by alpha-adrenergic blocking agents injected ventral to anterior septum but not into lateral septum, *J. Comp. Physiol. Psychol.*, 91, 1977, 886–896.

Albert, D.J., R.C.K. Wong, K.N. Brayley, H.C. Fibiger, Evaluation of adrenergic, cholinergic and dopaminergic involvement in the inhibition of hyperreactivity and interanimal aggression by the medial hypothalamus in the rat, *Pharmacol. Biochem. Behav.*, 11, 1979, 1–10.

Allikmets, L.H., Cholinergic mechanisms in aggressive behaviour, *Med. Biol.*, 52, 1974, 19–30.

Babor, T.F., H.R. Kranzler, R.J. Lauerman, Early detection of harmful alcohol consumption: Comparison of clinical laboratory and self report screening procedures, *Addict. Behav.*, 14, 1989, 139–157.

Bandler, R. and R. Halliday, Lateralized loss of biting attack patterned reflexes following induction of contralateral sensory neglect in the cat: a possible role for the striatum in centrally elicited aggressive behaviour, *Brain Res.*, 242, 1982, 165–177.

Barnett, A., J.B. Malick, R.I. Taber, Effects of antihistamines on isolation-induced fighting in mice, *Psychopharmacologia*, 19, 1971, 359–365.

Barr, G.A., J.L. Gibbons, W.H. Bridger, A comparison of the effects of acute and subacute administration of beta-phenylethylamine and d-amphetamine on mouse killing behavior of rats, *Pharmacol. Biochem. Behav.*, 11, 1979, 419–422.

Barrett, J., H. Edinger, A. Siegel, Intrahypothalamic injections of norepinephrine facilitate feline affective aggression via alpha-2 adrenoceptors, *Brain Res.*, 525, 1990, 285–293.

Barrett, J.A., M.B. Shaikh, H. Edinger, A. Siegel, The effects of intrahypothalamic injections of norepinephrine upon affective defense behavior in the cat, *Brain Res.*, 426, 1987, 381–384.

Baxter, B.L. The effect of chlordiazepoxide on the hissing response elicited via hypothalamic stimulation, *Life Sci.*, 3, 1964, 531–537.

Baxter, B.L., Elicitation of emotional behavior by electrical or chemical stimulation applied at the same loci in cat mesencephalon, *Exp. Neurol.*, 21, 1968a, 1–10.

Baxter, B.L., The effect of selected drugs on the "emotional" behavior elicited via hypothalamic stimulation, *Int. J. Neuropharmacol.*, 7, 1968b, 47–54.

Beart, P.M., L.S. Nicolopoulos, D.C. West, P.M. Headley, An excitatory amino acid projection from ventromedial hypothalamus to periaqueductal gray in the rat: autoradiographic and electrophysiological evidence, *Neurosci. Lett.*, 85, 1988, 205–211.

Beart, P.M., R.J. Summers, J.A. Stephenson, C.J. Cook, M.J. Christie, Excitatory amino acid projections to the periaqueductal gray in the rat: a retrograde transport study utilizing d(3H)aspartate and, 3H,GABA, *Neuroscience*, 34, 1990, 163–176.

Beitz, A.J., Possible origin of glutamatergic projections to the midbrain periaqueductal gray and deep layer of the superior colliculus of the rat, *Brain Res. Bull.*, 23, 1989, 25–35.

Beleslin, D.B. and R. Samardzic, Effects of parachlorophenylalanine and 5,6-dihydroxytryptamine on aggressive behaviour evoked by cholinomimetics and anticholinesterases injected into the cerebral ventricles of conscious cats, *Neuropharmacology*, 18, 1979, 251–257.

Beleslin, D.B., D. Jovanovic-Micic, N. Japundzic, A.M. Terzic, R. Samardzic, Behavioral, autonomic and motor effects of neuroleptic drugs in cats: motor impairment and aggression, *Brain Res. Bull.*, 15, 1985, 353–356.

Benton, D. and R. Smoothy, The relationship between blood alcohol levels and aggression in mice, *Physiol. Behav.*, 33, 1984, 757–760.1.

Berntson, G.G., M.S. Beattie, J.M. Walker, Effects of nicotinic and muscarinic compounds on biting attack in the cat, *Pharmacol. Biochem. Behav.*, 5, 1976, 235–239.

Berntson, G.G. and S.F. Leibowitz, Biting attack in cats: evidence for central muscarinic mediation, *Brain Res.*, 51, 1973, 366–370.

Blanchard, R., Differential effects of benzodiazepines and 5-HT 1A agonists on defensive patterns in wild rats, in P. Bevan, A. Cools, T. Archer, Eds., *Behavioural Pharmacology of 5-HT*, Lawrence Erlbaum Associates, Hillsdale, N.J., 1989, pp. 145–148.

Blanchard, R.J., C.D. Blanchard, K.J. Flannelly, K. Hori, Ethanol changes patterns of defensive behavior in wild rats, *Physiol. Behav.*, 38, 1986, 645–650.

Blanchard, R.J., K. Hori, C. Blanchard, Ethanol effects on aggression of rats selected for different levels of aggressiveness, *Pharmacol. Biochem. Behav.*, 27, 1987, 641–644.

Blanchard, R.J., E.B. Yudko, C.D. Blanchard, Alcohol, aggression, and the stress of subordination, *J. Stud. Alcohol.*, 11, 1993a, 146–155.

Blanchard, R.J., E.B. Yudko, R.J. Rodgers, D.C. Blanchard, Defense system psychopharmacology: an ethological approach to the pharmacology of fear and anxiety, *Behav. Brain Res.*, 58, 1993b, 155–165.

Block, C.H., A. Siegel, H. Edinger, Effects of amygdaloid stimulation upon trigeminal sensory fields established during hypothalamically-elicited quiet biting attack in the cat, *Brain Res.*, 197, 1980a, 39–55.

Block, C.H., A. Siegel, H.M. Edinger, Effects of stimulation of the substantia innominata upon attack behavior elicited from the hypothalamus in the cat, *Brain Res.*, 197, 1980b, 57–74.

Broderick, P.A. and W.H. Bridger, A comparative study of the effect of L-tryptophan and its acetylated derivative N-acetyl-L-tryptophan on rat muricidal behavior, *Biol. Psychiatry*, 19, 1984, 89–94.

Brudzynski, S.M., Carbachol-induced agonistic behavior in cats: aggressive or defensive response? *Acta Neurobiol. Exp.*, 41, 1981a, 15–32.

Brudzynski, S.M., Growling component of vocalization as a quantitative index of carbachol-induced emotional-defensive response in cats, *Acta Neurobiol. Exp.*, 41, 1981b, 33–51.

Brudzynski, S.M. and F. Bihara, Ultrasonic vocalization in rats by cholinergic stimulation of the brain, *Neurosci. Lett.*, 109, 1990, 222–226.

Brudzynski, S.M. and B. Eckersdorf, Vocalization accompanying emotional-aversive response induced by carbachol in the cat, *Neuropsychopharmacology*, 1, 1988, 311–320.

Brudzynski, S.M., B. Eckersdorf, H. Golebiewski, Evidence for involvement of endogenous acetylcholine in emotional-aversive response in the cat, *Prog. Neuro-Psychopharmacol., Biol. Psychiatry*, 14, 1990, 807–812.

Brudzynski, S.M., B. Eckersdorf, H. Golebiewski, Emotional-aversive nature of the behavioral response induced by carbachol in cats, *J. Psychiatr. Neurosci.*, 18, 1, 1993, 38–45.

Brudzynski, S.M., Ultrasonic vocalization induced by intracerebral carbachol in rats: localization and a dose–response study, *Behav. Brain Res.*, 63, 1994, 133–143.

Brutus, M., M.B. Shaikh, A. Siegel, H. Edinger, Effects of experimental temporal lobe seizures upon hypothalamically elicited aggressive behavior in the cat, *Brain Res.*, 366, 1986, 53–63.

Brutus, M. and A. Siegel, Effects of the opiate antagonist naloxone upon hypothalamically elicited affective defense behavior in the cat, *Behav. Brain Res.*, 33, 1989, 23–32.

Brutus, M., S. Zuabi, A. Siegel, Effects of D-ala2-met5-enkephalinamide microinjections placed into the bed nucleus of the stria terminalis upon affective defense behavior in the cat, *Brain Res.*, 473, 1988, 147–152.

Brutus, M., S. Zuabi, A. Siegel, Microinjection of D-ala2-met5-enkephalinamide placed into the nucleus accumbens suppresses feline affective defense behavior, *Exp. Neurol.*, 104, 1989, 55–61.

Bushman, B.J. and H.M. Cooper, Effects of alcohol on human aggression: an integrative research review, *Psychol. Bull.*, 107, 1990, 341–354.

Cabib, S., F.R. D'Amato, S. Puglisi-Allegra, D. Maestripieri, Behavioral and mesocorticolimbic dopamine responses to nonaggressive social interactions depend on previous social experiences and on the opponent's sex, *Behav. Brain Res.*, 112, 2000, 13–22.

Cabib, S. and S. Puglisi-Allegra, Genotype-dependent modulation of LY 171555-induced defensive behavior in the mouse, *Psychopharmacology (Berlin)*, 97, 1989, 166–168.

Chance, M.R.A., R.A.J.H. Mackintosh, A.K. Dixon, The effects of ethyl alcohol on social encounters between mice, *J. Alcoholism*, 8, 1973, 90–93.

Cheu, J.W. and A. Siegel, GABA receptor mediated suppression of defensive rage behavior elicited from the medial hypothalamus of the cat: role of the lateral hypothalamus, *Brain Res.*, 783, 1998, 293–304.

Chiavegatto, S., V.L. Dawsons, L.A. Mamounas, V.E. Koliatsos, T.M. Dawson, R.J. Nelson, Brain serotonin dysfunction accounts for aggression in male mice lacking neuronal nitric oxide synthase, *Proc. Natl. Acad. Sci. U.S.A.*, 98, 2001, 1277–1281.

Coccaro, E.F., Central serotonin and impulsive aggression, *Br. J. Psychiatr.*, 155 (Suppl. 8), 1989, 52–62.

Coccaro, E.F., S. Gabriel, L.J. Siever, Buspirone challenge: Preliminary evidence for a role for central 5HT–1a receptor function in impulsive aggressive behavior in humans, *Psychopharmacol. Bull.*, 26, 1990, 405.

Coccaro, E.F., R.J. Kavoussi, T.B. Cooper, R.L. Hauger, Central serotonin activity and aggression: Inverse relationship with prolactin response to *d*-fenfluramine, but not CSF 5-HIAA concentration, in human subjects, *Am. J. Psychiatr.*, 154, 1997, 1430–1435.

Cologer-Clifford, A., N.G. Simon, S.F. Lu, S.A. Smoluk, Serotonin agonist-induced decreases in intermale aggression are dependent on brain region and receptor subtype, *Pharmacol. Biochem. Behav.*, 58, 1997, 425–430.

Cologer-Clifford, A., S.A. Smoluk, N.G. Simon, Effects of serotonergic 1A and 1B agonists in androgenic versus estrogenic systems for aggression, *Ann. N.Y. Acad. Sci.*, 794, 1996, 339–342.

Crowley, T.J., A.J. Stynes, M. Hydinger, I.C. Kaufman, Ethanol, methamphetamine, pentobarbital, morphine, and monkey social behavior, *Arch. Gen. Psychiatry*, 31, 1974, 829–838.

Davis, W.M. and J.H. Khalsa, Increased shock induced aggression during morphine withdrawal, part 1, *Life Sci.*, 10, 1971, 1321–1327.

De Boer, S.F., M. Lesourd, E. Mocaër, J.M. Koolhaas, Somatodendritic 5-HT_{1A} autoreceptors mediate the anti-aggressive actions of 5-HT_{1A} receptor agonists in rats: an ethopharmacological study with S-15535, alnespirone, and WAY-100635, *Neuropsychopharmacology*, 23, 2000, 20–33.

Delgado, J.M.R., Antiaggressive effects of chlordiazepoxide, in S. Garattini, E. Mussini, L.O. Randall, Eds., *The Benzodiazepines*, Raven Press, New York, 1973, pp. 419–432.

Dominic, J.A. and K.E. Moore, Behavior and brain contents of catecholamines in mice during chronic administration of methyldopa, *Neuropharmacology*, 10, 1971, 565–570.

Dubinsky, B. and M.E. Goldberg, The effect of imipramine and selected drugs on attack elicited by hypothalamic stimulation in the cat, *Neuropharmacology*, 10, 1971, 537–545.

Dubinsky, B., J.K. Karpowicz, M.E. Goldberg, Effects of tricyclic antidepessants on attack elicited by hypothalamic stimulation: Relation to brain biogenic amines, *J. Pharmacol. Exp. Ther.*, 187, 1973, 550–557.

Dubinsky, B., K. Karpowicz, M.E. Goldberg, Effects of desipramine, imipramine and chlorimipramine on hypothalamically-induced attack in cats, *Pharmacologist*, 13, 1971, 1.

Eichelmann, B., The catecholamines and aggressive behavior, *Neurosci. Res.*, 5, 1973, 109–129.

Fanselow, M.S., R. Sigmundi, R.C. Bolles, Naloxone pretreatment enhances shock-elicited aggression, *Physiol. Psychol.*, 8, 1980, 369–371.

Ferguson, J., S. Henriksen, H. Cohen, G. Mitchell, J. Barchas, W. Dement, "Hypersexuality" and behavioral changes in cats caused by administration of p-chlorophenylalanine, *Science*, 168, 1970, 499–501.

File, S.E. and J.F.W. Deakin, Chemical lesions of both dorsal and median raphe nuclei and changes in social and aggressive behaviour in rats, *Pharmacol. Biochem. Behav.*, 12, 1980, 855–859.

Fukuda, T. and T. Tsumagari, Effects of psychotropic drugs on the rage responses induced by electrical stimulation of the medial hypothalamus in cats, *Jpn. J. Pharmacol.*, 33, 1983, 885–890.

Funderburk, W.H., M.H. Foxwell, M.W. Hakala, Effects of psychotherapeutic drugs on hypothalamic-induced hissing in cats, *Neuropharmacology*, 9, 1970, 1–7.

Gallagher, A.W., L.A. Chahl, A.M. Lynch, Distribution of substance P-like immunoreactivity in guinea pig central nervous system, *Brain Res. Bull.*, 29, 1992, 199–207.

Garmendia, L., J.R. Sánchez, A. Azpiroz, P.F. Brain, V.M. Simón, Clozapine: strong antiaggressive effects with minimal motor impairment, *Physiol. Behav.*, 51, 1992, 51–54.

Gay, P.E. and R.C. Leaf, Rat strain differences in pilocarpine-induced mouse killing, *Physiol. Psychol.*, 4, 1976, 28–32.

Gay, P.E., R.C. Leaf, F.B. Arble, Inhibitory effects of pre- and posttest drugs on mouse-killing by rats, *Pharmacol. Biochem. Behav.*, 3, 1975, 33–45.

Gay, P.E., L.S. Potter, J.A. Consalvi, R.C. Leaf, The effects of d-amphetamine on prey killing and prey eating in the rat and mouse, *Bull. Psychonom. Soc.*, 10, 1977, 385–388.

George, R., W.L. Haslett, D.J. Jenden, The central action of a metabolite of tremorine, *Life Sci.*, 8, 1962, 361–363.

Geyer, M. and D. Segal, Shock-induced aggression: opposite effects of intraventricularly infused dopamine and norepinephrine, *Behav. Biol.*, 10, 1974, 99–104.

Gianutsos, G., M.D. Hynes, H. Lal, Enhancement of morphine-withdrawal and apomorphine-induced aggression by clonidine, *Psychopharm. Comm.*, 2, 1976, 165–171.

Gianutsos, G. and H. Lal, Blockade of apomorphine-induced aggression by morphine or neuroleptics: differential alteration by antimuscarinics and naloxone, *Pharmacol. Biochem. Behav.*, 4, 1976, 639–642.

Gianutsos, G. and H. Lal, Narcotic analgesics and aggression, *Mod. Prob. Pharmacopsych.*, 13, 1978, 114–138.

Golebiewski, H. and A. Romaniuk, The participation of serotoninergic system in the regulation of emotional-defensive behavior evoked by intrahypothalamic carbachol injections in the cat, *Acta Neurobiol. Exp.*, 45, 1985, 25–36.

Hadfield, M.G., Mesocortical vs nigrostriatal dopamine uptake in isolated fighting mice, *Brain Res.*, 222, 1981, 172–176.

Hadfield, M.G., Dopamine: mesocortical vs nigrostriatal uptake in isolated fighting mice and controls, *Behav. Brain Res.*, 7, 1983, 269–281.

Han, Y., M.B. Shaikh, A. Siegel, Medial amygdaloid suppression of predatory attack behavior in the cat: I. Role of a substance P pathway from the medial amygdala to the medial hypothalamus, *Brain Res.*, 716, 1996a, 59–71.

Han, Y., M.B. Shaikh, A. Siegel, Medial amygdaloid suppression of predatory attack behavior in the cat: II. Role of a GABAergic pathway from the medial to the lateral hypothalamus, *Brain Res.*, 716, 1996b, 72–83.

Hasselager, E., Z. Rolinski, A. Randrup, Specific antagonism by dopamine inhibitors of items of amphetamine induced aggressive behaviour, *Psychopharmacologia*, 24, 1972, 485–495.

Higley, J.D., S.T. King, M.F. Hasert, M. Champoux, S.J. Suomi, M. Linnoila, Stability of interindividual differences in serotonin function and its relationship to severe aggression and competent social behavior in rhesus macaque females, *Neuropsychopharmacology*, 14, 1996a, 67–76.

Higley, J.D., P.T. Mehlman, S.B. Higley, B. Fernald, J. Vickers, S.G. Lindell, D.M. Taub, S.J. Suomi, M. Linnoila, Excessive mortality in young free-ranging male nonhuman primates with low cerebrospinal fluid 5-hydroxyindoleacetic acid concentrations, *Arch. Gen. Psychol.*, 53, 1996b, 537–543.

Higley, J.D., P.T. Mehlman, R.E. Poland, D.M. Taub, J. Vickers, S.J. Suomi, M. Linnoila, CSF testosterone and 5-HIAA correlate with different types of aggressive behaviors, *Biol. Psychiatry*, 40, 1996c, 1067–1082.

Igic, R., P. Stern, E. Basagic, Changes in emotional behaviour after application of cholinesterase inhibitor in the septal and amygdala region, *Neuropharmacology*, 9, 1970, 73–75.

Jacobs, B.L. and A. Cohen, Differential behavioral effects of lesions of the median or dorsal raphe nuclei in rats: open field and pain-elicited aggression, *J. Comp. Physiol. Psychol.*, 90, 1976, 102–108.

Kantak, K.M., L.R. Hegstrand, B. Eichelman, Dietary tryptophan modulation and aggressive behavior in mice, *Pharmacol. Biochem. Behav.*, 12, 1979a, 675–679.

Kantak, K.M., L.R. Hegstrand, J. Whitman, B. Eichelman, Effects of dietary supplements and a tryptophan-free diet on aggressive behavior in rats, *Pharmacol. Biochem. Behav.*, 12, 1979b, 173–179.

Kantak, K.M. and K.A. Miczek, Aggression during morphine withdrawal: effects of method of withdrawal, fighting experience, and social role, *Psychopharmacology (Berlin)*, 90, 1986, 451–456.

Katz, R.J. and E. Thomas, Effects of scopolamine and alpha-methylparatyrosine upon predatory attack in cats, *Psychopharmacologia*, 42, 1975, 153–157.

Katz, R.J. and E. Thomas, Effects of para-chlorophenylalanine upon brain stimulated affective attack in the cat, *Pharmacol. Biochem. Behav.*, 5, 1976, 391–394.

Kemble, E.D., M. Behrens, J.M. Rawleigh, B.M. Gibson, Effects of yohimbine on isolation-induced aggression, social attraction, and conspecific odor preference in mice, *Pharmacol. Biochem. Behav.*, 40, 1991, 781–785.

Knutson, J.F., N.L. Kane, A.J. Schlosberg, D.J. Fordyce, K.J. Simansky, Influence of PCPA, shock level, and home-cage conditions on shock-induced aggression, *Physiol. Behav.*, 23, 1979, 897–907.

Konig, M., A.M. Zimmer, H. Steiner, P.V. Holmes, J.N. Crawley, M.J. Brownstein, A. Zimmer, Pain responses, anxiety and aggression in mice deficient in pre-proenkephalin, *Nature*, 383, 1996, 535–538.

Koprowska, M. and A. Romaniuk, Behavioral and biochemical alterations in median and dorsal raphe nuclei lesioned cats, *Pharmacol. Biochem. Behav.*, 56, 1997, 529–540.

Korte, S.M., O.C. Meijer, E.R. De Kloet, B. Buwalda, J. Keijser, F. Sluyter, G. Van Oortmerssen, B. Bohus, Enhanced 5-HT_{1A} receptor expression in forebrain regions of aggressive house mice, *Brain Res.*, 736, 1996, 338–343.

Korzan, W.J., T.R. Summers, C.H. Summers, Monoaminergic activities of limbic regions are elevated during aggression: influence of sympathetic social signaling, *Brain Res.*, 870, 2000, 170–178.

Kostowski, W., O. Pucilowski, A. Plaznik, Effect of stimulation of brain serotonergic system on mouse-killing behavior in rats, *Physiol. Behav.*, 25, 1980, 161–165.

Kostowski, W. and L. Valzelli, Biochemical and behavioral effects of lesions of raphe nuclei in aggressive mice, *Pharmacol. Biochem. Behav.*, 2, 1974, 277–280.

Kromer, W. and J.E. Dum, Mouse-killing in rats induces a naloxone-blockable increase in nociceptive threshold, *Eur. J. Pharmacol.*, 63, 1980, 195–198.

Krsiak, M., Effect of ethanol on aggression and timidity in mice, *Psychopharmacology*, 51, 1976, 75–80.1.

Krsiak, M. and M. Borgesova, Effect of alcohol on behaviour of pairs of rats, *Psychopharmacologia*, 32, 1973, 201–209.

Kruk, M.R., Ethology and pharmacology of hypothalamic aggression in the rat, *Neurosci. Biobehav. Rev.*, 15, 1991, 527–538.

Kruk, M.R., A.M. Van der Poel, T.P. De Vos-Frerichs, The induction of aggressive behaviour by electrical stimulation in the hypothalamus of male rats, *Behaviour*, 70, 1979, 292–322.

Kruk, M.R., A.M. Van der Poel, J.H.C.M. Lammers, T. Hagg, A.M.D.M. de Hey, S. Oostwegel, Ethopharmacology of hypothalamic aggression in the rat, in B. Olivier, J. Mos, P.F. Brain, Eds., *Ethopharmacology of Agonistic Behaviour in Animals and Humans*, Martinus Nijhoff Publishers, Dordrecht, 1987, pp. 35–45.

Lagerspetz, K.M.J. and K. Ekqvist, Failure to induce aggression in inhibited and in genetically nonaggressive mice through injections of ethyl alcohol, *Aggress. Behav.*, 4, 1978, 105–113.

Lal, H., J. O'Brien, S.K. Puri, Morphine-withdrawal aggression: sensitization by amphetamines, *Psychopharmacologia*, 22, 1971, 217–223.

Lamprecht, F., B. Eichelman, N.B. Thoa, R.B. Williams, I.J. Kopin, Rat fighting behavior: serum dopamine-beta-hydroxylase and hypothalamic tyrosine hydroxylase, *Science*, 177, 1972, 1214–1215.

Lanaud, P., T. Popovici, E. Normand, C. Lemoine, B. Bloch, B.P. Roques, Distribution of CCK mRNA in particular region, hippocampus, periaqueductal grey and thalamus, of the rat by in situ hybridization, *Neurosci. Lett.*, 104, 1989, 38–42.

Langevin, H. and P.C. Emson, Distribution of substance P, somatostatin and neurotensin in the human hypothalamus, *Brain Res.*, 246, 1982, 65–69.

Leaf, R.C., L. Lerner, Z.P. Horovitz, Role of the amygdala in the pharmacological and endocrinological manipulation of aggression, in S. Garattini and E.B. Sigg, Eds., *Aggressive Behavior*, Wiley, New York, 1969, pp. 120–131.

Leslie, G.B., Central stimulant properties of compounds with peripheral muscarinic properties, *Nature*, 208, 1965, 1291–1293.

M.H. Lewis, J.-L. Gariépy, P. Gendreau, D.E. Nichols, R.B. Mailman, Social reactivity and D_1 dopamine receptors: studies in mice selectively bred for high and low levels of aggression, *Neuropsychopharmacology*, 10, 1994, 115–122.

Lindqvist, C.U., Criminal homicide in northern Sweden 1970–1981: alcohol intoxication, alcohol abuse and mental disease, *Int. J. Law Psychiatry*, 8, 1986, 19–37.

Liu, H., S. Chandler, A.J. Beitz, M.T. Shipley, M.M. Behbehani, Characterization of the effect of cholecystokinin, CCK, on neurons in the periaqueductal gray of the rat: immunocytochemical and *in vivo* and *in vitro* electrophysiological studies, *Brain Res.*, 642, 1994, 83–94.

Lu, C.-L., M.B. Shaikh, A. Siegel, Role of NMDA receptors in hypothalamic facilitation of feline defensive rage elicited from the midbrain periaqueductal gray, *Brain Res.*, 581, 1992, 123–132.

Luo, B., J.W. Cheu, A. Siegel, Cholecystokinin B receptors in the periaqueductal gray potentiate defensive rage behavior elicited from the medial hypothalamus of the cat, *Brain Res.*, 796, 1998, 27–37.

Lyons, W.E., L.A. Mamounas, G.A. Ricaurte, V. Coppola, S.W. Reid, S.H. Bora, C. Wihler, V.E. Koliatsos, L. Tessarollo, Brain-derived neurotrophic factor-deficient mice develop aggressiveness and hyperphagia in conjunction with brain serotonergic abnormalities, *Proc. Natl. Acad. Sci. U.S.A.*, 96, 1999, 15239–15244.

MacDonnell, M.F., L. Fessock, S.H. Brown, Aggression and associated neural events in cats. Effects of p-chlorophenylalanine compared with alcohol, *Q. J. Stud. Alcohol.*, 32, 1971a, 748–763.

MacDonnell, M.F., L. Fessock, S.H. Brown, Ethanol and the neural substrate for affective defense in the cat, *Q. J. Stud. Alcohol.*, 32, 1971b, 406–419.

Maeda, H., Effects of psychotropic drugs upon the hypothalamic rage response in cats, *Folia Psychiatr. Neurol. Jpn.*, 30, 1976, 539–546.

Maeda, H. and S. Maki, Dopaminergic facilitation of recovery from amygdaloid lesions which affect hypothalamic defensive attack in cats, *Brain Res.*, 363, 1986, 135–140.

Maeda, H., T. Sato, S. Maki, Effects of dopamine agonists on hypothalamic defensive attack in cats, *Physiol. Behav.*, 35, 1985, 89–92.

Malick, J.B., Effects of selected drugs on stimulus-bound emotional behavior elicited by hypothalamic stimulation in the cat, *Arch. Int. Pharmacodyn. Ther.*, 186, 1970, 137–141.

Malick, J.B., Differential effects of d- and l-amphetamine on mouse-killing behavior in rats, *Pharmacol. Biochem. Behav.*, 3, 1975, 697–699.

Malick, J.B. and A. Barnett, The role of serotonergic pathways in isolation-induced aggression in mice, *Pharmacol. Biochem. Behav.*, 5, 1976, 55–61.

Marini, J.L., J.K. Walters, M.H. Sheard, Effects of d- and l-amphetamine on hypothalamically-elicited movement and attack in the cat, *Agressologie*, 20, 1979, 155–160.

Mark, V.H., I. Takada, H. Tsutsumi, H. Takamatsu, E. Toth, D.B. Mark, Effect of exogenous catecholamines in the amygdala of a 'rage' cat, *Appl. Neurophysiol.*, 38, 1975, 61–72.

Marshall, J.F., Somatosensory inattention after dopamine-depleting intracerebral 6-OHDA injections: spontaneous recovery and pharmacological control, *Brain Res.*, 177, 1979, 311–324.

Mayer, M.L. and N.K. MacLeod, The excitatory action of substance P and stimulation of the stria terminalis bed nucleus on preoptic neurones, *Brain Res.*, 166, 1979, 206–210.

Mercer, L.D., P.M. Beart, M.K. Horne, D.I. Finkelstein, P. Carrive, G. Paxinos, On the distribution of cholecystokinin B receptors in monkey brain, *Brain Res.*, 738, 1996, 313–318.

Messeri, P., B.E. Eleftheriou, A. Oliverio, Dominance behavior: a phylogenetic analysis in the mouse, *Physiol. Behav.*, 14, 1975, 53–58.

Miczek, K.A., J.L. Altman, J.B. Appel, W.O. Boggan, Para-chlorophenylalanine, serotonin and killing behavior, *Pharmacol. Biochem. Behav.*, 3, 1975, 355–361.

Miczek, K.A., J.F. DeBold, M. Haney, J. Tidey, J. Vivian, E. Weerts, Alcohol, drugs of abuse, aggression, and violence, in A.J. Reiss and J.A. Roth, Eds., *Understanding and Preventing Violence*, vol. 3, National Academy Press, Washington, D.C., 1994a, pp. 377–570.

Miczek, K.A., J.F. DeBold, A.M.M. Van Erp, Neuropharmacological characteristics of individual differences in alcohol effects on aggression in rodents and primates, *Behav. Pharmacol.*, 5, 1994b, 407–421.

Miczek, K.A., M. Haney, J. Tidey, J. Vivian, E. Weerts, Neurochemistry and pharmacotherapeutic management of aggression and violence, in A. Reiss, K. Miczek, J. Roth, Eds., *Understanding and Preventing Violence*, vol. 2, National Academy Press, Washington, D.C., 1994c, pp. 245–514.

Miczek, K.A. and J.M. O'Donnell, Alcohol and chlordiazepoxide increase suppressed aggression in mice, *Psychopharmacology (Berlin)*, 69, 1980, 39–44.

Miyachi, Y.H., Kasai, H. Ohyama, T. Yamada, Changes of aggressive behavior and brain serotonin turnover after very low-dose X-irradiation of mice, *Neurosci. Lett.*, 175, 1994, 92–94.

Moore, T.M., A. Scarpa, A. Raine, A meta-analysis of serotonin metabolite 5-HIAA and antisocial behavior, *Aggress. Behav.*, 28, 2002, 299–316.

Mos, J. and B. Olivier, Differential effects of selected psychoactive drugs on dominant and subordinate male rats housed in a colony, *Neurosci. Res. Commun.*, 2, 1988, 29–36.

Mos, J., B. Olivier, M. Poth, R. Van Oorschot, H. Van Aken, The effects of dorsal raphe administration of eltoprazine, TFMPP and 8-OH-DPAT on resident intruder aggression in the rat, *Eur. J. Pharmacol.*, 238, 1993, 411–415.

Moyer, K.E., and J.M. Crabtree, Sex differences in fighting and defense induced in rats by shock and d-amphetamine during morphine abstinence, *Physiol. Behav.*, 11, 1973, 337–343.

Murdoch, D., R.O. Pihl, D. Ross, Alcohol and crimes of violence: present issues, *Int. J. Addict.*, 25, 1990, 1059–1075.

Myers, R.D., Emotional and autonomic responses following hypothalamic chemical stimulation, *Can. J. Psychol.*, 18, 1964, 6–14.

Nakamura, K. and H. Thoenen, Increased irritability: a permanent change induced in the rat by intraventricular administration of 6-hydroxydopamine, *Psychopharmacologia*, 24, 1972, 359–372.

Nakao, H., N. Tashiro, R. Kono, R. Araki, Effects of GABA and glycine on aggressive-defense reaction produced by electrical stimulation of the ventromedial hypothalamus in cats, in M. Itoh, Ed., *Integrative Control Functions of the Brain*, Elsevier, Amsterdam/Tokyo, 1979, pp. 332–334.

Navarro, J.J., E. Maldonado, D. Beltrán, Effects of methyl-chlorophenylpiperazine, m-CPP, on isolation-induced aggression in male mice, *Med. Sci. Res.*, 27, 1999, 661–662.

Olivier, B., J. Mos, D.L. Rasmussen, Behavioural pharmacology of the serenic, eltoprazine, in M. Raghoebar, B. Olivier, D.L. Rasmussen, J. Mos, Eds., *Drug Metabolism and Drug Interactions*, Freund, London, 1990a, pp. 31–83.

Olivier, B., J. Mos, M. Tulp, J. Schipper, S. DenDaas, G. Van Oortmerssen, Serotonergic involvement in aggressive behaviour in animals, in H.M. Van Praag, Ed., *Monoaminergic Regulation of Aggression and Impulse Control*, ACNP Proceedings, Washington, D.C., 1990b, pp. 79–137.

Otsuka, M. and K. Yoshioka, Neurotransmitter functions of mammalian tachykinins, *Physiol. Rev.*, 73, 1993, 229–308.

Panksepp, J,.Drugs and stimulus-bound attack, *Physiol. Behav.*, 6, 1971, 317–320.

Parmigiani, S. and P. Palanza, Fluprazine inhibits intermale attack and infanticide, but not predation, in male mice, *Neurosci. Biobehav. Rev.*, 15, 1991, 511–513.

Penaloza-Rojas, J.H., G. Bach-Y-Rita, H.F. Rubio-Chevannier, R. Hernandez-Peon, Effects of imipramine upon hypothalamic and amygdaloid excitability, *Exp. Neurol.*, 4, 1961, 205–213.

Pernanen, K., *Alcohol in Human Violence*, Gilford Press, New York, 1991.

Pott, C., S. Kramer, A. Siegel, Naloxone sensitivity of stimulation sites in the central gray effective in modulating feline aggression, *Physiol. Behav.*, 40, 1986, 207–213.

Pradhan, S.N., Aggression and central neurotransmitters, *Intl. Rev. Neurobiol.*, 18, 1973, 213–262.

Pucilowski, O., B. Eichelman, D.H. Overstreet, A.H. Rezvani, D.S. Janowsky, Enhanced affective aggression in genetically bred hypercholinergic rats, *Neuropsychobiology*, 24, 1991, 37–41.

Pucilowski, O., A. Plaznik, W. Kostowski, Aggressive behavior inhibition by serotonin and quipazine injected into the amygdala in the rat, *Behav. Neural Biol.*, 43, 1985, 58–68.

Puglisi-Allegra, S. and S. Cabib, Pharmacological evidence for a role of D2 dopamine receptors in the defensive behavior of the mouse, *Behav. Neural Biol.*, 50, 1988, 98–111.

Puglisi-Allegra, S., A. Oliverio, P. Mandel, Effects of opiate antagonists on social and aggressive behavior of isolated mice, *Pharmacol. Biochem. Behav.*, 17, 1982, 691–694.

Rao, R., M. Yamano, S. Shiosaka, A. Shinohara, M. Tohyama, Origin of leucine-enkephalin fibers and their two main afferent pathways in the bed nucleus of the stria terminalis in the rat, *Exp. Brain Res.*, 65, 1987, 411–420.

Reichling, D.B. and A.I. Basbaum, Contribution of brainstem GABAergic circuitry to descending antinocicieptive controls: I. GABA-immunoreactive projection neurons in the periaqueductal gray and nucleus raphe magnus, *J. Comp. Neurol.*, 302, 1990, 370–377.

Reichling, D.B. and A.I. Basbaum, Collateralization of periaqueductal gray neurons to forebrain or diencephalon and to the medullary nucleus raphe magnus in the rat, *Neuroscience*, 42, 1991, 183–200.

Reis, D.J., Central neurotransmitters in aggression, *Res. Publ. Assoc. Res. Neuro. Ment. Dis.*, 52, 1974, 119–148.

Reis, D.J. and K. Fuxe, Depletion of noradrenaline in brainstem neurons during sham rage behaviour produced by acute brainstem transection in cat, *Brain Res.*, 7, 1968, 448–451.

Reis, D.J. and K. Fuxe, Brain norepinephrine: evidence that neuronal release is essential for sham rage behavior following brainstem transection in cat, *Proc. Natl. Acad. Sci. U.S.A.*, 64, 1969, 108–112.

Reis, D.J. and L.-M. Gunne, Brain catecholamines: Relation to the defense reaction evoked by amygdaloid stimulation in cat, *Science*, 149, 1965, 450–451.

Rodgers, R.J., The medial amygdala: serotonergic inhibition of shock-induced aggression and pain sensitivity in rats, *Aggress. Behav.*, 3, 1977, 277–288.

Rodgers, R.J., D.C. Blanchard, L.K. Wong, R.J. Blanchard, Effects of scopolamine on anti-predator defense reactions in wild and laboratory rats, *Pharmacol. Biochem. Behav.*, 36, 1990, 575–583.

Rodgers, R.J. and K. Brown, Amygdaloid function in the central cholinergic mediation of shock-induced aggression in the rat, *Aggress. Behav.*, 2, 1976, 131–152.

Rodgers, R.J. and C.A. Hendrie, Agonistic behaviour in rats: evidence for non-involvement of opioid mechanisms, *Physiol. Behav.*, 29, 1982, 85–90.

Roeling, T.A.P., M.R. Kruk, R. Schuurmans, J.G. Veening, Behavioral responses of bicuculline methiodide injections into the ventral hypothalamus of freely moving, socially interacting rats, *Brain Res.*, 615, 1993, 121–127.

Romaniuk, A., S. Brudzynski, J. Gronska, The effect of chemical blockade of hypothalamic cholinergic system on defensive reactions in cats, *Acta Physiol. Pol.*, 24, 1973, 809–816.

Romaniuk, A., S. Brudzynski, J. Gronska, The effects of intrahypothalamic injections of cholinergic and adrenergic agents on defensive behavior in cats, *Acta Physiol. Pol.*, 25, 1974, 297–305.

Romaniuk, A. and H. Golebiewski, Adrenergic modulation of the hypothalamic cholinergic mechanism in the control of emotional-defensive behavior in the cat, *Acta Neurobiol. Exp.*, 39, 1979, 313–326.

Sánchez, C., J. Arnt, J. Hyttel, E.K. Moltzen, The role of serotonergic mechanisms in inhibition of isolation-induced aggression in male mice, *Psychopharmacology (Berlin),* 110, 1993, 53–59.

Saudou, F., D.A. Amara, A. Dierich, M. Lemeur, S. Ramboz, L. Segu, M.C. Buhot, R. Hen, Enhanced aggressive behavior in mice lacking the 5ht1b receptor, *Science*, 265, 1994, 1875–1878.

Schubert, K., M.B. Shaikh, Y. Han, L. Pohorecky, A. Siegel, Differential effects of ethanol on feline rage and predatory attack behavior: an underlying neural mechanism, *Alcohol. Clin. Exp. Res.*, 20, 1996a, 882–889.

Schubert, K., M.B. Shaikh, A. Siegel, NMDA receptors in the midbrain periaqueductal gray mediate hypothalamically evoked hissing behavior in the cat, *Brain Res.*, 726, 1996b, 80–90.

Shaikh, M.B., M. Brutus, A. Siegel, H.E. Siegel, Regulation of feline aggression by the bed nucleus of stria terminalis, *Brain Res. Bull.*, 16, 1986, 179–182.

Shaikh, M.B., N.C. De Lanerolle, A. Siegel, Serotonin 5-HT $_{1A}$ and 5-HT$_{2/1C}$ receptors in the midbrain periaqueductal gray differentially modulate defensive rage behavior elicited from the medial hypothalamus of the cat, *Brain Res.*, 765, 1997, 198–207.

Shaikh, M.B., H. Edinger, A. Siegel, Carbamazepine regulates feline aggression elicited from the midbrain periaqueductal gray, *Pharmacol. Biochem. Behav.*, 30, 1987, 409–415.

Shaikh, M.B., C.-L. Lu, A. Siegel, Affective defense behavior elicited from the feline midbrain periaqueductal gray is regulated by mu- and delta-opioid receptors, *Brain Res.*, 557, 1991a, 344–348.

Shaikh, M.B., C.-L. Lu, A. Siegel, D-2 dopamine receptor involvement in the regulation of quiet biting attack behavior in the cat, *Brain Res. Bull.*, 27, 1991b, 725–730.

Shaikh, M.B., C.-L. Lu, A. Siegel, An enkephalinergic mechanism involved in amygdaloid suppression of affective defense behavior elicited from the midbain periaqueductal gray in the cat, *Brain Res.*, 559, 1991c, 109–117.

Shaikh, M.B., K. Schubert, A. Siegel, Basal amygdaloid facilitation of midbrain periaqueductal gray elicited defensive rage behavior in the cat is mediated through NMDA receptors, *Brain Res.*, 635, 1994, 187–195.

Shaikh, M.B., A.B. Shaikh, A. Siegel, Opioid peptides within the midbrain periaqueductal gray suppress affective defense behavior in the cat, *Peptides*, 9, 1988, 999–1004.

Shaikh, M.B. and A. Siegel, Naloxone-induced modulation of feline aggression elicited from midbrain periaqueductal gray, *Pharmacol. Biochem. Behav.*, 31, 1989, 791–796.

Shaikh, M.B. and A. Siegel, GABA-mediated regulation of feline aggression elicited from midbrain periaqueductal gray upon affective defense behavior in the cat, *Brain Res.*, 507, 1990, 51–56.

Shaikh, M.B., A. Steinberg, A. Siegel, Evidence that substance P is utilized in medial amygdaloid facilitation of defensive rage behavior in the cat, *Brain Res.*, 625, 1993, 283–294.

Sheard, M.H., The effects of amphetamine on attack behavior in the cat, *Brain Res.*, 5, 1967, 330–338.

Sheard, M.H., The effect of p-chlorophenylalanine on behavior in rats: relation to brain serotonin and 5-hydroxyindoleacetic acid, *Brain Res.*, 15, 1969, 524–528.

Sheard, M.H., Behavioral effects of p-chlorophenylalanine in rats: inhibition by lithium, *Commun. Behav. Biol.*, 5, 1970, 71–73.

Sheard, M.H., Brain serotonin depletion by p-chlorophenylalanine of lesions of raphe neurons in rats, *Physiol. Behav.*, 10, 1973, 809–811.

Siegel, A., T.A.P. Roeling, T.R. Gregg, M.R. Kruk, Neuropharmacology of brain-stimulation-evoked aggression, *Neurosci. Biobehav. Rev.*, 23, 1999, 359–389.

Smith, D.E., M.B. King, B.G. Hoebel, Lateral hypothalamic control of killing: evidence for a cholinoceptive mechanism, *Science*, 167, 1970, 900–901.

Smoothy, R., N.J. Bowden, M.S. Berry, Ethanol and social behaviour in naive Swiss mice, *Aggress. Behav.*, 8, 1982, 204–207.

Stoddard-Apter, S.L. and M.F. MacDonnell, Septal and amygdalar efferents to the hypothalamus which facilitate hypothalamically elicited intraspecific aggression and associated hissing in the cat: an autoradiographic study, *Brain Res.*, 193, 1980, 19–32.

Sweidan, S., H. Edinger, A. Siegel, The role of D1 and D2 receptors in dopamine agonist-induced modulation of affective defense behavior in the cat, *Physiol. Biochem. Behav.*, 36, 1990, 491–499.

Sweidan, S., H. Edinger, A. Siegel, D2 dopamine receptor-mediated mechanisms in the medial preoptic-anterior hypothalamus regulate affective defense behavior in the cat, *Brain Res.*, 549, 1991, 127–137.

Tazi, A., R. Dantzer, P. Mormede, M. Le Moal, Effects of post-trial administration of naloxone and beta-endorphin on shock-induced fighting in rats, *Behav. Neural Biol.*, 39, 1983, 192–202.

Thoa, N.B., B. Eichelman, L.K.Y. Ng, Shock-induced aggression: effects of 6-hydroxydopamine and other pharmacological agents, *Brain Res.*, 43, 1972, 467–475.

Thor, D., Amphetamine induced fighting during morphine withdrawal, *J. Gen. Psychol.*, 84, 1971, 245–250.

Tizabi, Y., N.B. Thoa, G.D. Maengwyn-Davies, I.J. Kopin, D.M. Jacobowitz, Behavioral correlation of catecholamine concentration and turnover in discrete brain areas of three strains of mice, *Brain Res.*, 166, 1979, 199–205.

Torda, C., Effects of catecholamines on behavior, *J. Neurosci. Res.*, 2, 1976, 193–202.

Ueki, S. and H. Yoshimura, Regional changes in brain norepinephrine content in relation to mouse-killing behavior by rats, *Brain Res. Bull.*, 7, 1981, 151–155.

Uhl, G.R., M.J. Kuhar, S.H. Synder, Enkephalin-containing pathway: amygdaloid efferents in the stria terminalis, *Brain Res.*, 149, 1978, 223–228.

Valzelli, L., S. Garattini, S. Bernasconi, A. Sala, Neurochemical correlates of muricidal behavior in rats, *Neuropsychobiology*, 7, 1981, 172–178.

Valzelli, L., E. Giacalone, S. Garattini, Pharmacological control of aggressive behavior in mice, *Eur. J. Pharmacol.*, 2, 1967, 144–146.

Van der Poel, A.M. and M. Remmelts, The effect of anticholinergics on the behaviour of the rat in a solitary and in a social situation, *Arch. Int. Pharmacodyn. Ther.*, 189, 1971, 394–396.

Van Erp, A.M.M. and K.A. Miczek, Aggressive behavior, increased accumbal dopamine, and decreased cortical serotonin in rats, *J. Neurosci.*, 20, 2000, 9320–9325.

Vergnes, M., Induction du comportement d'agression rat-souris par la p-chlorophenylalanine: Role de L'amygdale, *Physiol. Behav.*, 25, 1980, 353–356.

Vergnes, M. and E. Kempf, Tryptophan deprivation: effects on mouse-killing and reactivity in the rat, *Physiol. Behav.*, 1981, 1–5.

Vergnes, M. and E. Kempf, Effect of hypothalamic injections of 5,7-dihydroxytryptamine on elicitation of mouse-killing in rats, *Behav. Brain Res.*, 5, 1982, 387–397.

Vergnes, M., E. Kempf, G. Mack, Controle inhibiteur du comportement d'agression interspecifique du rat: systeme serotoninergique du raphe et afferences olfactives, *Brain Res.*, 70, 1974, 481–491.

Vergnes, M., G. Mack, E. Kempf, Lesions du raphe et reaction d'agression interspecifique rat-souris: effets comportementaux et biochimiques, *Brain Res.*, 57, 1973, 67–74.

Vivian, J.A. and K.A. Miczek, Diazepam and gepirone selectively attenuate either 20–32 or 32–64 kHz ultrasonic vocalizations during aggressive encounters, *Psychopharmacology (Berlin),* 112, 1993, 66–73.

Walsh, M.L., R. White, D.J. Albert, Dissociation of prey killing and prey eating by naloxone in the rat, *Pharmacol. Biochem. Behav.*, 21, 1984, 5–7.

Warden, M.K. and W. S. Young, Distribution of cells containing mRNA's encoding substance P and neurokinin B in the rat central nervous system, *J. Comp. Neurol.*, 272, 1988, 90–113.

Weerts, E.M., W. Tornatzky, K.A. Miczek, Prevention of pro-aggressive effects of alcohol in rats and squirrel monkeys by benzodiazepine receptor antagonists, *Psychopharmacology (Berlin),* 111, 1993, 144–152.

Weiner, S., M.B. Shaikh, A.B. Shaikh, A. Siegel, Enkephalinergic involvement in periaqueductal gray control of hypothalamically elicited predatory attack in the cat, *Physiol. Behav.*, 49, 1991, 1099–1105.

Welch, A.S. and B.L. Welch, Effect of stress and para-chlorophenylalanine upon brain serotonin, 5-hydroxyindoleacetic acid and catecholamines in grouped and isolated mice, *Biochem. Pharmacol.*, 17, 1968, 699–708.

Welch, B.L. and A.S. Welch, Effect of grouping on the level of brain norepinephrine in white Swiss mice, *Life Sci.*, 4, 1965, 1011–1018.

White, S.M., R.F. Kucharik, J.A. Moyer, Effects of serotonergic agents on isolation-induced aggression, *Pharmacol. Biochem. Behav.*, 39, 1991, 729–736.

Winslow, J.T., J. Ellingboe, K.A. Miczek, Effects of alcohol on aggressive behavior in squirrel monkeys: influence of testosterone and social context, *Psychopharmacology*, 95, 1988, 356–363.

Yao, R.H., P. Rameshwar, R.J. Donnelly, A. Siegel, Neurokinin–1 expression and co-localization with glutamate and GABA in the hypothalamus of the cat, *Brain Res. Mol. Brain Res.*, 71, 1999, 149–158.

Yao, R.H., P. Rameshwar, T. Gregg, A. Siegel, Co-localization of NK_1-receptor mRNA with glutamate immunopositivity in cat hypothalamic neurons by the combination of in situ hybridization and immunohistochemistry, *Brain Res. Brain Res. Protoc.*, 7, 2001, 154–161.

Zablocka, B. and D. Esplin, Central excitatory and neuromuscular paralyzant effects of pilocarpine in cats, *Arch. Int. Pharmacodyn. Thér.*, 147, 1964, 490–496.

8 Aggression and Hormonal Status

M. Demetrikopoulos and A. Siegel

It is well established that various hormonal systems are critically involved in regulating aggressive behavior. Numerous review chapters (Siegel and Demetrikopoulos, 1993; Barfield, 1984; Bouissou, 1983; Brain, 1977, 1978; Coe and Levine, 1983) as well as a monograph (Svare, 1983) were written on this subject before the early 1990s. In this chapter, our review is limited to recent literature. As in our previous review (Siegel and Demetrikopoulos, 1993), we will focus principally on studies involving androgens and estrogens at different levels over the phylogenetic scale, and also include a brief review of experiments involving adrenal steroids. We explore how aggression is affected by hormonal status by reviewing the experimental manipulation of hormones, seasonal variance of hormones, and the role that hormones play in development in order to understand the underlying mechanisms by which hormones modulate aggressive responses.

TESTOSTERONE AND AGGRESSIVE BEHAVIOR

In the following section, we examine the relationship of testosterone to aggressive behavior in various animal species, and discuss how seasonal changes and development may affect both testosterone levels as well as aggressive behavior. Additionally, we assess the relationship between testosterone and aggressive behavior at the human level.

EXPERIMENTAL MANIPULATION OF TESTOSTERONE LEVELS

Experimental manipulation of testosterone levels can be accomplished through systemic administration of testosterone, castration, or intracerebral administration. The effects of testosterone on aggression are related to the gender and neuroendocrine status of participants as well as to environmental conditions. For example, attack on a lactating female is considered to be a female-typical form of aggressive behavior, while attack on an olfactory bulbectomized male is considered to be a male-typical form of aggressive behavior. Castration of male mice leads to an increase in attacks on lactating females (i.e., female-typical aggression) (Haug and Brain, 1983; Whalen and Johnson, 1987; Haug et al., 1986), which is suppressed by testosterone treatment (Haug and Brain, 1983; Whalen and Johnson, 1987). In contrast, testosterone treatment of female ovariectomized mice increases attacks on lactating female targets (Whalen and Johnson, 1988). Castration of male mice leads to decreased attack of olfactory bulbectomized targets (i.e., male-typical aggression), and testosterone treatment leads to an increase in this type of attack (Whalen and Johnson, 1987). Furthermore, ovariectomized females given testosterone display increased male-typical aggression toward olfactory bulbectomized males (Simon et al., 1984; Simon and Masters, 1987; Whalen and Johnson, 1988). Thus, the effects of testosterone on these two models of aggression are governed by both the target stimulus under study and the gender of the resident. Specifically, testosterone increases male-typical aggression in both male and female residents, while it decreases female-typical aggression in male residents and increases female-typical aggression in female residents.

More recent studies have demonstrated that testosterone treatment not only increases aggressive behavior by the subjects given testosterone, but also increases attacks on treated subjects. A series of studies by Rhen and Crews (1999, 2000) and Rhen et al. (1999) examined the effects of administering testosterone to gonadectomized adult leopard geckos. They demonstrated that testosterone treatment of both genders elicited more attacks and increased aggressive behavior in male geckos. The result with female geckos was more complex since the aggressive displays were generally shorter and depended on the duration of testosterone treatment. Similar results were found using prepubertal male lambs (Ruiz-de-la-Torre and Manteca, 1999). They found that testosterone-treated animals were more likely to be attacked in a familiar flock, and were more likely to attack unfamiliar animals in a mixed flock. They suggested that the testosterone was important for dominance development. This fits in well with a previous study by Albert et al. (1989), which demonstrated that when rats were subjected to a competitive experience for palatable food, testosterone-dependent aggressive behaviors ensued that were characterized by lateral attacks and piloerection.

Modification of gonadotropin-releasing hormone (GnRH) is an indirect method of manipulating testosterone levels. European ground squirrels were injected with GnRH before, during, and after mating season to determine behavioral effects, and to document changes in testosterone secretion. Testosterone and aggressive behavior were increased prior to female emergence by GnRH, but were not induced after mating (Millesi et al., 2002). In a related study, male bulls were immunized against GnRH at 4 months of age in an attempt to chemically castrate and manage the animals. At 16 months, immunocastrated bulls demonstrated reduced aggression levels that were typical of steers (Price et al., 2003). Thus, modulation of the GnRH system was able to release context-dependent behaviors in the squirrels, and served as an alternative to surgical castration in beef cattle, which further suggests hypothalamic involvement in testosterone-induced aggressive behaviors.

Since peripherally administered testosterone presumably acts through receptors present within the central nervous system (CNS), a number of studies have explored the central administration of testosterone. Albert et al. (1987a) demonstrated increased aggression toward a male intruder when testosterone was implanted into the medial hypothalamus of castrated male hooded rats. This effect was site specific, and was not obtained when testosterone was implanted either dorsally or anterior to this site, nor when cholesterol was implanted into the medial hypothalamus. Similarly, Albert et al. (1987b) demonstrated that male hooded rats with electrolytic lesions of the medial hypothalamus showed lowered aggressiveness as measured by deficits in attack, biting, and piloerection than sham-operated controls when presented with an unfamiliar male intruder. Additionally, Bermond et al. (1982) demonstrated that the threshold current for electrical stimulation of hypothalamic-induced biting behavior was higher in castrated subjects, which lack androgens in their general circulation, than normal subjects. In addition, this increase in threshold was lowered following intramuscular testosterone propionate treatment. Collectively, these studies suggest that the hypothalamus is a central site critical for the development of aggressive behavior, and that testosterone acts as a neuromodulator in this region. The effects of experimental manipulation of testosterone on aggression in animals are summarized in Table 8.1.

TABLE 8.1
Effects of Experimental Manipulation of Testosterone on Aggression in Animals

Citation	Species	Results Obtained
Haug and Brain, 1983 Whalen and Johnson, 1987 Haug et al., 1986	Male mice	Castration increased female-typical aggression
Haug and Brain, 1983 Whalen and Johnson, 1987	Male mice	T suppressed castration-induced aggression
Whalen and Johnson, 1988	Female mice	T increased female-typical aggression
Whalen and Johnson, 1987	Male mice	Castration decreased male-typical aggression T increased male-typical aggression
Simon and Masters, 1987 Whalen and Johnson, 1988	Female mice	T increased male-typical aggression
Albert et al., 1987a	Male rat	Hypothalamic implant increased aggression
Albert et al., 1987b	Male rat	Hypothalamic lesion decreased aggression
Albert et al., 1989	Male rat	T increased aggression
Bermond et al., 1982	Male rat	Hypothalamic stimulation increased aggression; castration increased and T decreased hypothalamic current thresholds for aggression
Rhen and Crews, 1999	Leopard gecko	T increased aggressive displays in females; T and DHT increased aggressive displays in males
Ruiz-de-la-Torre and Manteca, 1999	Male lambs	T increased aggression and caused evoked attacks
Rhen and Crews, 2000	Male leopard geckos	T increased aggression and caused evoked attacks
Rhen et al., 1999	Female leopard geckos	T increased aggression and caused evoked attacks
Price et al., 2003	Male bulls	GnRH immunization decreased aggression
Millesi et al., 2002	Male squirrels	GnRH increased pre-mating season aggression

Note: T, testosterone.

Seasonal Variations in Testosterone Levels

Seasonal variations of aggressive behavior are evident in vertebrate classes from fish through mammals (Table 8.2). Tricas et al. (2000) demonstrated that male Atlantic stingrays have elevated androgen production and aggressive reproductive behavior during their very long mating season. Seasonal variations of testosterone and aggression have been more extensively studied in lizards. Testosterone levels in male mountain spiny lizards correlate well with changes in aggressive territorial behavior such that when testosterone levels are low, territorial aggression levels are low, and when testosterone levels are high, territorial aggression levels are

TABLE 8.2
Effects of Seasonal Variations in Testosterone Levels on Aggression in Animals

Citation	Species	Results Obtained
Tricas et al., 2000	Male stingrays	Androgen levels positively correlated with aggression
Moore, 1986		
Moore and Marler, 1987	Lizard	T levels positively correlated with aggression
Moore, 1988		
Klukowski and Nelson, 1998		
Smith and John-Alder, 1999		
Watt et al., 2003	Lizard	T levels positively correlated with aggression and high displayers had increased corticosterone
Schlinger, 1987		
Archawaranon and Wiley, 1988	White-throated sparrow	T levels positively correlated with aggression
Wingfield, 1984		
Soma et al., 1999a	Song sparrow	Aromatase inhibitor and androgen antagonist decreased aggression
Soma and Wingfield, 2001	Territorial all year	May have local conversion of T precursors in brain
Hau et al., 2000	Antbird, territorial all year	T increased during social challenge and T increases aggression
Meddle et al., 2002	Arctic sparrow	T and aggression increased only at beginning of breeding season — no effect later
Soma et al., 1999b	Arctic longspur	Aromatase activity decreased in rostral hypothalamus late in breeding season
Caldwell et al., 1984	Male wood rats	T levels not correlated with aggression
Michael and Zumpe, 1981	Male rhesus monkeys	T castration not related to aggression
Barrett et al., 2002	Male macaques	T correlated with noncontact aggression but not contact aggression

Note: T, testosterone.

high (Moore, 1986). Moore and Marler (1987) also demonstrated that while testosterone implants in the nonbreeding season led to an increase in aggression, the aggression level was lower than that observed during the breeding season despite the fact that the testosterone levels were equivalent. When testosterone implants were given to castrated subjects during the breeding season, normal breeding season levels of territorial aggression were observed (Moore, 1988). Similar seasonal relationships between testosterone and aggression have been obtained with northern fence lizards (Smith and John-Alder, 1999; and Klukowski and Nelson, 1998) and male Jacky dragons (Watt et al., 2003). Watt et al. demonstrated that elevated testosterone levels were the best predictor of territorial display and that subjects with high levels of display demonstrated an acute rise in corticosterone. Thus, territorial aggression during the breeding season is controlled by testosterone, and affected by a number of additional mechanisms. It is likely that during the breeding season, other neuroendocrine factors interact with increased levels of testosterone to yield high levels of territorial aggression.

Several investigators have examined the effects of testosterone levels on aggression in birds. Free-living birds that received testosterone capsule implants which maintained breeding level testosterone levels showed greater injury and had lower survival rates to the following year (Dufty, 1989). It was suggested that prolonged, elevated testosterone levels produced a risk factor by maintaining aggressive intramale interactions that are normally evident only during the breeding season. An earlier study by Wingfield (1984) suggests that this conclusion may be true. Free-living adult male song sparrows were subcutaneously implanted with testosterone to maintain plasma testosterone at the springtime peak. Sparrows with the testosterone implants showed higher levels of aggression than controls when challenged with tape recordings of conspecifics intruding into their territory.

Depending on the climatic conditions, birds demonstrate great variability in their seasonal behavior. Seasonal variations in aggression are evident in the white-throated sparrow, and these can be correlated with androgen levels (Schlinger, 1987) such that both androgen levels and aggression are increased in November but not in January or March. Researchers have suggested that the increases in both androgen and aggression help establish flock formation and dominance hierarchies. Archawaranon and Wiley (1988) examined whether the effects of androgens on aggression and dominance in these sparrows were due to testosterone or to its metabolites. The subjects were subcutaneously implanted with testosterone, that is, its metabolites — androstenedione, 5α-dihydrotestosterone (DHT), androsterone, or estradiol. Testosterone administration, or the combined treatment of its metabolites, produced the most potent effect on aggression and dominance. Further, when the conversion of testosterone to both DHT and estradiol was blocked, aggression and dominance scores were lower compared to blockade of the conversion of testosterone to *either* DHT or estradiol, despite the fact that blockade of a single metabolite produced elevated testosterone levels. Thus, androgenic effects on aggression are significantly dependent on the conversion of testosterone to *both* androgenic and estrogenic metabolites.

Some birds, such as the male song sparrows, fail to show seasonal behavioral changes and demonstrate territorial behavior throughout the year. However, their nonbreeding-season aggressive behaviors appear to be testosterone independent, since they are unaffected by castration, and plasma testosterone levels are very low during this period. However, birds treated with an aromatase inhibitor and an androgen receptor antagonist demonstrated decreased aggressive behaviors (Soma et al., 1999a). Interestingly, plasma dehydroepiandrosterone is high in these subjects, and may be converted locally in the brain into active sex steroids (Soma and Wingfield, 2001). Similar mechanisms may be involved in the control of territorial aggression in the tropical spotted antbird, which also displays year-long defensive behaviors despite low plasma testosterone levels (Hau et al., 2000). During periods of social challenge, these birds show elevated testosterone levels, and testosterone implants increase aggressive behaviors. Additionally, birds treated with an aromatase inhibitor and an androgen receptor antagonist demonstrated decreased aggressive behaviors just as the song sparrows did. Thus, these birds may also be converting testosterone precursors directly at their site of action or else use discrete testosterone secretion to avoid chronically elevated testosterone.

Due to extreme weather conditions, Arctic birds such as the Gambel's white-crowned sparrow have a brief breeding season (unlike temperate birds), and only demonstrate high testosterone and luteinizing hormone early in the breeding season (Meddle et al., 2002). During incubation, simulated intrusions increased LH but not testosterone, although testosterone could be elevated by injection of gonadotropin-releasing hormone, and testosterone implantation did not increase aggression in males. However, corticosterone implants decreased aggression in response to a simulated intrusion during incubation. It is thought that modifications in the typical effects of these hormones are due to the fact that in such a harsh climate, interruptions in parental care may adversely affect reproductive success. Results by Soma et al. (1999b) with another Arctic bird, the Lapland longspur, may provide a possible mechanism for this lack of effect of testosterone, since they demonstrated that aromatase activity was decreased in the rostral hypothalamus in the late breeding season. A recent review of hormonal behavioral interactions in Arctic birds explores a variety of unique interrelationships of testosterone and aggression (Wingfield and Hunt, 2002).

With mammals, the relationship among seasonal effects, testosterone, and aggression is more dynamic. Caldwell et al. (1984) castrated male wood rats after puberty, and demonstrated that seasonal increases in aggression could be independent of testosterone since the castrated subjects maintained their seasonal increases in aggression. This finding essentially replicates an earlier result with rhesus monkeys in which both castrated and intact subjects maintained their annual changes in aggression (Michael and Zumpe, 1981). The relationship between testosterone and aggression can also be more nuanced with higher animalsm as demonstrated by testosterone being correlated with noncontact female-directed aggression, but not with contact aggression in male Japanese macaques (Barrett et al., 2002).

Testosterone Levels during Development

In mammals, sexual differentiation begins before birth and continues through puberty. Androgens are thought to play an important role in inducing prenatal

TABLE 8.3
Developmental Effects of Testosterone Levels on Aggression in Animals

Citation	Species	Results Obtained
Gandelman and Graham, 1986		
Mann and Svare, 1983	Female mice	T *in utero* increased adult aggression
Rines and Vom Saal, 1984		
Simon et al., 1984	Female mice	Neonatal testosterone increased adult aggression
Freeman et al., 1998	Female shrews	
Frank et al., 1991	Hyenas	Neonatal testosterone correlates with fatal sibling fighting
Shrenker et al., 1985	Male mice	Castration at 30 days postnatally decreased aggression, while castration at 50 days was not effective

changes that occur in the structure and function of various tissues including the gonads and brain producing masculinization and defeminization in males. Androgens continue to be important through puberty and may affect long-term physiological and behavioral manifestations (Table 8.3).

In rodents, *in utero* position is critical in determining their hormonal status since it establishes the microenvironment for the developing fetus and governs its later propensity to display aggressive behavior. Rodent fetuses are positioned randomly *in utero*. Fetuses that develop between two male fetuses have higher blood testosterone levels during gestation than do fetuses that develop between two females (Perrigo et al., 1989); fetuses adjacent to a male fetus receive *in utero* androgen stimulation from that male fetus. Therefore, a subject that develops between two females *in utero* would be endocrinologically distinct from one that develops between two males. Furthermore, due to the microenvironment present during gestation, a subject that develops alone *in utero* would be expected to be endocrinologically distinct from those that develop with a litter. Ovariectomized adult female mice with an *in utero* position between two females attacked olfactory bulbectomized males when treated with testosterone, compared to female subjects that developed as a single fetus due to the removal of their littermates during gestation (Gandelman and Graham, 1986). In fact, the subject that developed alone *in utero* did not show aggressive behavior as an adult and was unresponsive to hormonal treatment. Thus, in intact litters containing both male and female pups, the presence of male fetuses causes some masculinization of female fetuses to occur *in utero* even if they are not contiguous. The importance of *in utero* testosterone in adult aggression was similarly demonstrated when female subjects, whose mothers received testosterone during the gestational period, showed increased aggression toward male intruders (Mann and Svare, 1983).

Rines and Vom Saal (1984) controlled for possible differential levels of hormones postnatally by ovariectomizing subjects at birth and giving hormone replacement. Young females that were in an *in utero* position between two males (2M females) were more aggressive than young females that were in an *in utero* position between zero males (OM females). However, old OM females were aggressive, suggesting an age-related difference where OM females developed sensitivity to testosterone later in life. Similarly, Perrigo et al. (1989) demonstrated that male mice that developed between two females *in utero*, and thus had lower circulating testosterone levels during gestation, had increased infanticide both before and after mating. Thus, these studies indicate the importance of developmental factors that contribute to the organization of aggressive behavior during the postnatal period. A more thorough discussion of the subtleties of this area is beyond the scope of this chapter, but a recent review of this area has been conducted by Ryan and Vandenbergh (2002).

In addition to the critical time period during prenatal development, during postnatal development there is a temporal relationship between castration and the tendency for expression of aggressive behavior. Shrenker et al. (1985) demonstrated that male mice castrated 30 days after birth were less aggressive than those castrated at 50 days after birth. In fact, subjects castrated at 50 days showed levels of aggression similar to sham-operated controls. Similarly, female mice given testosterone at birth behaved more aggressively when treated with testosterone as adults than did subjects given estradiol neonatally and similarly tested with testosterone as adults (Simon et al., 1984). Neonatal testosterone treatment of female musk shrews also produced more aggressive adults (Freeman et al., 1998). Additionally, early postnatal effects of testosterone are important in the early expression of fatal aggressive behavior in neonatal spotted hyenas. During the first month following birth, testosterone remained higher in males than females, while androstenedione, which was elevated in both sexes at birth, remained high in females and fell in males. Sibling fighting was most pronounced during this time period, and was directly related to high androgen levels (Frank et al., 1991).

There are critical time periods during development when circulating testosterone must be present for appropriate interactions with neural structures associated with the expression of aggressive reactions to take place, and thus ensure that such responses will remain present in the adult organism. Although this would seem to contradict studies discussed earlier that used adult subjects in which testosterone was shown to be effective in modulating aggression, once this critical developmental period has passed, the presence or absence of circulating testosterone becomes less important. However, while there are critical periods during which changes have the most dramatic and long-lasting effects, further manipulation of the system can occur at a later time as well. Whether such changes can significantly alter brain function in adulthood has yet to be determined.

Testosterone's Role in Human Aggression

Various studies have been designed to assess the relationship between testosterone and aggressive behavior in humans (Table 8.4). Several methodological difficulties are apparent, including differences in the (1) nature of the populations sampled, (2) ages of the sample, (3) size of the sample considered, (4) measurement of testosterone, and (5) kinds of behavioral processes investigated that are interpreted as being aggressive in nature. A recent review of this area and its potential shortcomings was conducted by Zitzmann and Nieschlag (2001). Nevertheless, a pattern appears to emerge from these studies that, while not entirely consistent, is supportive of the findings obtained from the animal literature that demonstrate a positive correlation between testosterone levels and the presence of aggressive behavior.

Perhaps the most convincing data are derived from studies involving subjects incarcerated in either a prison or mental institution. In an early study, Ehrenkranz et al. (1974) categorized prison inmates on the basis of their level of aggressiveness (as determined from a battery of psychological tests) and were thus able to distinguish a chronically aggressive group from socially dominant and nonaggressive individuals. Testosterone levels were consistently higher in individuals displaying chronic aggressive behavior than in those inmates who were classified as either socially dominant or nonaggressive. Similarly, Dabbs et al. (1987) categorized prison inmates in terms of whether they had committed violent crimes. Positive correlations were obtained between individuals displaying high free salivary testosterone levels and those who committed violent crimes. Moreover, individuals who had committed nonviolent crimes had the lowest testosterone levels. Prisoners who were rated as being "tougher" by their peers as well as individuals who had committed nonviolent crimes but had received punishment for disciplinary infractions also had high testosterone levels. In a related study (Rada, 1976; Rada et al., 1983), the authors classified patients at the Atascadero State Hospital in California according to the degree of violence exhibited during the commission of a rape and according to their scores on the Buss-Durkee and Megargee Overcontrolled hostility inventories. Consistent with the findings described above, rapists who were judged to be most violent (i.e., those who beat their victims during the rape) had significantly higher plasma testosterone levels than child molesters or normal subjects. Although individual correlations between testosterone levels and hostility ratings were not demonstrated, hostility rating scores were also higher for rapists than for normals. A more recent study also demonstrated a correlation between plasma testosterone and aggressive acts in prisoners (Dolan et al., 2001). They also demonstrated that primary psychopaths, as defined by the Special Hospital Assessment of Personality and Socialization, had lower cortisol and higher testosterone concentrations. Similar findings have been reported in elderly demented patients such that free plasma testosterone was positively correlated with aggression, while estrogen levels were negatively correlated with aggression (Orengo et al., 2002).

Similar positive correlations between aggressive behavior and high testosterone levels were obtained from a number of studies on normal adolescents or young adults. Testosterone levels were higher in college students who scored high for hostility and aggression as measured on a hostility inventory scale (Persky et al., 1971); serum testosterone levels were positively correlated with self-ratings of spontaneous aggression and dominance in young adult males given standardized and projective tests (Christiansen and Knussmann, 1987); and adolescent males who responded more vigorously to provocations and threats when given the Olweus Aggression Inventory scale had high testosterone levels and were more irritable and more likely to engage in aggressive behavior (Olweus et al., 1980, 1988). Furthermore, Scaramella and Brown (1978) demonstrated that testosterone levels were positively correlated with the response to threat in college hockey players; and Booth

TABLE 8.4
Human Studies Correlating Testosterone Levels and Aggressive Behavior

Citation	Population Sample	Results Obtained
Studies Indicating a Positive Relationship between T Levels and Aggression		
Dabbs et al., 1987	Prison inmates	High T levels associated with violent criminal acts, and low T values associated with nonviolent crimes
Ehrenkranz et al., 1974	18- to 45-year-old prisoners	T levels were higher in a chronically aggressive inmate group relative to a socially dominant or nonaggressive group
Rada et al., 1983	Rapists, child molesters	T levels were higher in rapists than in child molesters or nonviolent control subjects
Rada, 1976	Rapists	T values were higher in rapists who were violent than those who raped but who were not otherwise violent
Dolan et al., 2001	Prisoners with personality disorders	T levels correlated with aggression disorders
Orengo et al., 2002	Demented elderly men	T levels correlated with aggression
Persky et al., 1971	College students	In younger men but not in older men, T levels were higher for Buss–Durkee Hostility Inventory and aggression
Christiansen and Knussmann, 1987	20- to 30-year-old males	Subjects with high T showed high self-ratings of spontaneous aggression
Olweus et al., 1980, 1988	Adolescent males	Subjects with high T levels showed increased readiness to respond to threat and increased propensity to engage in aggressive destructive behavior (Olweus Aggression Inventory)
Scaramella and Brown, 1978	Hockey players	T levels were higher in response to threat, but other measures of aggression were not positively correlated with T levels
Booth et al., 1989	Male college tennis players	T levels rose before a tennis match and were highest in the winners
Finkelstein et al., 1997	Hypogonadal adolescents	Administration of T and estrogen increased aggression
Gerra et al., 1997	18- to 19-year-old normal males	High normal aggressiveness correlated with T
Studies That Do Not Show a Positive Relationship between T Levels and Aggression		
Bradford and McLean, 1984	Sex offenders	No correlation observed between high levels of violence and high levels of T
Kreuz and Rose, 1972	Young prisoners	Little correlation observed between T levels of verbally aggressive people who frequently fight and nonaggressive individuals

TABLE 8.4 (CONTINUED)
Human Studies Correlating Testosterone Levels and Aggressive Behavior

Citation	Population Sample	Results Obtained
Worthman and Konner, 1987	!Kung San men	Pattern of change in T levels is associated with prolonged exercise rather than the success or failure of the hunt
Sourial. and Fenton, 1988	Klinefelter's syndrome	Case study — aggressive and sexual fantasies and impulses declined with T treatment in individual with low endogenous T levels
Raboch et al., 1987	Klinefelter's syndrome	Case study — low T levels associated with sexually motivated murder

Note: T, testosterone.

et al. (1989) reported that testosterone levels rose just prior to a tennis match and were highest in the winners of the matches.

Since correlational studies have numerous limitations, more recent studies have manipulated testosterone and aggression levels in human subjects. Hypogonadal adolescents were treated with sex hormones at early, middle, and late pubertal doses. Both testosterone (in boys) and estrogen (in girls) increased physical aggressive behaviors and aggressive impulses but not verbal aggressive behaviors as measured by the Olweus Multifaceted Aggression Inventory (Finkelstein et al., 1997). Gerra et al. (1997) experimentally induced aggressiveness in psychophysically healthy male teenagers. Subjects with high-normal basal aggressiveness demonstrated elevated basal testosterone levels and elevated norepinephrine, growth hormone, and cortisol in response to the experimental induction of aggression.

A smaller number of studies have failed to provide data in support of a positive relationship between high testosterone levels and aggressive behavior. Males incarcerated for sex offenses that ranged from exhibitionism, fetishism, and pedophilia to more violent behavior (i.e., rape) did not demonstrate a significant correlation between sexual violence and testosterone levels (Bradford and McLean, 1984). However, this result may not conflict with Rada (1976) and Rada et al. (1983) described above since these authors demonstrated the dependence of other violent acts being committed on the victim, and Bradford and McLean did not mention whether the rapists in their study committed other violent acts on their victims. Kreuz and Rose (1972) examined young prisoners and observed little correlation between testosterone levels of verbally aggressive people who frequently fought and those who were nonaggressive. Worthman and Konner (1987) followed the pattern of change in testosterone levels of !Kung San men during the course of their hunt and reported changes in testosterone levels were correlated more closely with the length of exercise rather than with the success or failure of the hunt. Finally, investigators provided case study reports of individuals diagnosed with Klinefelter's syndrome, which is characterized, in part, by lower levels of testosterone and by the relative absence of secondary sex characteristics. Testosterone treatment resulted in a decline of aggressive impulses and sexual fantasies in one individual, which may have been due to the psychological benefit of secondary sex characteristic development following testosterone treatment (Sourial and Fenton, 1988). Raboch et al. (1987) reported that another individual with Klinefelter's syndrome committed a sexually motivated murder despite relatively low testosterone levels.

In spite of some negative data, the overall observations obtained from human studies support the presence of a relationship between endogenous testosterone levels and the propensity to commit aggressive acts. However, other variables such as environmental conditions, the psychological state of the individual, and the type of aggressive act committed appear to play more critical roles concerning whether a positive correlation will be obtained in humans.

ESTROGEN AND AGGRESSIVE BEHAVIOR

While the role of testosterone on aggressive behavior has been more widely studied, other hormones have also been shown to play an important part in the expression of aggressive behavior. In the following section, the main findings of the effects of estrogen on aggressive behavior will be summarized. It should be noted that the effects of estrogen on aggressive behavior are much more inconsistent than the effects of testosterone.

Experimental Manipulation of Estrogen Levels

A number of studies have failed to demonstrate a direct effect of estrogen on aggression. For example, DeBold and Miczek (1984) found that residents are more likely to attack (biting toward the neck, back, or flank, or a nip to the snout or face) intruders of the same sex than those of the opposite sex. This form of aggressive behavior is gonadally dependent in males but not in females since castration of the male intruder led to a decrease in attack from male residents that could be reversed by testosterone replacement, while ovariectomy of female intruders did not change the frequency of attack by female residents. Similarly, although baseline levels of aggression were relatively high, chronic estradiol treatment failed to alter attack responses in female resident hamsters that were ovariectomized, treated with varying doses of estradiol, and retested 3, 7, 10, and 14 days post-implantation (Meisel et al., 1988). However, when subsequently injected with progesterone, estradiol-treated subjects showed a reduction in aggression compared to cholesterol-treated controls that remained aggressive after progesterone treatment. Thus, inhibition of aggression in female subjects is likely to be dependent on the combined effects of estradiol and progesterone.

More recent studies have demonstrated mixed effects of estrogen on aggressive behavior. For example, Stavisky et al. (1999) demonstrated that ovariectomized female cynomolgus macaques had a two- to three-fold increase in aggressive behavior compared to sham or hormonally replaced subjects while affiliative behaviors were unaffected. This effect is opposite that seen in spiny lizards that display reduced aggression following ovariectomy (Woodley and Moore, 1999).

Since it had been suggested that estrogen acts via the CNS to affect aggressive reactions, Takahashi et al. (1985) implanted estradiol into the ventromedial hypothalamus and medial preoptic area or anterior hypothalamus where estrogen receptors are believed to be located. Controls had cholesterol injected into the ventromedial nucleus, estradiol placed into the anterior hypothalamus or preoptic area, or cholesterol injected into both sites. Decreased agonistic responses toward a male partner were observed following both single and dual estradiol implantation suggesting that estrogen might act at receptor sites within the anterior hypothalamus or preoptic region to modulate agonistic behavior.

Naturally Occurring Variations in Estrogen Levels

A large source of estrogen variance occurs in response to pregnancy and parturition. A recent review of the hormonal control of maternal aggression was conducted by Lonstein and Gammie (2002). Mayer and Rosenblatt (1987) examined hormonal factors that mediate maternal care as potential mediators of maternal aggression. Resident subjects were either pregnant or virgin female rats. All subjects were hysterectomized, ovariectomized, injected with estrogen or sham operated, and then exposed to pups continuously. Nonfamiliar male intruders were introduced at several different times during the induction of maternal behavior. Prior to their sensitization with the pups, aggressive behavior was evident in groups with elevated estrogen levels. Once maternal behavior was initiated, pregnant and pregnancy-terminated females showed increased aggression, but females that had never been impregnated did not. Thus, while elevated estrogen levels may have some modulatory effects on nonmaternal aggression, factors other than estrogen may be critical for the expression of maternal aggression. Conception itself, or the cascade of physiological processes that follow conception, may cause long-lasting neuroendocrine changes that affect ensuing response patterns, including maternal aggression.

Mayer et al. (1990) employed nonpregnant, ovariectomized rats treated with levels of estrogen and progesterone normally seen during pregnancy, which caused a short latency maternal behavior in the nonpregnant subjects. Subjects given hormonal treatment at pregnancy levels displayed aggressive behavior even if they did not express maternal behavior, while vehicle controls were not aggressive. In a second experiment, some of the subjects were additionally hypophysectomized. However, hypophysectomy had no effect on this response, and so it does not seem to be pituitary dependent.

The effect of varying estrogen levels across the estrous cycle on agonistic behavior of a female hamster toward a sexually active male was explored. The pair was acclimated in separate halves of a two-compartment test chamber for 2 days, and then the partition was removed and the frequency of attack and chase were recorded for 15 minutes. Higher levels of attack–chase were found when endogenous estrogen levels were low in comparison to the response pattern present when estrogen levels were naturally high in the estrous cycle (Lisk and Nachtigall, 1988). Further, ovariectomy produced an increase in attack–chase levels, and the addition of exogenous estrogen decreased them. Thus, it appears that estrogen can play an inhibitory role in the regulation of aggressive behavior.

ESTROGEN LEVELS DURING DEVELOPMENT

As previously indicated, sexual differentiation begins during fetal life and continues through puberty in mammals. In addition to androgens, estrogens are also important in inducing changes in the structure and function of various tissues, such as the brain and gonads, which may be important in aggressive behavior. Males that developed between two female fetuses *in utero* had higher levels of estradiol in their amniotic fluid than males located between two male fetuses and were less aggressive as adults (Vom Saal et al., 1983). Possible differences in postnatal exposure to gonadal hormones were controlled by castrating the subjects at birth and giving hormone replacement treatment. Variations in levels of aggression observed were the result of developmental differences in brain tissue that occurred prenatally, due, in part, to the action of estrogen. However, prenatal exposure to estrogen — administered to the mother during days 12 through 18 of gestation — did not appear to alter the expression of aggressive behavior to chronic testosterone treatment given as adults (Gandelman et al., 1982). It can be assumed from the testosterone studies that during gestation a critical time period exists for the development of aggression, and this study may have failed to administer estrogen during the critical time window for this hormone to be effective. Additionally, the *in utero* position findings are not likely to be attributed solely to differences in estrogen levels, but are probably the result of the interaction of the constellation of endocrine micro-changes. Such effects would also include changes in androgen levels, as well as any neuroendocrine factors that interact with estrogens and androgens.

Female mice ovariectomized and administered androgen 90 to 140 days post-natally demonstrated increased aggression in neonatally treated controls, while post-natal estrogen administration was not effective (Simon and Whalen, 1987). This might have been predicted since the authors only tested aggressive behaviors by biting or chasing of an olfactory bulbectomized male intruder, which is a male typical form of aggression. Neonatal androgen treatment led to enhanced effectiveness of subsequent androgen, but not estrogen, treatment in eliciting aggression as adults. Similarly, neonatal estrogen exposure resulted in the expression of aggression when the subjects were subsequently given estrogenic, but not androgenic, stimulation as adults. Thus, responses to these compounds in adult subjects are dependent on the entire neuroendocrine history of the subject. Moreover, it should be further noted that the differential effects of perinatal androgens and estrogens on aggression in adulthood are confounded by the fact that androgens such as testosterone are metabolically converted (aromatized) to estradiol in the developing and adult mammalian brain (Naftolin et al., 1975). Factors that affect this conversion have been shown to alter the organization (McEwen et al., 1977) as well as the activation (Harding, 1986) of sexual and sexually dimorphic behavior.

PROGESTERONE AND AGGRESSIVE BEHAVIOR

The following section briefly reviews the relationship of progesterone to aggressive behaviors. As with the experiments exploring the effects of estrogens on aggressive behavior, this literature is not as consistent as the testosterone literature.

Experimental Manipulation of Progesterone Levels

Through the course of pregnancy, both circulating levels of progesterone and maternal aggressive behavior such as attack on a male intruder change in magnitude (Mann and Svare, 1982). Early in pregnancy, progesterone levels were somewhat correlated with aggression in which both variables were low on days 6 and 10 and high on day 14 (Mann et al., 1984). However, by day 18 in pregnancy, there was dissociation between progesterone levels and aggressive behavior, with progesterone values at their nadir while aggressive behavior was still high. Although supplemental administration of progesterone did not yield an early onset of aggression, progesterone treatment in virgin mice with intact ovaries yielded an increase in aggressive behavior that was weaker than that evident in pregnant females. Furthermore, when progesterone levels were reduced, aggressiveness in virgin mice was attenuated, while the levels of aggressive behavior remained high under this hormonal condition in pregnant mice.

Hysterectomy on day 15 decreased aggression toward male intruders and this decrease could be attenuated by progesterone implanted under the nape of the neck (Svare et al., 1986). Additionally, estradiol inhibited progesterone treatment induced aggression in hysterectomized females even though estradiol treatment by itself did not affect aggression in hysterectomized mice. This supports the data of Mann et al. (1984), demonstrating that while progesterone is associated with enhancement of aggressive encounters, the overall aggressive reaction is further modified by the interactions with other hormones such as estrogen as well as other aspects of the internal milieu of the pregnant animal.

This conclusion may be quite limited since the effects of progesterone on pregnancy-induced aggression are strain specific (Svare, 1988). Female DBA/2J mice were more likely to show pregnancy-induced aggressive behaviors than C57BL6J mice. Furthermore, virgin DBA/2J mice that received subcutaneously implanted progesterone exhibited aggressive behavior, while such responses were not evident in C57BL/6J mice. Similar levels of circulating progesterone were observed in both strains during pregnancy and following implantation, which suggested that genotypic differences result from variations in CNS tissue sensitivity to progesterone.

While progesterone has been shown, under some conditions, to increase pregnancy-induced aggression, it may inhibit other forms of aggression. In a same sex resident–intruder model, castrated male hamsters showed decreased aggressive behavior after progesterone treatment, and ovariectomized female hamsters treated with progesterone showed decreased frequency of attack (Fraile et al., 1987). However, ovariectomized female hamsters treated with estradiol and then given subcutaneous progesterone injections two and three days later showed higher levels of aggression than control subjects (Meisel and Sterner, 1990). A site-specific central effect of progesterone was found. Decreased aggression by female hamsters with progesterone implants in the ventromedial nucleus and medial preoptic area, but not in the anterior hypothalamus, was expressed toward a sexually experienced male (Takahashi and Lisk, 1985). A more recent study has suggested that progesterone can inhibit testosterone-induced aggression, but not aggression induced by testosterone metabolites (Gravance et al., 1996).

Progesterone Levels during Development

Female subjects that received progesterone prenatally during days 12 through 16 of gestation showed increased postpartum aggression as adults (Wagner et al., 1986). The authors suggested that prenatal exposure affected brain differentiation, which led to a masculinization of the brain as shown by an increase in aggression.

ADRENAL STEROIDS AND AGGRESSIVE BEHAVIOR

A number of investigators have attempted to examine the possible role that adrenal steroids may play during the process of aggression. Authors have primarily focused on how corticosterone levels are altered in an animal defeated after fighting rather than on the aggressor. The data concerning how adrenal steroids may alter the propensity to commit aggressive reactions is more pertinent to the present chapter, and will be briefly summarized here.

Studies that have attempted to determine how adrenal steroids affect aggressive reactions have been reported in the lizard, rat, and human. Tokarz (1987) reported that lizards pretreated with corticosterone pellets showed a reduction in aggressive tendencies such as approach and biting behavior as well as aggressive postural responses. A more recent study suggests a mechanism for this by demonstrating that systemic corticosterone administration in male *Anolis carolinenis* enhanced hippocampal and medial amygdala serotonergic turnover, and that testosterone enhanced serotonergic turnover in the hippocampus (Summers et al., 2000). Although no changes were found in circulating levels of corticosterone collected during aggressive encounters in lizards (Moore, 1987), a more recent study by this group has demonstrated both elevated plasma corticosterone levels and a lowered ratio of forebrain:brainstem serotonin in both male and female mountain spiny lizards following an aggressive interaction. Thus, at the level of the lizard along the phylogenetic scale, it is not possible to draw any clear-cut conclusions concerning whether aggressive reactions are mediated by changes in circulating levels of corticosterone.

Neonate Norway rat pups treated with corticosterone within the first 4 days of life eventually fought less frequently than did control animals. Moreover, corticosterone treatment at a later period (day 9 or 10) had little effect on fighting, thus suggesting that a critical period exists with respect to the time when corticosterone treatment can be effective in modulating aggressive behavior (Meaney et al., 1982). Similarly, administration of low doses of the corticosterone precursor, deoxycorticosterone, resulted in a decrease in fighting, as well as in an increase in the current threshold (delivered to an electrified grid) required to initiate fighting behavior in a shock-induced model of aggression (Severyanova, 1988).

Microinjections of corticotropin-releasing factor (CRF) administered intracerebroventricularly facilitated the occurrence of shock-induced fighting (Tazi et al., 1987), and delivery of the CRF antagonist, alpha-helical CRF-(9-41), could block shock-induced fighting. It is difficult to relate this finding to the previous results where corticosterone was employed for several reasons. One possibility is that CRF, in addition to its actions on the anterior pituitary, may act as a transmitter at sites in the brain that are associated with the expression of aggressive behavior. Second, CRF is likely to have widespread effects on the adrenal system and alter the release of other hormones. These might include aldosterone as well as androgens that could alter aggressive reactions in a manner different from that of corticosterone. Additionally, CRFs effects are likely to be receptor subtype specific. For example, Jasnow et al. (1999) demonstrated that a nonspecific CRF1/CRF2 receptor antagonist reduced the expression of conditioned defeat while a specific nonpeptide CRF1 receptor antagonist failed to do so. Furthermore, while ACTH administration can lead to an increase in aggressive behavior possibly by acting directly on brain mechanisms governing this response, or by acting through a corticosterone mechanism (Brain and Evans, 1977), ACTH treatment can also reduce aggressiveness in both intact mice as well as in mice with controlled corticosterone levels (Leshner, 1975). Thus, our present level of understanding of the role of adrenal cortical steroids on aggression in rodents remains unclear.

Several studies conducted with humans used either male violent offenders or substance users and attempted to correlate cortisol levels with aggression, impulsivity, or other antisocial behaviors. Again, these studies were inconsistent. For example, one investigation reported low cortisol levels among habitually violent offenders who maintained antisocial personalities in comparison to a control group of people who had antisocial personalities but who did not display violent behavior (Virkkunen, 1985). In contrast, individuals with higher levels of aggressiveness and impulsivity following administration of the indirect serotonin agonist fenfluramine had higher cortisol levels (Fishbein et al., 1989).

Thus, at different levels along the phylogenetic scale, identification of a relationship between the pituitary–adrenal axis and aggressive behavior has been inconsistent and does not permit clear-cut conclusions to be drawn. These data suggest that adrenal steroids function primarily in response to acts of violence or other forms of stress, and it should also be noted that many investigators have sought to examine how the adrenocortical system responds when the organism is subjected to aggression. For example, in studies involving subjects from mice (File, 1984) to monkeys (Martensz et al., 1987; Scallet et al., 1981), cortisol levels were reported as elevated following an attack (or defeat) by an aggressor of the same species. It has also been demonstrated that 3 α, 5 α-tetrahydroxycorticosterone, a metabolite of the corticosterone precursor deoxycorticosterone, administered peripherally to intruder mice hastens the time for defeat to occur (Kavaliers, 1988). The data suggest that this naturally occurring steroid can have a potent effect on how the organism reacts to aggressive encounters.

MODULATION OF HORMONAL-AGGRESSION INTERACTIONS BY ETHANOL AND ILLICIT DRUGS

The possible relationship of ethanol and drugs of abuse to aggressive behavior was considered in the previous chapter. However, we have chosen to briefly review this subject again because of their interactions with endocrine functions with respect to aggression and rage.

Alcohol may serve to induce aggressive behavior, and aggressive responses to alcohol such as attack bites, offensive sideways threats, and tail rattles, are thought to be dose dependent. In fact, at higher doses, there is a possibility that alcohol decreases aggression. Male subjects receiving high doses of testosterone in connection with a low alcohol dosage displayed elevated levels of aggression that could be reduced somewhat by the administration of a higher dose of alcohol (DeBold and Miczek, 1985). With higher levels of testosterone, more alcohol was needed to reduce aggressive behavior. Additionally, sham castrated males had increased levels of aggression following alcohol administration, while neither androgenized females nor neonatally gonadectomized males showed this response (Lisciotto et al., 1990). Therefore, it is not just the level of circulating testosterone but also the gender that is important as a determinant of aggressive behavior. Estrogen has also been shown to increase ethanol intake and aggression in male mice, while low estrogen levels increase ethanol intake in females (Hilakivi-Clarke, 1996). Ethanol also has modulating effects on the hormone–aggression interaction in humans. A recent study demonstrated lower outward-directed aggressiveness and increased plasma cortisol and prolactin levels in adult children of alcoholic fathers during experimental induction of aggression (Gerra et al., 1999). In a similar study, men with a history of alcohol-related aggression displayed higher aggression, while control subjects demonstrated a positive correlation between testosterone (and DHT) and anger as well as a negative correlation between cortisol and anger (Von der et al., 2002). Furthermore, the Michigan Alcoholism Screening Test correlated positively with anger in both groups. Similarly, Gerra et al. (2001) demonstrated that heroin addicts have higher outward-directed aggression and a blunted response of ACTH, cortisol, and growth hormone to an experimentally induced aggressive situation. This blunted neuroendocrine response is similar to that seen in highly aggressive men to ipsapirone (5-HT_{1A}) challenge, and may be due to impaired serotonergic (1A) receptor function that can be associated with increased aggressivity.

COMMENTS AND CONCLUSIONS

The studies reviewed in this chapter provide substantive evidence that sex hormones play significant roles in the expression of aggressive behavior. However, this relationship is highly complex and depends on interactions with such factors as gender, species, and model of aggressive behavior under consideration. The diversity of aggression models is part of the difficulty in evaluating this literature. In fact, the various models of aggression may reflect different behaviors rather than a unitary process. However, from the literature that we have reviewed, the following conclusions are drawn. The studies conducted with animal subjects have, for the most part, demonstrated that testosterone is the most important hormonal variable involved in the induction of aggressive behavior. This response may be gender specific and selective to the particular form of aggression measured, suggesting that the effects may not be totally generalizeable. Furthermore, it has been demonstrated that there are specific critical times developmentally that are important for the manifestation of the effects of testosterone on the induction of aggressive behavior. The human data are less consistent, but still suggest that testosterone affects aggressive behavior. It is possible that at different levels along the phylogenetic scale, other psychogenic variables begin to override the relationship between testosterone and aggression.

With respect to other sex hormones, estrogen appears to oppose the effects of testosterone and may inhibit aggressive responses, while the effects of progesterone are less clear-cut and may be dependent on the interaction with other neuroendocrine factors. The adrenal steroid literature fails to demonstrate that these hormones play a conclusive role in the regulation of aggressive behavior. Instead, adrenal steroid hormones may serve a more significant function with respect to processes associated with the defeat of an organism following an aggressive bout.

The possible mechanism by which sex hormones control aggressive reactions likely involves the action of hormones that pass through the blood–brain barrier on selective neurons within the brain. These hormones will presumably modify the release of specific neurotransmitters and may possibly affect the activity of second messenger systems as well. Since the overwhelming majority of the studies considered involve models that may be classified as defensive aggression, it would seem that two of the most critical neuroanatomical structures involved are the medial hypothalamus and midbrain periaqueductal gray (PAG). These structures are significant because they are regions from which this form of aggression can be readily elicited by electrical or chemical stimulation (Fuchs et al., 1985a, 1985b; Shaikh et al., 1987; Siegel and Pott, 1988; Siegel and Brutus, 1990). The medial hypothalamus plays a unique role in this process since (1) implantation of hormones into this structure can clearly modify the propensity for the occurrence of an attack response; (2) lesions of the medial hypothalamus result in an attenuation of aggressive behavior such as maternal aggression (Hansen, 1989), which has been shown to be modified by hormonal treatment; and (3) sex hormone-concentrating cells can be found in high densities within the medial hypothalamus (Pfaff, 1968a, 1968b; Pfaff and Keiner, 1973). Although the medial hypothalamus and possibly the PAG would appear to be primary candidates where hormonal actions would most effectively modify aggressive behavior, other structures should also be considered. These include the limbic system structures — amygdala, hippocampal formation, septal area, and prefrontal and cingulate cortices — which are characterized by their modulatory actions on aggressive behavior (Siegel and Edinger, 1981, 1983; Siegel and Brutus, 1990). It is suggested that sex hormone modulation of aggressive behavior may likely result from its combined effects on both the medial hypothalamus as well as limbic structures. This notion is depicted in Figure 8.1.

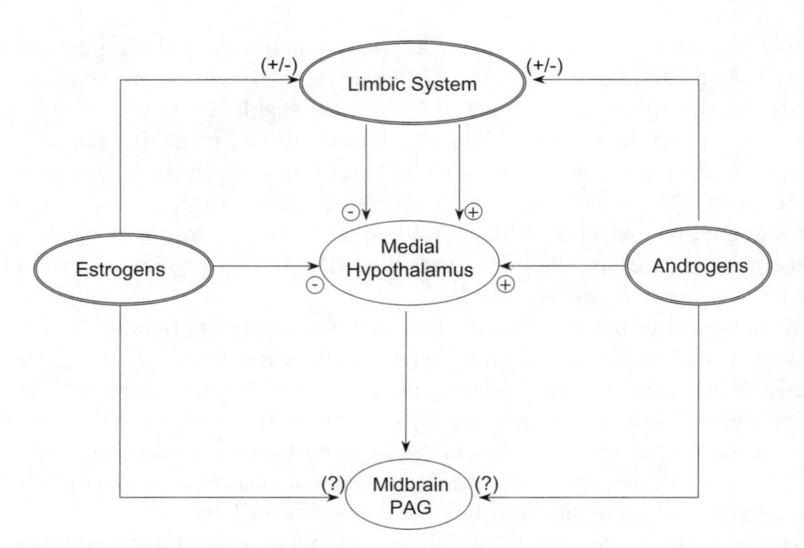

FIGURE 8.1 Schematic diagram illustrating possible mechanism by which sex hormones may regulate aggressive behavior. The expressions of aggressive responses that are defensive in nature are likely to be mediated via a pathway from the medial hypothalamus into the midbrain periaqueductal gray (PAG). Neurons from the PAG then pass caudally to lower regions of the brainstem and ultimately to the spinal cord somatomotor and autonomic cell groups. This represents the final common paths for the elicitation of aggressive responses. Output neurons in the medial hypothalamus and PAG are also directly modulated by various component nuclei of the limbic system. The discharge patterns of neurons within the medial hypothalamus, limbic system, and possibly the midbrain PAG may be altered by circulating levels of androgens and estrogens, which can easily pass through the blood–brain barrier. This modification of the firing patterns results in a change in the propensity for the occurrence of aggressive reactions. Evidence suggests that androgens normally serve to facilitate positive aggressive responses, while the role of estrogens is less clear, but they appear to play an inhibitory, negative role. That these hormones may act directly on neurons within the PAG remains a possibility that has yet to be experimentally tested.

REFERENCES

Albert, D.J., E.M. Dyson, M.L. Walsh, Intermale social aggression: reinstatement in castrated rats by implants of testosterone propionate in the medial hypothalamus, *Physiol. Behav.*, 39, 1987a, 555–560.

Albert, D.J., E.M. Dyson, M.L. Walsh, B.B. Gorzalka, Intermale social aggression in rats: suppression by medial hypothalamic lesions independently of enhanced defensiveness of decreased testicular testosterone, *Physiol. Behav.*, 39, 1987b, 693–698.

Albert, D.J., D.M. Petrovic, M.L. Walsh, Competitive experience activates testosterone-dependent social aggression toward unfamiliar males, *Physiol. Behav.*, 45, 1989, 723–727.

Archawaranon, M. and R.H. Wiley, Control of aggression and dominance in white-throated sparrows by testosterone and its metabolites, *Horm. Behav.*, 22, 1988, 497–517.

Barfield, R.J., Reproductive hormones and aggressive behavior, *Prog. Clin. Bio. Res.*, 169, 1984, 105–134.

Barrett, G.M., K. Shimizu, M. Bardi, S. Asaba, A. Mori, Endocrine correlates of rank, reproduction, and female-directed aggression in male Japanese macaques, *Macaca fuscata*, *Horm. Behav.*, 42, 2002, 85–96.

Bermond, B., J. Mos, W. Meelis, A.M. VanDerPoel, M.R. Kruk, Aggression induced by stimulation of the hypothalamus: effects of androgens, *Pharmacol. Biochem. Behav.*, 16, 1982, 41–45.

Booth, A., G. Shelley, A. Mazur, G. Tharp, R. Kittok, Testosterone, and winning and losing in human competition, *Horm. Behav.*, 23, 1989, 556–571.

Bouissou, M.F., Androgens, aggressive behaviour and social relationships in higher mammals, *Horm. Res.*, 18, 1983, 43–61.

Bradford, J. and D. McLean, Sexual offenders, violence and testosterone: a clinical study, *Can. J. Psychiatry*, 29, 1984, 335–343.

Brain, P.F., *Hormones and aggression*, Eden Press, Montreal, 1977.

Brain, P.F., *Hormones and aggression*, Eden Press, Montreal, 1978.

Brain, P.F. and A.E. Evans, Acute influences of some ACTH-related peptides on fighting and adrenocortical activity in male laboratory mice, *Pharmacol. Biochem. Behav.*, 7, 1977, 425–433.

Caldwell, G.S., S.E. Glickman, E.R. Smith. Seasonal aggression independent of seasonal testosterone in wood rats, *Proc. Natl. Acad. Sci. U.S.A.*, 81, 1984, 5255–5257.

Christiansen, K. and R. Knussmann, Androgen levels and components of aggressive behavior in men, *Horm. Behav.*, 21, 1987, 170–180.

Coe, C.L. and S. Levine, Biology of aggression, *Bull. Am. Acad. Psychiatry Law*, 11, 1983, 131–148.

Dabbs, J.M., R.L. Frady, T.S. Carr, N.F. Besch, Saliva, testosterone and criminal violence in young adult prison inmates, *Psychosom. Med.*, 49, 1987, 174–182.

DeBold, L.F. and K.A. Miczek, Aggression persists after ovariectomy in female rats, *Horm. Behav.*, 18, 1984, 177–190.

DeBold, J.F. and K.A. Miczek, Testosterone modulates the effects of ethanol on male mouse aggression, *Psychopharmacology*, 86, 1985, 286–290.

Dolan, M., I.M. Anderson, J.F. Deakin, Relationship between 5-HT function and impulsivity and aggression in male offenders with personality disorders, *Br. J. Psychiatr.*, 178, 2001, 352–359.

Dufty, A.M., Testosterone and survival: a cost of aggressiveness? *Horm. Behav.*, 23, 1989, 185–193.

Ehrenkranz, J., E. Bliss, M.H. Sheard, Plasma testosterone: correlation with aggressive behavior and social dominance in man, *Psychosom. Med.*, 36, 1974, 469–475.

File, S.E., The stress of intruding: reduction by chlordiazepoxide, *Physiol. Behav.*, 33, 1984, 345–347.

Finkelstein, J.W., E.J. Susman, V.M. Chinchilli, S.J. Kunselman, M.R. D'Arcangelo, J. Schwab, LM. Demers, L.S. Liben, G. Lookingbill, H.E. Kulin, Estrogen or testosterone increases self-reported aggressive behaviors in hypogonadal adolescents, *J. Clin. Endocrin. Metab.*, 82, 1997, 2433–2438.

Fishbein, D.H., D. Lozovsky, J.H. Jaffe, Impulsivity, aggression, and neuroendocrine responses to serotonergic stimulation in substance abusers, *Biol. Psychiatry*, 25, 1989, 1049–1066.

Fraile, I.G., B.S. McEwen, D.W. Pfaff, Progesterone inhibition of aggressive behaviors in hamsters, *Physiol. Behav.*, 39, 1987, 225–229.

Frank, L.G., S.E. Glickman, P. Light, Fatal sibling aggression, precocial development, and androgens in neonatal spotted hyenas, *Science*, 252, 1991, 702–704.

Freeman, L.M., T. Arora, E.F. Rissman, Neonatal androgen affects copulatory behavior in the female musk shrew, *Horm. Behav.*, 34, 1998, 231–238.

Fuchs, S.A.G., H.M. Edinger, A. Siegel, The role of the anterior hypothalamus in affective defense behavior elicited from the ventromedial hypothalamus of the cat, *Brain Res.*, 330, 1985a, 93–107.

Fuchs, S.A.G., H.M. Edinger, A. Siegel, The organization of the hypothalamic pathways mediating affective defense behavior in the cat, *Brain Res.*, 330, 1985b, 77–92.

Gandelman, R. and S. Graham, Singleton female mouse fetuses are subsequently unresponsive to the aggression-activating property of testosterone, *Physiol. Behav.*, 37, 1986, 465–467.

Gandelman, R, C. Peterson, H. Hauser, Mice: fetal estrogen exposure does not facilitate later activation of fighting by testosterone, *Physiol. Behav.*, 29, 1982, 397–399.

Gerra, G., A. Zaimovic, A.P. Avanzini, B. Chittolini, G. Giucastro, R. Caccavari, M. Palladino, D. Maestri, C. Monica, R. Delsignore, F. Brambilla, Neurotransmitter-neuroendocrine responses to experimentally induced aggression in humans: influence of personality variable, *Psychiatr. Res.*, 66, 1997, 33–43.

Gerra, G., A. Zaimovic, M.A. Raggi, F. Giusti, R. Delsignore, S. Bertacca, F. Brambilla, Aggressive responding of male heroin addicts under methadone treatment: psychometric and neuroendocrine correlates, *Drug Alcohol Depend.*, 65, 2001, 85–95.

Gerra, G., A. Zaimovic, R. Sartori, M.A. Raggi, C. Bocchi, U. Zambelli, M. Timpano, V. Sanichelli, R. Delsignore, F. Brambilla, Experimentally induced aggressiveness in adult children of alchoholics, ACOAs: preliminary behavioral and neuroendocrine findings, *J. Stud. Alcohol*, 60, 1999, 776–783.

Gravance, C.G., P.J. Casey, M.J. Erpino, Progesterone does not inhibit aggression induced by testosterone metabolites in castrated male mice, *Horm. Behav.*, 30, 1996, 22–25.

Hansen, S., Medial hypothalamic involvement in maternal aggression of rats, *Behav. Neurosci.*, 103, 1989, 1035–1046.

Harding, C., The role of androgen metabolism in the activation of male behavior, *Ann. N.Y. Acad. Sci.*, 4, 1986, pp. 371–378.

Hau, M., M. Wikelski, K.K. Soma, J.C. Wingfield, Testosterone and year-round territorial aggression in a tropical bird, *Gen. Comp. Endocrin.*, 117, 2000, 20–33.

Hau, M. and P.F. Brain, The effects of differential housing, castration and steroidal hormone replacement on attacks directed by resident mice towards lactating intruders, *Physiol. Behav.*, 30, 1983, 557–560.

Haug, M., L.F. Spetz, M.L. Ouss-Schlegel, D. Benton, P.F. Brain, Effects of gender, gonadectomy and social status on attack directed towards female intruders by resident mice, *Physiol. Behav.*, 37, 1986, 533–537.

Hilakivi-Clarke, L., Role of estradiol in alcohol intake and alcohol-related behaviors, *J. Stud. Alcohol*, 57, 1996, 162–170.

Jasnow, A.M., M.C. Banks, E.C. Owens, K.L. Huhman, Differential effects of two corticotropin-releasing factor antagonists on conditioned defeat in male Syrian hamsters, *Mesocricetus auratus*, *Brain Res.*, 846, 1999, 122–128.

Kavaliers, M., Inhibitory influences of the adrenal steroid, 3alpha, 5alpha-tetrahydroxycorticosterone on aggression and defeat-induced analgesia in mice, *Psychopharmacology*, 95, 1988, 488–492.

Klukowski, M. and C.E. Nelson, The challenge hypothesis and seasonal changes in aggression and steroids in male northern fence lizards, *Sceloporus undulatus hyacinthinus*, *Horm. Behav.*, 33, 1998, 197–204.

Kreuz, L.E. and R.M. Rose, Assessment of aggressive behaviour and plasma testosterone in a young criminal population, *Psychosom. Med.*, 34, 1972, 321–332.

Leshner, A.I., A model of hormones and agonistic behavior, *Physiol. Behav.*, 15, 1975, 225–235.

Lisciotto, C.A., J.F. DeBold, K.A. Miczek, Sexual differentiation and the effects of alcohol on aggressive behavior in mice, *Pharmacol. Biochem. Behav.*, 35, 1990, 357–362.

Lisk, R.D. and M.J. Nachtigall, Estrogen regulation of agonistic and proceptive responses in the golden hamster, *Horm. Behav.*, 22, 1988, 35–48.

Lonstein, J.S. and S.C. Gammie, Sensory, hormonal, and neural control of maternal aggression in laboratory rodents, *Neurosci. Biobehav. Rev.*, 26, 2002, 869–888.

Mann, M.A., C. Konen, B. Svare, The role of progesterone in pregnancy-induced aggression in mice, *Horm. Behav.*, 18, 1984, 140–160.

Mann, M.A. and B. Svare, Factors influencing pregnancy-induced aggression in mice, *Behav. Neural Biol.*, 36, 1982, 242–258.

Mann, M.A. and B. Svare, Prenatal testosterone exposure elevates maternal aggression in mice, *Physiol. Behav.*, 30, 1983, 503–507.

Martensz, N.D., S.V. Vellucci, L.M. Fuller, B.J. Everitt, E.B. Keverne, J. Herbert, Relation between aggressive behaviour and circadian rhythms in cortisol and testosterone in social groups of talapoin monkeys, *J. Endocrin.*, 115, 1987, 107–120.

Mayer, A.D., H.B. Ahdieh, J.S. Rosenblatt, Effects of prolonged estrogen–progesterone treatment and hypophysectomy on the stimulation of short-latency maternal behavior and aggression in female rats, *Horm. Behav.*, 24, 1990, 152–173.

Mayer, A.D. and J.S. Rosenblatt, Hormonal factors influence the onset of maternal aggression in laboratory rats, *Horm. Behav.*, 21, 1987, 253–267.

McEwen, B.S., I. Lieberburg, C. Chaptal, L.C. Krey, Aromatization: important for sexual differentiation of the neonatal rat brain, *Horm. Behav.*, 9, 1977, 249–263.

Meaney, M.J., J. Stewart, W.W. Beatty, The influence of glucocorticoids during the neonatal period on the development of play-fighting in Norway rat pups, *Horm. Behav.*, 16, 1982, 475–491.

Meddle, S.L., L.M. Romero, L.B. Astheimer, W.A. Buttemer, I.T. Moore, J.C. Wingfield, Steroid hormone interrelationships with territorial aggression in an Arctic-breeding songbird, *Horm. Behav.*, 42, 2002, 212–221.

Meisel, R.L. and M.R. Sterner, Progesterone inhibition of sexual behavior is accompanied by an activation of aggression in female Syrian hamsters, *Physiol. Behav.*, 47, 1990, 415–417.

Meisel, R.L., M.R. Sterner, M.A. Diekman, Differential hormonal control of aggression and sexual behavior in female Syrian hamsters, *Horm. Behav.*, 22, 1988, 453–466.

Michael, R.P. and D. Zumpe, Relation between the seasonal changes in aggression, plasma testosterone and the photoperiod in male rhesus monkeys, *Psychoneuroendocrinology*, 6, 1981, 145–158.

Millesi, E., I.E. Hoffmann, S. Steurer, M. Metwaly, J.P. Dittami, Vernal changes in the behavioral and endocrine responses to GnRH application in male European ground squirrels, *Horm. Behav.*, 41, 2002, 51–58.

Moore, M.C., Elevated testosterone levels during nonbreeding season territoriality in a fall-breeding lizard, *Sceloporus jarrovi, J. Comp. Physiol.*, 158, 1986, 159–163.

Moore, M.C., Circulating steroid hormones during rapid aggressive responses of territorial male mountain spiny lizards, *Sceloporus jarrovi, Horm. Behav.*, 21, 1987, 511–521.

Moore, M.C., Testosterone control of territorial behavior: tonic-release implants fully restore seasonal and short-term aggressive responses in free-living castrated lizards, *Gen. Comp. Endocrinol.*, 70, 1988, 450–459.

Moore, M.C. and C.A. Marler, Effects of testosterone manipulations on nonbreeding season territorial aggression in free-living male lizards, *Sceloporus jarrovi, Gen. Comp. Endocrinol.*, 65, 1987, 225–232.

Naftolin, F., K.J. Ryan, I.J. Davies, V.V. Reddy, F. Flores, Z.I. Petro, M. Kuhn, The formation of estrogens by central neuroendocrine tissues, *Rec. Prog. Horm. Res.*, 31, 1975, 295–315.

Olweus, D., A. Mattsson, D. Schalling, H. Low, Testosterone, aggression, physical, and personality dimensions in normal adolescent males, *Psychosom. Med.*, 42, 1980, 253–269.

Olweus, D., A. Mattsson, D. Schalling, H. Low, Circulating testosterone levels and aggression in adolescent males: a causal analysis, *Psychosom. Med.*, 50, 1988, 261–272.

Orengo, C., M.E. Kunik, V. Molinari, K. Wristers, S.C. Yodofsky, Do testosterone levels relate to aggression in elderly men with dementia? *J. Neuropsychiatr. Clin. Neurol.*, 14, 2002, 161–166.

Perrigo, G., W.C. Bryant, F.S. Vom Saal, Fetal, hormonal and experiential factors influencing the mating-induced regulation of infanticide in male house mice, *Physiol. Behav.*, 46, 1989, 121–128.

Persky, H., K.D. Smith, G.K. Basu, Relation of psychologic measures of aggression and hostility to testosterone production in man, *Psychosom. Med.*, 33, 1971, 265–277.

Pfaff, D.W., Autoradiographic localization of radioactivity in the rat brain after injection of tritiated sex hormones, *Science*, 161, 1968a, 1355–1356.

Pfaff, D.W., Uptake of estradiol-17β-H3 in the female rat brain: An autoradiographic study, *Endocrinology*, 82, 1968b, 1149–1155.

Pfaff, D.W. and M. Keiner. Atlas of estradiol-concentrating cells in the central nervous system of the female rat, *J. Comp. Neurol.*, 151, 1973, 121–158.

Price, E.O., T.E. Adams, C.C. Huxsoll, R.E. Borgwardt, Aggressive behavior is reduced in bulls actively immunized against gonadotropin-releasing hormone, *J. Anim. Sci.*, 81, 2003, 411–415.

Raboch, J., H. Cerna, P. Zemek, Sexual aggressivity and androgens, *Br. J. Psychiatry*, 151, 1987, 398–400.

Rada, R.T., Alcoholism and the child molester, *Ann. N.Y. Acad. Sci.*, 273, 1976, 492–496.

Rada, R.T., D.R. Laws, R. Kellner, L. Stivastava, G. Peake, Plasma androgens in violent and nonviolent sex offenders, *Bull. Am. Acad. Psychiatry Law*, 11, 1983, 149–158.

Rhen, T. and D. Crews, Embryonic temperature and gonadal sex organize male-typical sexual and aggressive behavior in a lizard with temperature-dependent sex determination, *Endocrinology*, 140., 1999, 4501–4508.

Rhen, T. and D. Crews, Organization and activation of sexual and agonistic behavior in the leopard gecko, *Neuroendocrinology*, 71, 2000, 252–261.

Rhen, T., J. Ross, D. Crews, Effects of testosterone on sexual behavior and morphology in adult female leopard geckos *Eublepharis macularius*, *Horm. Behav.*, 36, 1999, 119–128.

Rines, J.P. and F.S. Vom Saal, Fetal effects on sexual behavior and aggression in young and old female mice treated with estrogen and testosterone, *Horm. Behav.*, 18, 1984, 117–129.

Ruiz-de-la-Torre, J.L. and X. Manteca, Effects of testosterone on aggressive behavior after social mixing in male lambs, *Physiol. Behav.*, 68, 1999, 1–15.

Ryan, B.C. and J.G. Vandenbergh, Intrauterine position effects, *Neurosci. Biobehav. Rev.*, 26, 2002, 665–678.

Scallet, A.C., S.J. Suomi, R.E. Bowman, Sex differences in adrenocortical response to controlled agonistic encounters in rhesus monkeys, *Physiol. Behav.*, 26, 1981, 385–390.

Scaramella, T.J. and W.A. Brown, Serum testosterone and aggressiveness in hockey players, *Psychosom. Med.*, 40, 1978, 262–265.

Schlinger, B.A., Plasma androgens and aggressiveness in captive winter white-throated sparrows, *Zonotrichia albicollis, Horm. Behav.*, 21, 1987, 203–210.

Severyanova, L.A., Neuromodulator mechanism of the inhibitory influence of deoxycorticosterone on the aggressive-defensive behavior of rats, *Neurosci. Behav. Physiol.*, 18, 1988, 486–492.

Shaikh, M.B., J.A. Barrett, A. Siegel, The pathways mediating affective defense and quiet biting attack behavior from the midbrain central gray of the cat: an autoradiographic study, *Brain Res.*, 437, 1987, 9–25.

Shrenker, P., S.C. Maxson, B.E. Ginsburg, The role of postnatal testosterone in the development of sexually dimorphic behaviors in DBA/lBg mice, *Physiol. Behav.*, 35, 1985, 757–762.

Siegel, A. and M. Brutus, Neurosubstrates of aggression and rage in the cat, in A.N. Epstein and A.R. Morrison, Eds., *Progress in Psychobiology and Physiological Psychology*, Academic Press, San Diego, Calif., 1990, pp. 135–233.

Siegel, A. and M.K. Demetrikopoulos, Hormones and aggression, in J. Shulkin, Ed., *Hormonally Induced Changes in Mind and Brain*, Academic Press, San Diego, Calif., 1993, pp. 99–127.

Siegel, A. and H. Edinger, Neural control of aggression and rage, in P.J. Morgane and J. Panksepp, Eds., *Handbook of the Hypothalamus*, Marcel Dekker, New York, 1981, pp. 203–240.

Siegel, A. and H.M. Edinger, Role of the limbic system in hypothalamically elicited attack behavior, *Neurosi. Biobehav. Rev.*, 7, 1983, 395–407.

Siegel, A. and C.B. Pott, Neural substrate of aggression and flight in the cat, *Prog. Neurobiol.*, 31, 1988, 261–283.

Simon, N.B., R. Gandelman, L.L. Gray, Endocrine induction of intermale aggression in mice: a comparison of hormonal regimens and their relationship to naturally occurring behavior, *Physiol. Behav.*, 33, 1984, 379–383.

Simon, N.G. and D.B. Masters, Activation of male-typical aggression by testosterone but not its metabolites in C57BL/6J female mice, *Physiol. Behav.*, 41, 1987, 405–407.

Simon, N.G. and R.E. Whalen, Sexual differentiation of androgen-sensitive and estrogen-sensitive regulatory system for aggressive behavior, *Horm. Behav.*, 21, 1987, 493–500.

Smith, L.C. and H.B. John-Alder, Seasonal specificity of hormonal, behavioral, and coloration responses to within- and between-sex encounters in male lizards, *Sceloporus undulatus, Horm. Behav.*, 36, 1999, 39–52.

Soma, K.K., R.K. Bindra, J. Gee, J.C. Wingfield, B.A. Schlinger, Androgen-metabolizing enzymes show region-specific changes across the breeding season in the brain of a wild songbird, *J. Neurobiol.*, 41, 1999b, 176–188.

Soma, K.K., K. Sullivan, J. Wingfield, Combined aromatase inhibitor and antiandrogen treatment decreases territorial aggression in a wild songbird during the nonbreeding season, *Gen. Comp. Endocrinol.*, 115, 1999a, 442–453.

Soma, K.K. and J.C. Wingfield, Dehydroepiandrosterone in songbird plasma: seasonal regulation and relationship to territorial aggression, *Gen. Comp. Endocrinol.*, 123, 2001, 144–155.

Sourial, N. and F. Fenton, Testosterone treatment of an XXYY male presenting with aggression: a case report, *Can. J. Psychol.*, 33, 1988, 846–850.

Stavisky, R.C., T.C. Register, S.L. Watson, D.S. Weaver, J.R. Kaplan, Behavioral responses to ovariectomy and chronic anabolic steroid treatment in female cynomolgus macaques, *Physiol. Behav.*, 66, 1999, 95–100.

Summers, C.H., E.T. Larson, P.J. Ronan, P.M. Hofmann, A.J. Emerson, K.J. Renner, Serotonergic responses to corticosterone and testosterone in the limbic system, *Gen. Comp. Endocrinol.*, 117, 2000, 151–159.

Svare, B., Genotype modulates the aggression-promoting quality of progesterone in pregnant mice, *Horm. Behav.*, 22, 1988, 90–99.

Svare, B., J. Miele, C. Kinsley, Mice: Progesterone stimulates aggression in pregnancy-terminated females, *Horm. Behav.*, 20, 1986, 194–200.

Svare, B.B., Ed., *Hormones and Aggressive Behavior*, Plenum Press, New York, 1983.

Takahashi, L.K. and R.D. Lisk, Diencephalic sites of progesterone action for inhibiting aggression and facilitating sexual receptivity in estrogen-primed golden hamsters, *Endocrinology*, 116, 1985, 2393–2399.

Takahashi, L.K., R.D. Lisk, A.L. Burnett, Dual estradiol action in diencephalon and the regulation of sociosexual behavior in female golden hamsters, *Brain Res.*, 359, 1985, 194–207.

Tazi, A., R. Dantzer, M. Le Moal, J. Rivier, W. Vale, G.F. Koob, Corticotropin-releasing factor antagonist blocks stress-induced fighting in rats, *Regul. Peptides,* 18, 1987, 37–42.

Tokarz, R.R., Effects of corticosterone treatment on male aggressive behavior in a lizard, *Anolis sagrei, Horm. Behav.*, 21, 1987, 358–370.

Tricas, T.C., K.P. Maruska, L.E. Rasmussen, Annual cycles of steroid hormone production, gonad development, and reproductive behavior in the Atlantic stingray, *Gen. Comp. Endocrinol.*, 118, 2000, 209–225.

Virkkunen, M., Urinary free cortisol secretion in habitually violent offenders, *Acta Psychiatr. Scand.*, 72, 1985, 40–44.

Vom Saal, F.S., W.M. Grant, C.W. McMullen, K.S. Laves, High fetal estrogen concentrations: correlation with increased adult sexual activity and decreased aggression in male mice, *Science*, 220, 1983, 1306–1308.

Von der, P.B., T. Sarkola, K. Seppa, C.J. Eriksson, Testosterone, 5 alpha-dihydrotestosterone and cortisol in men with and without alcohol-related aggression, *J. Stud. Alcohol*, 63, 2002, 518–526.

Wagner, C.K., C. Kinsley, B. Svare, Mice: postpartum aggression is elevated following prenatal progesterone exposure, *Horm. Behav.*, 20, 1986, 212–221.

Watt, M.J., G.L. Forster, J.M. Joss, Steroid correlates of territorial behavior in male jacky dragons, *Amphibolurus muricatus*, *Brain Behav. Evol.*, 61, 2003, 184–194.

Whalen, R.E. and F. Johnson, Individual differences in the attack behavior of male mice: a function of attack stimulus and hormonal state, *Horm. Behav.*, 21, 1987, 223–233.

Whalen, R.E. and F. Johnson, Aggression in adult female mice: chronic testosterone treatment induces attack against olfactory bulbectomized male and lactating female mice, *Physiol. Behav.*, 43, 1988, 17–20.

Wingfield, J.C., Environmental and endocrine control of reproduction in the song sparrow, *Melospiza melodia*. 11. Agonistic interactions as environmental information stimulating secretion of testosterone, *Gen. Comp. Endocrinol.*, 56, 1984, 417–424.

Wingfield, J.C. and K.E. Hunt, Arctic spring: hormone-behavior interactions in a severe environment, *Comp. Biochem. Physiol.*, 132, 2002, 275–286.

Woodley, S.K. and M.C. Moore, Ovarian hormones influence territorial aggression in free-living female mountain spiny lizards, *Horm. Behav.*, 35, 1999, 205–214.

Worthman, C.M. and M.J. Konner, Testosterone levels change with subsistence hunting effort in !Kung San men, *Psychoneuroendocrinology*, 12, 1987, 449–458.

Zitzmann, M. and E. Nieschlag, Tesototsterone levels in healthy men and the relation to behavioral and physical characteristics: facts and constructs, *Eur. J. Endocrinol.*, 144, 2001, 183–197.

9 Aggression and Immune Function

M. Demetrikopoulos and A. Siegel

The current literature suggests that there is a reciprocal interaction between aggression and immune function such that aggressive behaviors affect immune system functioning and immune events contribute to aggressive behavior. Thus, the interplay of aggression and immune function has been studied from a variety of vantage points, including (1) the role of social status on immune function, (2) the effect of an aggressive act on the recipient, (3) the effect of an aggressive act on the perpetrator, and (4) the effect of immune events on aggressive behaviors. Rather than providing an exhaustive review of these interactions, this chapter will use illustrative examples that represent the larger body of knowledge. The objective of this chapter is to review the interplay of aggression with immune function with particular attention to these vantage points. While the discussion of interspecific aggression may add further dimensions to this interaction, this chapter will specifically focus on intraspecific aggression. Following review of the interaction between aggression and immunity, we briefly review the converging neuronal circuitry that may account for this phenomenon.

SOCIAL STATUS EFFECTS ON IMMUNITY

The literature describing the effects of social status on immunity serves as a general introduction to this discussion since these social issues are closely related to intraspecific aggressive interactions. Many mammals live in colonies, packs, troops, herds, tribes, or other social structures. Within these social structures there is often an order such that animals within the group are part of a social hierarchy. Often, aggressive interactions are involved in the development and continuation of these social hierarchies. While a thorough review of this area is beyond the scope of this chapter, several representative examples are explored in order to examine the effect of social status on immunity within a stable social hierarchy, and as a result of hierarchy formation in an unstable social situation. A comparison of results for these two types of studies will provide suggestive evidence for the role that aggressive behaviors play in social status effects on immune function.

STABLE HIERARCHICAL GROUPS

One approach to examining the effects of social status on immune function involves an experimentally developed stable social hierarchy in the laboratory. Once the hierarchy is well established, differences due to rank can be examined. Vekovishcheva et al. (2000) examined the role of dominance in white mice on a variety of physiological functions, including pain threshold, thymus and spleen weights, and primary antibody production, as well as behavioral measures. Through selective manipulations of the groups, they were able to form a number of linear hierarchies whereby subject A dominated over subject B and C, and subject B dominated over subject C. Despite differences in motor activity, there were no statistically significant differences in the basic physiology of dominant and subordinate mice in this stable linear hierarchy. However, when more specific immunological measures were examined, Stefanski et al. (2001) demonstrated that subdominant male rats had impaired immune systems. This study used a despotic hierarchy with five pairs of subjects. In a despotic hierarchy, one subject dominates the others while the latter are relatively equal and do not demonstrate aggression toward one another. In this model, subdominant males had decreased numbers of CD4 and CD8 T cells, as well as decreased functional capacity of the T cells as measured by their proliferative capacity to the mitogen ConA. In addition to changes in T cells, subdominant subjects showed a marked enhancement of granulocyte numbers, which the authors suggested was of adaptive value if wounding was likely. However, in this experimental paradigm wounding was not likely since dominance was largely maintained by threat. It is likely that the immunological changes were related to the elevated plasma epinephrine and norepinephrine in subdominant subjects, which may suggest that these animals were somewhat stressed. The potential clinical significance of these changes was apparent in a study by Hessing et al. (1994) that demonstrated that social status in pigs was critical in predicting morbidity and mortality following challenge by Aujeszky disease virus (also known as pseudorabies virus, which is found mainly in swine). They found that morbidity and mortality were highest for subordinate pigs in a stable social structure.

The experiments described thus far were conducted after the formation of a stable hierarchy within the laboratory. Other researchers have used a naturalistic model of observing behavior in the wild. Naturalistic observation ensures that findings about dominance are not an artifact of experimentally manipulated social hierarchy. In one such study, Sapolsky and Spencer (1997) studied a troop of wild baboons in the Masai Mara National Reserve in Kenya. This troop had been habituated to observation for 4 years before the start of the study. The study examined the role of dominance hierarchies on anabolic hormone insulin-like growth factor-I (IGF-I), and found a suppression of IGF-I in subordinate male baboons. This finding is germane to the current discussion since IGF-I has been implicated as a possible mediator in a variety of physiological and immunological functions.

A third model for studying the effect of dominance rank on immune function uses a comparison of baseline values to the response evoked by competition for limited resources. It has been postulated that an important function of dominance hierarchies is their use in determining access to resources. Boccia et al. (1992) used a competitive water test, whereby the group was given access to a single waterspout following 24 hours of deprivation. Subjects were observed after the water was turned back on to determine the order in which the subjects drank, length of time they drank, and displacement of subjects from the spout. In addition to comparing the baseline values to the competitive water test condition, this study used pigtail and bonnet macaques, which exhibit differences in social behavior. Interestingly, dominance did not affect baseline immune functioning in either species, although there were immunological differences between the two species such that pigtails showed higher natural killer (NK) cell function and lower lymphocyte activation than the bonnets. As expected, the pigtail macaques showed more agonistic behaviors than did the bonnet macaques irrespective of the competition condition. During the competition task, the behavior of the two species varied dramatically. The bonnet monkeys formed a line by dominance rank in order to wait for their turn to drink. In contrast, the pigtail monkeys competed for access to the water through the increased use of agonistic social interactions. Additionally, the pigtail macaques demonstrated an increase in NK activity, and a decrease in lymphocyte activity following the water competition task. These finding for the pigtail macaques are similar to those found in the literature on aggression and immune function in humans (see Aggression, Hostility, and Marital Conflict Effects on Immunity in Humans section).

Hierarchy Formation

In addition to studying already established social hierarchies, a number of investigators have studied the immunological consequences of active hierarchy formation. Results from these studies have demonstrated that change in the social structure, and the subsequent aggressivity resulting from social pressure to reestablish a hierarchy, produce a variety of immune changes. For example, male guinea pigs put into social situations with unstable dominance had lower complement activity and engaged in frequent stand-threat interactions (Stefanski and Hendrichs, 1996). Generally, guinea pigs are able to form stable dominance hierarchies within a very brief period of time, after which these stand-threat interactions disappear. Although plasma cortisol was not correlated with the changes in complement activity, the authors hypothesized that the protracted aggressive interactions were evidence of a stressful environment.

Other investigators have demonstrated an initial concomitant rise in cortisol with a decrease in immune function in response to the establishment of a dominance hierarchy (Gust et al., 1991). Within 48 hours of the formation of a new social group, female rhesus monkeys were able to establish a dominance hierarchy with only minor fighting and no wounding or trauma. While the effects on cortisol levels was transitory, the effects on CD4 and CD8 T cells persisted for more than 2 months, with the low ranking subjects showing significantly reduced numbers of T cells. Similar results were found in a more chronic model of social hierarchy disruption using male cynomolgus monkey subjects (Cohen et al., 1997). Subjects were put into an unstable social situation for 15 months whereby they were reorganized into new groups every month. Following 8 months of reorganization, the subjects were exposed to an influenza virus and then subsequently infected with adenovirus after 15 months of reorganization. Subjects exposed to the instability manipulation demonstrated increased aggressive behaviors. However, low social status, irrespective of social reorganization, was associated with enhanced probability of infection and less aggressive behaviors. Therefore, the reorganization *per se* was less important in terms of immune competence than was social rank. Interestingly, subjects with lower rank showed enhanced cortisol responses to social reorganization, but these elevated hormonal levels did not account for the changes in infection rates. The differences in infection rates due to social rank may have clinical implications. Capitanio et al. (1998) studied this more directly by examining survival rates of male rhesus macaques in unstable social environments that were exposed to simian immunodeficiency virus (SIV). Animals with a disrupted social hierarchy survived significantly shorter periods of time and engaged in more aggressive behaviors, while subsequently engaging in less affiliative behaviors such as grooming. Furthermore, when the data were examined across social disruption conditions, subjects receiving more social threats had higher plasma SIV levels. These results correspond well with the prior experiment: since the authors chose the receipt of threats as the agonistic behavior to study, it could be postulated that these were not the dominant subjects.

An interesting complication to the relationship of aggression, dominance, and immunity was explored by Barnard et al. (1994). Contrary to the studies just described, these authors demonstrated that dominant male mice were infected with the parasite *Babesia microti* sooner and were slower to clear the infection. They suggested that this effect was due to the immunosuppressive effects of the elevated testosterone evident in the dominant mice.

Summary

Because so many mammals demonstrate hierarchical group formation, it is tempting to postulate that there must be an advantage to this social structure and that maintenance of a stable group should not be detrimental to the individuals within said group. However, even within a stable hierarchy, most studies demonstrate that socially subordinate animals have suboptimal immune functioning (Table 9.1), which may be partially related to their rank imposing a chronic stress. During the formation of the hierarchy, socially subordinate animals also have suboptimal immune functioning, which puts them at a higher risk of opportunistic infections. Thus, subjects that engage in less aggressive behaviors tend to have lower immune competencies.

EFFECTS OF AGGRESSION ON RECIPIENT'S IMMUNE FUNCTION

When examining the effect of aggression on immune function, it is useful to distinguish between the recipient and perpetrator of the aggressive act. This factor is often related to various other factors including the dominance status of the individual, social context, and hormonal and neuroendocrine status at the time of the confrontation. A large body of literature focuses on the effect of aggression on the recipient's immune function. While several models have been used to study this phenomenon, this review is limited to the social disruption model, whereby an aggressive intruder is placed into an established cage of subjects, and to the social confrontation model, whereby the subject serves as the intruder into an aggressive resident animal's home cage. However, a large proportion of this literature frames the findings in terms of the aggressive encounter being either a short-term or long-term stressor. Many of these papers do not focus on aggression *per se* but rather view it as a naturalist stressor. Nevertheless, a few selected

TABLE 9.1
Social Status Effect on Immune Function

Citation	Model	Species	Outcome
Vekovishcheva et al., 2000	Stable linear hierarchies	White mice	No effect of rank on spleen or thymus weight or primary immune response
Stefanski et al., 2001	Stable despotic hierarchies	Long Evans rats	Subdominant decreased T cells
Hessing et al., 1994	Stable hierarchy	Pigs	Subordinate increased morbidity and mortality to Aujeszky disease
Sapolsky and Spencer, 1997	Dominance hierarchy in wild	Baboons	Subordinate decreased insulin-like growth factor
Boccia et al., 1992	Competition for limited access to water	Pigtail macaques	Baseline, no effect of dominance Postcompetition, increased agonistic behavior and natural killer activity
Boccia et al., 1992	Competition for limited access to water	Bonnet macaques	Baseline, no effect of dominance Postcompetition, no agonistic behavior or immune effects
Stefanski and Hendrichs, 1996	Unstable dominance	Guinea pigs	Decreased complement activity
Gust et al., 1991	New group formation	Rhesus monkeys	Low-ranking subjects, decreased T cells
Cohen et al., 1997	Social reorganization	Cynomolgus monkeys	Low social status, increase probability of infection
Capitanio et al., 1998	Disrupted social hierarchy	Rhesus macaques	Shorter survival, simian immunodeficiency virus
Barnard et al., 1994	Disrupted social hierarchy	CFLP mice	Dominant, slower to clear infection

papers from this area are reviewed in order to gain some understanding of this aspect of the interplay between aggression and immunity.

Social Disruption

In this model, a dominant animal is repeatedly transferred into cages with an established social hierarchy. This aggressive animal attacks the home cage subjects that typically demonstrate submissive behaviors. If a resident animal attacks an intruder, that intruder will be replaced with another aggressive intruder so that in all cases the resident animals will become submissive. Experiments by Avitsur et al. (2001) and Stark et al. (2001) have begun to examine how this disruption, and subsequent elevation of glucocorticoids, affects an immunological cascade of events. They have proposed that the elevated glucocorticoids leads to glucocorticoid resistance in subordinate animals that may be important for wound healing and be related to changes seen in humans during depression. This effect was particularly pronounced in macrophages, and was likely to develop in animals that showed low levels of social exploration prior to attack and high levels of submission following attack. Thus, the behavioral profile of the animals, prior to their interaction in an agonistic encounter, was able to predict their behavior in this encounter as well as predict the physiological consequences of the agonistic encounter. Hence, there appears to be a social disposition that predisposes an individual animal to respond in a particular physiological way to an aggressive encounter.

Social Confrontation

This model is often termed resident–intruder confrontation and has been widely used as a social stress model to study a variety of physiological processes, including changes in immune function, cardiovascular issues, and stress-related digestive issues. It has been likened to the naturalistic stressors that territorial animals would encounter in the wild. In this model, the subject acts as an intruder into an aggressive animal's home cage. In general, the intruder animal is attacked and is often defeated by the resident animal. The intruder may become submissive and cower away from the resident animal or it may become subdominant and show enhanced motor activity. Due to the complex nature of this manipulation, a variety of control groups are necessary for this paradigm, including a group that experiences a novel environment.

While it is beyond the scope of this chapter to present a thorough review of this literature, a few papers have been selected that demonstrate some of the complexity of responses and variables that contribute to the richness of this model. For example, Dreau et al. (1999) used this model to examine the effect of social confrontation on immune function. Subjects exposed to an aggressive resident animal five times a day for 2 days demonstrated decreases in thymocyte numbers, decreased *in vitro* response to LPS, and decreases in CD4 and CD8 T cells. Although not statistically significant, the subjects also had increased *in vivo Escherichia coli* bacterial growth.

Stefanski and Engler (1998) sought to distinguish the effects of acute and chronic social confrontation on cellular immunity. The subjects were exposed to the aggressive resident for 2 hours or for 48 hours with either exposure time resulting in aggressive encounters to establish dominance. Many of the immunological variables examined were affected similary in both the chronic and acute exposure times. However, some of the impairment present following 2 hours of agonistic behavior had been resolved in the 48-hour group. Interestingly, the correlation that was present at 2 hours between immune function and level of submissive behavior expressed by losers was not present at 48 hours.

In order to better characterize immune status following social confrontation, Stefanski and Ben-Eliyahu (1996) examined the development of syngeneic tumor metastasis. Subjects experienced 7 hours of confrontation and received mammary tumor cells 1 hour into the confrontation. The intruders had increased tumor retention that was related to the level of submissive behaviors displayed. Furthermore, pretreatment with butoxamine, a beta-adrenergic antagonist, reduced this effect by half and adrenal demedullation essentially eliminated it. Interestingly, neither of these treatments modified the social interaction. In the studies described above, not all of the intruder subjects displayed submissive behaviors, despite the fact that they were all attacked by aggressive residents. Instead of the more typical submissive posture, some of the subjects displayed subdominant behaviors characterized by enhanced activity or routines that may be independent of environmental stimuli. The submissive display or subordinate losers in the confrontation model are generally passive and avoid the aggressive animal. The behavior of the subordinate loser is often likened to defeat. Stefanski (1998) sought to examine the immune changes in these two types of losers in the social confrontation model. The immune changes following social confrontation significantly differed in these two groups and, in some cases, were in opposite directions, which suggested that these subjects may be experiencing vastly different neuroendocrine/neuroimmune cascade responses to the same aggressive encounter.

SUMMARY

The primary literature has examined the effects of aggression on the recipient, but has used this approach mainly as a model of stress. Consequently, manipulations of social variables and descriptions of social interactions are generally not thorough enough to give a clear understanding of the nuances involved. Despite these limitations, these findings do generally demonstrate that submissive behaviors often result in immunosuppression (Table 9.2).

TABLE 9.2
Effect of Aggression on Recipient's Immune Function

Citation	Model	Species	Outcome
Avitsur et al., 2001	Social disruption	C57BL/6 mice	Subordinate, glucocorticoid-resistant splenocytes
Stark et al., 2001	Social disruption	C57BL/6 mice	Subordinate, glucocorticoid-resistant macrophages and increased IL-6
Dreau et al., 1999	Social confrontation	DBA/2 mice	Exposure to aggressive resident yields decreased thymocytes and T cells
Stefanski and Engler, 1998	Social confrontation	Long Evans rats	Increased granulocytes; decreased lymphocytes, T cells, and B cells
Stefanski and Ben-Eliyahu, 1996	Social confrontation	Fischer 344 rats	Increased tumors in intruders blocked by beta-adrenergic antagonist
Stefanski, 1998	Social confrontation	Long Evans rats	Subdominant decreased T and B cells, submissive decreased T cells, increased B cells and granulocytes

EFFECTS OF AGGRESSION ON PERPETRATOR'S IMMUNE FUNCTION

While the previous section reviewed the effects of aggression on the recipient, the current section will review the effects of aggression on the perpetrator's immune system. I attempt to describe what happens to an individual's immune system as it engages in aggressive acts. This literature is quite broadly defined, and includes (1) a number of models of aggression such as social isolation and resident–intruder models; (2) the relationship of aggressive behaviors to strain differences in immune function; and (3) the effects of aggressive behaviors in humans, including hostility and marital conflict. Once again, rather than providing an exhaustive review of this area, the selected experiments highlight some important aspect of the larger literature.

SOCIAL ISOLATION–INDUCED FIGHTING

In this model of aggression, subjects are housed in isolation for some period of time prior to reintroduction to a neutral group setting. Often, following the isolation housing, subjects are kept in cages equipped with a transparent divider, so that they can see the other subject but not physically interact with it prior to the experimental encounter. These encounters generally involve an aggressive confrontation that ultimately leads to dominant and submissive behavioral development.

Gryazeva et al. (2001) examined the effect of multiple aggressive encounters on immune function. They found that all participants in the aggressive encounters increased the proportion of segmented neutrophils and CD4 and CD8 T cells in the spleen. Despite these general findings, there were additional immune changes that demonstrated divergence such that the winners and losers of these encounters had opposite immunological outcomes. This finding agrees with that of Hardy et al. (1990), who demonstrated that the direction of functional T-cell changes was dependent on the dominance status of the fighting subject. Submissive mice had lower T-cell proliferation compared to dominant and nonfighter controls. In a second experiment, they used a test that was less likely to produce wounding. In this experiment, they found that dominant mice had elevated T-cell proliferation compared to submissive and nonfought controls. Thus, the physical consequences of fighting had an effect independent of the psychological aspects of the aggressive encounter.

RESIDENT–INTRUDER INTERACTION

This model was described in the social confrontation subsection above, and is generally used to study the effects of aggression on the recipient's immune system since the model normally uses a highly aggressive resident. As previously mentioned, in this model an intruder rat is introduced into a resident's home cage. Kavelaars et al. (1999) used this model to study the effect of the attack latency of wild-type resident rats on the immune function of the resident. Animals with short attack latency were less resistant to experimental autoimmune encephalomyelitis (EAE). Thus, aggressive rats were more susceptible to EAE than nonaggressive rats. This finding somewhat contradicts much of the immune/aggression literature in that it demonstrates a detrimental outcome for the perpetrator of the aggressive act.

Gasparotto et al. (2002) modified the basic resident–intruder preparation further by having both the residents and intruders individually housed prior to the social confrontation. Following social isolation, half of the subjects were moved into a larger environment 3 days before the confrontation experience to become essentially a resident subject. This differs from the more classic resident–intruder model in that the resident is not very well established, and only obtains dominance in approximately 65% of the dyads, whereas in the classic resident–intruder model, the resident is generally dominant. Subjects were socially challenged for 2 or 3 weeks and immunologically challenged with sheep red blood cells (SRBC). The shorter chronic social challenge produced a depression in the primary immune response in submissive mice, while the longer chronic social challenge produced an enhancement of the primary immune response in dominant subjects. Thus, this distinction was similar to that seen in the less chronic confrontation that occurred in either the presence or absence of wounding. It is possible that in the longer chronic confrontation, the level of agonistic behavior was somewhat reduced or more stylized. However, this cannot be determined from this study, since the authors did not address the nature or severity of the agonistic behaviors in the two conditions.

Strain Differences in Behavior and Immune Function

In addition to individual differences, many species have strains that differ greatly in their levels of aggressive and submissive behaviors. In fact, some strains of mice have been selectively bred to produce high and low levels of aggressive behaviors. These selectively bred animals are models for understanding genetic contributions to individual differences. Petitto et al. (1993, 1994) examined a number of immunological factors in ICR mice that were selectively bred for differences in levels of social isolation–induced aggression. Selection was originally based on the frequency of attacks following contact with a randomly bred conspecific. Although the selection program was bidirectional, the high-aggressive line (NC900) does not differ from the unselected line. However, the low-aggressive line (NC100) exhibited enhanced freezing and immobility that were characterized as social inhibition. Importantly, these lines only appear to differ behaviorally in social contexts, and are quite similar in terms of their other behavioral responses, including their response to novelty and fearfulness. In addition to the differences in aggressive behaviors, these mice have significant and biologically relevant changes in immune function. These socially inhibited mice had reduced NK activity and increased susceptibility to tumor development (Petitto et al., 1993). They also had lower T-cell proliferation and lower IL-2 and gamma interferon production (Petitto et al., 1994). Similarly, Devoino et al. (1993) demonstrated that strains of mice which differed in social isolation induced aggression had correspondingly different immune responses following immunization with sheep red blood cells. The aggressive CBA strain had greater numbers of plaque-forming cells in response to aggressive encounters, while the plaque-forming cells decreased in the submissive C57BL mice compared to controls. Thus, association of the aggressive behavior with the primary immune response to SRBC is dependent on the subject's strain.

Aggression, Hostility, and Marital Conflict Effects on Immunity in Humans

Thus far, this chapter has examined the interaction of aggression and immune function in nonhuman subjects. The current section briefly examines the interplay of aggression and immunity in normal human subjects, rather than addressing the potential effects of pathological aggressive behaviors or aggressive behaviors committed by individuals with psychiatric or neurological disorders. Granger et al. (2000) demonstrated a positive curvilinear relationship between aggression and immune function, such that moderate levels of aggressive behavior were correlated with enhanced numbers of CD4 and B cells. This study used men who served in the U.S. Army in the 1965–1971 period as part of the Vietnam Experience Study, which was designed to examine the long-term health consequences of serving in Vietnam. Although the study was conducted on normal individuals, the aggression measure was derived from "antisocial personality" as defined in the DSM-III, and was designed to measure aggression and antisocial behavior. A high level of individual variability in relationship to these two variables among Vietnam War veterans may be expected. Christensen et al. (1996) examined the role that cynical hostility has on modifying the effect of self-disclosure of personal information about a stressful experience. Among those who participated in self-disclosure, high-hostility subjects had higher NK activity. Thus, in a potentially stressful or arousing condition, the personality variable of hostility predicts both the general status of arousal during the disclosure task and the immunological reaction to the task.

Hostility was measured in introductory psychology students by the Cook–Medley hostility scale, which identifies individuals who are suspicious, cynical, mistrust others, and are easily aroused by anger. In the studies mentioned above, underlying aggression or hostility was measured using standardized testing instruments. In a study by Kiecolt-Glaser et al. (1997), social conflict was encouraged. Older couples who had been married an average of 42 years were used in the study. Subjects who responded poorly to the functional immune assays displayed more negative behaviors in the conflict task.

Summary

Numerous studies have examined the effect of aggression on the perpetrator's immune system (Table 9.3). A variety of behavioral models are useful in understanding the physiological consequences of aggressive behaviors, and strain differences help to highlight the genetic contribution to these phenomena. The universality of aggressive behaviors is apparent, given that human studies have been conducted with veterans, college students, and elderly married couples. In many cases, subjects who score high on aggression also score high on immune function; thus, more aggressive subjects have enhanced immune function.

TABLE 9.3
Effects of Aggression on Perpetrator's Immune Function

Citation	Model	Species	Outcome
Gryazeva et al., 2001	Social isolation–induced fighting	C57B1/6J mice	Increased neutrophils, decreased T cells
Hardy et al., 1990	Social isolation–induced fighting	C3H/HeJ mice	Submissive, decreased T-cell proliferation; dominant, elevated T-cell proliferation
Kavelaars et al., 1999	Resident intruder	Wild-type rat	Short attack latency, less resistant to experimental autoimmune encephalomyelitis
Gasparotto et al., 2002	Isolated resident intruder	Swiss mice	Dominant, enhanced primary immune response to sheep red blood cells
Petitto et al., 1993	Genetically aggressive strain	ICR mice	Socially inhibited, decreased NK activity, increased tumor development
Petitto et al., 1994	Genetically aggressive strain	ICR mice	Socially inhibited, decreased T-cell prolifertion, lower IL-2 gamma interferon
Devoino et al., 1993	Genetically aggressive strain	CBA and C57BL mice	Aggressive strain, increased plaque-forming cells
Granger et al., 2000	Aggression scores in Vietnam veterans	Humans	Aggression increased CD4 and B cells
Christensen et al., 1996	Cynical hostility and self-disclosure	Humans	High hostility increased NK activity
Kiecolt-Glaser et al., 1997	Induced marital conflict	Humans	Negative behaviors decreased T- and B-cell activity

Note: NK, natural killer.

CAN IMMUNE EVENTS AFFECT AGGRESSION?

The preceding sections have explored the role that aggression has on the immune system. In this section, I briefly review some recent experimental evidence suggesting that this relationship is reciprocal and dynamic in that the immune system can affect change in aggressive behaviors.

Immune Challenge Effects

Granger et al. (1996) demonstrated that neonatal exposure to bacterial endotoxin increased socially reactive behaviors to conspecifics when tested as adults. These effects could be partially explained by changes in corticotropin-releasing factor levels during neonatal development that had long-term social consequences. In a similar study, Granger et al. (2001) explored the effect of early immune challenge on adult aggressive behaviors. When ICR neonatal mice were treated with endotoxin, adults from the high-aggressive line tended to show lower levels of aggressive behaviors. Thus, an acute immune challenge was capable of modifying a genetically determined behavioral expression.

Autoimmune Disorder Effects

Sakic et al. (1998) demonstrated that autoimmune, lupus-prone MRL-1pr mice showed reduced aggressiveness in the isolation-induced model of aggression. Several aspects of aggressive behavior, including longer attack latencies, fewer fights, and shorter duration of fights were manifest. However, it is unclear how this immune activation produced its effect since these subjects also had concomitant decreases in testosterone.

Potential Mediators

Brain cytokines may represent a potential mediator for the effects of immune variables on aggressive behaviors. This possibility was recently tested in a study by Hassanain et al. (2003a, 2003b) in which the role of interleukin-1 beta (IL-1β) in the hypothalamus on defensive rage behavior elicited from the periaqueductal gray (PAG) was assessed. IL-1β is derived from both the immune and nervous systems, and is endogenously present in the hypothalamus. It is thought to affect aggression by the induction of serotonin release. In the first part of this study, it was shown that microinjections of a 5-HT_{1A} agonist into the medial hypothalamus suppressed defensive rage behavior elicited from the PAG, while microinjections of a 5-HT_2 agonist into the same region of the medial hypothalamus facilitated PAG-elicited defensive rage. In the second part of the study, microinjections of an IL-1β agonist into the same region also facilitated PAG-elicited defensive rage. Of particular interest was the third part of the study, which demonstrated that pretreatment with a 5-HT_2 antagonist into the medial hypothalamus blocked the facilitatory effects of IL-1β. Thus, this finding was the first to demonstrate that IL-1β in the medial hypothalamus can potentiate defensive rage behavior elicited from the PAG, an effect that is mediated through a serotonergic mechanism.

Summary

There is a growing body of literature demonstrating the reciprocal nature of the interaction between aggression and immune functioning (Table 9.4). Thus, the findings of the effects of aggression on immunity and the effects of immunity on aggression must be placed within this larger framework. A careful examination of the effects of the immune system on aggression may suggest critical developmental stages for aggressive behavioral development, as well as pathways and neuroendocrine cascades for its expression. In addition, the discovery that cytokines in the brain can powerfully modulate aggression and rage, and that such effects are mediated through classical neurotransmitter mechanisms may open an entirely new direction of research governing the relationships among cytokines, aggression, and rage behavior.

TABLE 9.4
Immune Event Effects on Aggression

Citation	Model	Species	Outcome
Granger et al., 1996	Neonatal exposure to endotoxin	ICR mice	Increased socially reactive behaviors as adults
Granger et al., 2001	Neonatal exposure to endotoxin	ICR mice selectively bred	Decreased aggressive behavior of genetically aggressive adults
Sakic et al., 1998	Autoimmune disorder	Lupus-prone MRL-1pr mice	Reduced aggression in isolation-induced aggression
Hassanain et al., 2003a, 2003b	Midbrain periaqueductal gray–elicited defensive rage	Cats	Medial hypothalamus 5-HT_{1A} agonist suppressed rage; 5-HT_2 agonist or IL-1β facilitated rage; IL-1β effect blocked by 5-HT_2 antagonist

CONVERGING NEURONAL CIRCUITRY

ROLE OF CYTOKINES

The studies described above by Hassanain et al. (2003a, 2003b) have provided an important clue concerning how centrally mediated cytokines may function within existing neuronal circuits that govern defensive rage behavior. The discovery of the presence of IL-1β receptors in the medial hypothalamus, and the fact that its activation can facilitate defensive rage, suggest that the functional effects of these receptors are mediated through their actions on medial hypothalamic neurons. Recall from earlier chapters that the major descending pathway from the medial hypothalamus underlying the expression of defensive rage is one that projects to the PAG and whose functions are mediated principally by N-methyl-D-aspartate receptors in the PAG. As noted above, IL-1β receptor facilitation of defensive rage appears to be linked to 5-HT_2 receptors, whose activation was shown to also facilitate defensive rage.

A significant feature of these findings is that they provide a basis for understanding how other cytokines may modulate aggression and rage. Namely, we suggest that different cytokines in the hypothalamus and related regions probably act through classical neurotransmitter–receptors such as monoamines, amino acids, and neuropeptides to either excite or inhibit descending medial hypothalamic neurons, which project to the PAG and adjoining sites in the brainstem that mediate the rage response. In view of the fact that this line of research is in its infancy, there is a basic need for us to strengthen our understanding of how different cytokines may modulate aggression and rage and to establish the mechanisms underlying such effects.

Role of Periaqueductal Gray

As discussed in previous chapters, the PAG has a prominent role in the modulation of aggression, and there is a growing body of evidence which suggests that the PAG has an important role in immune function. For example, Demetrikopoulos et al. (1994) demonstrated that stimulation of the rostral-dorsal PAG suppressed NK cell function, and then Demetrikopoulos and Zhang (1998) showed that stimulation of the rostral-ventral PAG enhanced NK cell activity. The role of the PAG in aggression-induced effects on immunomodulation may be directly related to the neural circuitry underlying aggressive behaviors or may be due to the effects of secondary stress (reviewed in Demetrikopoulos et al., 1999).

Secondary Stress Effects

While a thorough review of the effects of stress on immune function are beyond the scope of this chapter, in this section I will briefly review the primary pathways that may be secondarily activated in response to aggressive encounters (Demetrikopoulos et al., 1999). The classic stress response involves the following cascade: corticotropin-releasing factor is released from the hypothalamus, which leads to release of adrenocorticotripic hormone from the anterior pituitary, which leads to release of glucocorticoid from the adrenal cortex. While it is tempting to suggest that this system's effects on immunity are simply due to the well-known immunosuppressive effect of glucocorticoid, an examination of this literature demonstrates that glucocorticoids constitute one of various factors that regulate immune function. Another aspect of this hypothalamic-mediated stress response includes release of luteinizing hormone from the pituitary, which stimulates the testis to produce testosterone (Tewes, 1999), which leads to glucose uptake in muscles. Testosterone is critically important for aggression (see Chapter 8) and can be immunosuppressive.

The sympathetic nervous system is also critically involved with the effects of stress on immunity and equally important in the study of aggression. It has been well documented that immune organs are directly innervated by the sympathetic nervous system and that many immune cells contain adrenergic receptors. Opiates and serotonin are other neurotransmitter systems that may be important for the convergence of aggression, stress, and immune function (Demetrikopoulos et al., 1999). For example, opiate-dependent immunosuppressive effects may be mediated through the μ receptors in the caudal-ventral PAG. Furthermore, it has been suggested that the raphe nucleus serotonergic system is immunosuppressive.

Summary

Various factors may be directly or indirectly involved in the effects of aggression on immunity. There is a clear convergence in the literature of neuroanatomical pathways, neurotransmitter systems, cytokines, and hormonal mediators that are common to the aggression and immune system literature.

OVERALL SUMMARY

As a general rule, aggressive behaviors are correlated to increased immune function in the perpetrator and decreased immune function in the losers. For the most part, in this chapter I did not attempt to look at more complex systems that would relate additional variables such as the interplay of aggression, sex hormones, and immune function. A number of these three-part interactions may help to explain some of the contradictions in the literature in this area. Future research may attempt to elucidate the potential mediators of these phenomena, as well as explore the converging neuronal circuitry that underlies the interaction between aggression and immune functions.

REFERENCES

Avitsur, R., J.L. Stark, J.F. Sheridan, Social stress induces glucocorticoid resistance in subordinate animals, *Horm. Behav.*, 39, 2001, 247–257.

Barnard, C.J., J.M. Behnke, J. Sewell, Social behaviour and susceptibility to infection in house mice, Mus musculus: effects of group size, aggressive behavior and status-related hormonal responses prior to infection on resistance to Babesia microti, *Parasitology*, 108, 1994, 487–96.

Boccia, M.L., M.L. Laudenslager, C.L. Broussard, A.S. Hijazi, Immune responses following competitive water tests in two species of macaques, *Brain Behav. Immun.*, 6, 1992, 201–213.

Capitanio, J.P., S.P. Mendoza, N.W. Lerche, W.A. Mason, Social stress results in altered glucocorticoid regulation and shorter survival in simian acquired immune deficiency syndrome, *Proc. Natl. Acad. Sci. U.S.A.*, 95, 1998, 4714–4719.

Christensen, A.J., D.L. Edwards, M.A. John, S. Wiebe, E.G. Benotsch, L. McKelvey, M. Andrews, D.M. Lubaroff, Effect of verbal self-disclosure on natural killer cell activity: moderating influence of cynical hostility, *Psychosom. Med.*, 58, 1996, 150–155.

Cohen, S., S. Line, S. Manuck, B.S. Rabin, E.R. Heise, J.R. Kaplan, Chronic social stress, social status, and susceptibility to upper respiratory infections in nonhuman primates, *Psychosom. Med.*, 59, 1997, 213–221.

Demetrikopoulos, M.K., S.E. Keller, S.J. Schleifer. Stress effects on immune function in rodents, in *Psychoneuroimmunology: an interdisciplinary introduction*, M. Schedlowski and U. Tewes, Eds., Kluwer Academic/Plenum Publishers, 1999, pp. 259–275.

Demetrikopoulos, M.K., A. Siegel, S.J. Schleifer, J. Obedi, S.E. Keller, Electrical stimulation of the dorsal midbrain periaqueductal gray suppresses peripheral blood natural killer cell activity, *Brain Behav. Immun.*, 8, 1994, 218–228.

Demetrikopoulos, M.K. and Z. Zhang, Rostal-ventral electrical stimulation of the midbrain periaqueductal gray limits tumor metastasis in rats, *Soc. Neurosci.*, 24, 1998, 737.

Devoino, L., E. Alperina, N. Kudryavrseva, N. Popova, Immune responses in male mice with aggressive and submissive behavior patterns: Strain differences, *Brain Behav. Immun.*, 7, 1993, 91–96.

Dreau, D., G. Sonnenfeld, N. Fowler, D.S. Morton, M. Lyte, Effects of social conflict on immune responses and E. coli growth within closed chambers in mice, *Physiol. Behav.*, 67., 1999, 133–140.

Gasparotto, O.C., Z.M. Ignacio, K. Lin, S. Goncalves, The effect of different psychological profiles and timings of stress exposure on humoral immune response, *Physiol. Behav.*, 76, 2002, 321–326.

Granger, D.A., A. Booth, D.R. Johnson, Human aggression and enumerative measures of immunity, *Psychosom. Med.*, 62, 2000, 583–590.

Granger, D.A., K.E. Hood, N.A. Dreschel, E. Sergeant, A. Likos, Developmental effects of early immune stress on aggressive, socially reactive, and inhibited behaviors, *Dev. Psychopathol.*, 13, 2001, 599–610.

Granger, D.A., K.E. Hood, S.C. Ikeda, C.L. Reed, M.L. Block, Neonatal endotoxin exposure alters the development of social behavior and the hypothalamic-pituitary-adrenal axis in selectively bred mice, *Brain Behav. Immun.*, 10, 1996, 249–259.

Gryazeva, N.I., A.V. Shurlygina, L.V. Verbitskaya, E.V. Melnikova, N.N. Kudryavtseva, V.A. Trifakin, Changes in various measures of immune status in mice subject to chronic social conflict, *Neurosci. Behav. Physiol.*, 31, 2001, 75–81.

Gust, D.A., T.P. Gordon, M.E. Wilson, A. Ahmed-Ansari, A.R. Brodie, H.M. McClure, Formation of a new social group of unfamiliar female rhesus monkeys affects the immune and pituitary adrenocortical systems, *Brain Behav. Immun.*, 5, 1991, 296–307.

Hardy, C., J. Quay, S. Livnat, R. Ader, Altered T-lymphocyte response following aggressive encounters in mice, *Physiol. Behav.*, 47, 1990, 1245–1251.

Hassanain, M., S. Bhatt, A. Siegel, Differential modulation of feline defensive rage behavior in the medial hypothalamus by 5-HT_{1A} and 5-HT_2 receptors, *Brain Res.*, 981, 2003a, 201–209.

Hassanain, M., S. Zalcman, S. Bhatt, A. Siegel, Interleukin-1 in the hypothalamus potentiates feline defensive rage: Role of 5-HT_2 receptors, *Neuroscience*, 120, 2003b, 227–233.

Hessing, M.J., C.J. Scheepens, W.G. Schouten, M.J. Tielen, P.R. Wiepkema, Social rank and disease susceptibility in pigs, *Vet. Immunol. Immunopathol.*, 43, 1994, 337–387.

Kavelaars, A., C.J. Heijnen, R. Tennekes, J.E. Bruggink, J.M. Koolhaas. Individual behavioral characteristics of wild-type rats predict susceptibility to experimental autoimmune encephalomyelitis, *Brain Behav. Immun.*, 13, 1999, 279–286.

Kiecolt-Glaser, J.K., R. Glaser, J.T. Cacioppo, R.C. MacCallum, M. Snydersmith, K. Cheongrag, W.B. Malarkey, Marital conflict in older adults: Endocrinological and immunological correlates, *Psychosom. Med.*, 59, 1997, 339–349.

Petitto, J.M., D.T. Lysle, J. Gariepy, P.H. Clubb, R.B. Cairns, M.H. Lewis, Genetic differences in social behavior: Relation to natural killer cell function and susceptibility to tumor development, *Neuropsychopharmacology*, 8, 1993, 35–43.

Petitto, J.M., D.T. Lysle, J. Gariepy, M.H. Lewis, Association of genetic differences in social behavior and cellular immune responsiveness: effects of social experience, *Brain Behav. Immun.*, 8, 1994, 111–122.

Sakic, B., L. Gurunlian, S.D. Denburg, Reduced aggressiveness and low testosterone levels in autoimmune MRL-1pr males, *Physiol. Behav.*, 63, 1998, 305–309.

Sapolsky, R.M. and E.M. Spencer, Insulin-like growth factor I is suppressed in socially subordinate male baboons, *Am. J. Physiol.*, 273, 1997, R1336–R1351.

Stark, J.L., R. Avitsur, D.A. Padgett, K.A. Campbell, F.M. Beck, J.F. Sheridan, Social stress induces glucocorticoid resistance in macrophages, *Am. J. Physiol.*, 280, 2001, R1799–R1805.

Stefanski, V., Social stress in loser rats: opposite immunological effects in submissive and subdominant males, *Physiol. Behav.*, 63, 1998, 605–613.

Stefanski, V. and S. Ben-Eliyahu, Social confrontation and tumor metastasis in rats: defeat and B-adrenergic mechanisms, *Physiol. Behav.*, 60, 1996, 277–282.

Stefanski, V. and H. Engler, Effects of acute and chronic social stress on blood cellular immunity in rats, *Physiol. Behav.*, 64, 1998, 733–741.

Stefanski, V. and H. Hendrichs, Social confrontation in male guinea pigs: behavior, experience, and complement activity, *Physiol. Behav.*, 60, 1996, 235–241.

Stefanski, V., G. Knopf, S. Schulz, Long-term colony housing in Long Evans rats: immunological, hormonal, and behavioral consequences, *J. Neuroimmunol.*, 114, 2001, 122–130.

Tewes, U., Concepts in psychology. In Psychoneuroimmunology: an interdisciplinary introduction, in M. Schedlowski and U. Tewes, Eds., Kluwer Academic/Plenum Publishers, Dordrecht, 1999, pp. 93–111.

Vekovishcheva, O.Y., I.A. Sukhotina, E.E. Zvartau, Co-housing in a stable hierarchical group is not aversive for dominant and subordinate individuals, *Neurosci. Behav. Physiol.*, 30, 2000, 195–200.

10 Genetics and Aggression

A frequently asked question is: What role does genetics or heredity play in the expression of behavior in animals and in humans? In effect, this question is addressing one of the oldest issues confronting students of ethology and related fields of animal behavior, the "nature" versus "nurture" debate, that is, the relative contributions of genetic and environmental factors to given behavioral processes. In the present context, this question relates directly to how genetic factors may influence rage and aggression. In this chapter, evidence is presented in support of genetic factors that contribute to these forms of emotional behavior. Moreover, attempts are made to relate the findings of these studies to the wider body of literature on the neurochemistry of rage and aggression.

STUDIES CONDUCTED IN ANIMALS

EARLY BEHAVIORAL STUDIES

Numerous studies conducted in the first half of the 20th century provided the first supporting data on the relationship between genetic factors and aggressive behavior. In perhaps the first study of its kind, Yerkes (1913) compared wild and tame rats and their F1 and F2 hybrids with respect to wildness, savageness, and timidity. Using behavioral measures including biting, exposing of the teeth, squealing, and jumping, it was concluded that savageness and timidity are heritable behavioral traits. Using a slightly different approach, Utsurikawa (1917) compared inbred and randomly bred strains of albino rats with respect to aggressive tendencies. Savageness was induced after the investigator scratched the floor of the cage with a copper wire, and the animals would bite viciously at the wire. The results showed that females surpassed males in vicious behavior, and that inbred animals were more savage than those that were bred randomly. A related approach was used by Coburn (1922) who applied Mendelian laws to determine the role of heredity in aggressive behavior. Here, the author used crosses in three generations of mice (i.e., tame vs. captured "wild" mice) and concluded that aggressive characteristics did have a hereditary basis. Following this line of research, Keller 1942) made use of the principle of "pleiotropy," which means that a single mutant gene can generate multiple, unrelated effects at clinical or phenotypical levels. In this particular study, Keller was able to show the relationship between coat color and temperament by contrasting gray "wild" rats with black "tame" rats.

Scott (1940) conducted a series of studies in order to delineate more systematically the behavioral characteristics of aggressive behavior in male mice. One of the major findings of these studies was that different strains of mice displayed extreme differences in reactions toward their partners with respect to response latency for initiating fights, thus supporting a general hereditary hypothesis for the expression of aggression. Lagerspetz (1961) studied other characteristics of genetically manipulated offspring of mice. In this study, mice selectively bred over three generations displayed motor activity that was greater in aggressive than in nonaggressive animals. In addition, it was noted that S3 descendants of aggressive mice were less emotionally reactive than controls.

Thus, the early behavioral studies described above provide evidence supporting the view that hereditary factors do, in fact, play a role in the propensity for expression of aggression and rage behavior.

RECENT STUDIES

In recent years, more sophisticated approaches to the study of genetics and emotional behavior and aggression have been employed. These approaches have focused on a variety of manipulations, including observations of the effects of chromosome alterations and mutations involving genetic deficiencies in specific neurotransmitter and related systems on aggressive behavior. The findings of these studies are described below.

Sluyter et al. (1995) sought to determine the effects of different parts of the Y chromosome of wild house mice on aggression. The strategy employed was to establish intercrosses between two selection lines for attack latency and their congenic strains. This procedure resulted in F1 hybrids that carried the same autosomes, but differed in their X chromosome and the two different parts of their Y chromosome (referred to as an autosomal and nonrecombining part). The authors concluded that both parts of the Y chromosome contribute to changes in aggression with the major change associated with the nonrecombining part of the aggressive parent. In a separate but related study, Sluyter et al. (1994) demonstrated that genetically more aggressive mice were much more sensitive to apomorphine with respect to stereotyped responses than nonaggressive mice.

The likely genetic disposition for aggressive behavior by specific strains of mice was used in a series of studies conducted by Sandnabba and colleagues (Sandnabba, 1994; Sandnabba et al., 1994; Sandnabba and Korpela, 1994). After demonstrating that various strains of mice can be selectively bred for high or low levels of aggression, these investigators sought to determine the effects of several variables on aggressive behavior. In one study (Kouri et al., 1995), it was noted that male mice exposed to mating early in life had higher levels of aggression, and that genetic disposition toward aggression was related to higher levels of sexual activity. In a parallel study (Schenberg et al., 1995), testosterone propionate given to a genetically aggressive strain of mice resulted in enhanced levels of aggression, and such levels were reduced in isolated male mice of the same line. In contrast, mice that exhibited low levels of aggression were relatively unaffected by isolation or testosterone propionate treatment. This finding suggests that when genes related to aggression are present, the male hormone has an enhancing effect on aggression.

Recently, Brodkin et al. (2002) provided an elegant analysis that identified the natural gene alleles that contribute to both individual and strain differences in aggression. These authors employed an outcross–backcross breeding protocol as well as a genome-wide scan, which enabled them to identify an aggressive behavior quantitative trait localized on distal chromosome 10 and the proximal chromosome X. The authors suggested that candidate genes for aggression included dicylycerol kinase α subunit gene and the glutamate receptor subunit AMP-3 gene. The significance of this finding is that it is the first of its kind to provide a genome-wide scan in a mammal, the net effect of which may lead to a greater clarification of the mouse alleles that affect aggression as well as alleles that could account for individual differences in aggression within the same and other mammalian species.

GENETIC EFFECTS ON NEUROANATOMIC CIRCUITRY AND NEUROTRANSMITTER–RECEPTOR SYSTEMS

Several of these studies were reviewed in Chapter 7, but are considered again because of their direct relevance to the current chapter.

A somewhat different approach was used by Guillot et al. (1994), who attempted to relate various strains of mice to alterations in aggressive behavior as well as to variations in selective neuroanatomic circuits. In this study, large strain differences with respect to aggressive tendencies were observed, which were strongly correlated to differences in the size of the hippocampal and infrapyramidal mossy-fiber terminal fields. In particular, the authors found a large negative correlation between the propensity to attack and the size of the hippocampal terminal field, leading the authors to conclude that differences in aggressive tendencies are related to neuroanatomic variations in specific brain structures, both of which are genetically determined. While these authors have focused their attention on intrahippocampal pathways, it would be of interest to determine whether similar changes might be observed in other neuronal groups that have also been classically associated with aggression and rage.

Numerous studies have been conducted that have tried to relate genetic manipulations affecting selective neurotransmitter systems with respect to aggression and rage. In one of the first such studies, Saudou et al. (1994) attempted to determine the effects of the role of 5-HT1B receptors in aggression in mice by producing mutant mice lacking this receptor. The authors reported that these animals showed no obvious developmental or behavioral deficits. With respect to aggressive behavior, the authors reported that when confronted with an intruder, these mutant mice attacked the intruder faster and more intensely than even wild-type mice. When the mice were administered the 5-HT1A/1B agonist, it eliminated only the hyperlocomotor effects observed in these mutants, leading the authors to conclude that the principal basis for enhanced aggressive responses was due to the absence of the 5-HT1B receptor.

In a related study, Chiavegatto et al. (2001) used mice lacking the specific genes for neuronal nitric oxide synthase, and the neural adhesion molecule, which is believed to affect 5-HT turnover and 5-HT receptor sensitivity. The results of this study were consistent with the findings described above in that mice lacking nitric oxide synthase were highly aggressive. The authors concluded that the heightened aggression was the result of a selective decrease in 5-HT turnover and a possible deficiency in the 5-HT1A or 5-HT1B receptor in brain regions regulating aggression.

Chen et al. (1994) provided another piece of evidence related to these studies. In this study, it was shown that mice deficient in α-calcium-calmodulin-dependent kinase II resulted in an increase in defensive aggression without inducing any known cognitive deficits. This author further reported that in these animals, there was a reduction in 5-HT release in serotonergic neurons of the dorsal raphe. Such an interpretation would clearly be consistent with the findings of the studies described above.

While the above studies are consistent with the large body of literature presented in Chapter 7, there is a caveat about the general design and organization of these studies that should be noted. On the one hand, the use of genetic "knockout" mice for the study of aggression and other processes may result in significant progress in this field. On the other hand, difficulties in interpretation of the data may result from nonspecific effects generated from this approach, and thus may lead to false conclusions. For example, it is possible that the absence of a given receptor–neurotransmitter may result in heightened aggression, but that the effect may be an indirect effect due, perhaps, to loss of sleep or a greater sensitization to pain stimuli. Second, because the loss of a given receptor–neurotransmitter would affect all parts of the central nervous system, interpretation of such findings due to the effects on all brain systems associated with the receptor in question is problematic. Nevertheless, and in spite of these problems, the novelty of the approach may provide new ways of studying the role of specific receptor systems in the regulation of aggression and rage.

STUDIES CONDUCTED IN HUMANS

The basic question raised from the studies conducted in animals is whether a genetic disposition for aggression in humans exists. A number of studies have been conducted over the past decade that provide evidence supporting the view of a genetic linkage to aggression and rage. Several of these findings are summarized here.

One line of investigation carried out by several investigators has been to use a co-twin approach in order to ascertain the presence or absence of a genetic disposition for aggression. Gottesman and Goldsmith (1994) reported a higher correlation between identical twins than between fraternal twins with respect to aggression, leading these investigators to conclude that both genetic as well as environmental factors exist that contribute to the organization of aggressive behavior. Grove et al. (1990) compared identical twins reared apart at an early age and drew similar conclusions, which were based on scores determined from childhood and adult antisocial behavior, alcohol- and drug-related problems. Their findings showed within-scale cohesion (heritability) as measured by Chronbach's coefficient α, in particular for drug and antisocial scales.

Another approach has been to attempt to correlate neurochemical measurements with antisocial behavior. In one study, Caspi et al. (2002) studied maltreated male children with a genotype of high levels of monamine oxidase A expression. It was observed that these children were more likely to develop antisocial behavior, suggesting that genotype can moderate sensitivity to environmental events. Lappalainen et al. (1998) observed that Finnish antisocial alcohol-dependent patients had a significantly higher HTRIB-861C allele frequency than other Finns. The paired linkage of antisocial alcoholism to this allele suggested to these authors that the predisposition to antisocial behavior and aggressiveness was clearly connected to this allele. More recently, Underwood and Mann (2003) reviewed the literature concerning genetics, aggression, and antisocial behavior and, in general, reached similar conclusions. However, a study by Nielsen et al. (1994) attempted to provide evidence of a linkage between the tryptophan hydroxylase (TPH) genotype and the presence of SCF 5-HIAA in impulsive and nonimpulsive alcoholic violent offenders and healthy controls. These authors observed a relationship between TPH and 5-HIAA in the cerebrospinal fluid, but failed to show a similar relationship between the TPH genotype and impulsive behavior.

Similar to animal studies, there are concerns associated with the design and organization of studies linking genetic disposition to human aggression. There are several key issues relevant to these studies. One is the relatively small sample sizes used in a number of these studies. Another concerns differences in the environments of identical versus fraternal twins or of other control groups used. However, in spite of these problems that are generic to these methods, the data obtained from such studies provide evidence consistent with the animal studies that reveal a linkage between differences in genotype and the propensity for elicitation of aggression and impulsive behavior.

SUMMARY

The studies summarized in this chapter provide reasonable support for the existence of a genetic basis for aggressive and impulsive behavior. As might be expected, the data obtained from animal studies, including those employing knockout mice, provide the most compelling evidence in support of a genetic connection. Likewise, the data obtained from the majority of human studies provide less compelling but, nevertheless, supporting data consistent with the findings from animal studies. Potential pitfalls associated with both animal and human studies were noted. In spite of these problems, these studies provide very promising data that could very well open new lines of investigation that could reveal the genetic predispositions for elicitation of aggressive behavior.

REFERENCES

Brodkin, E.S., S.A. Goforth, A.H. Keene, J.A. Fossella, L.M. Silver, Identification of quantitative trait loci that affect aggressive behavior in mice, *J. Neurosci.*, 22, 2002, 1165–1170.

Caspi, A., J. McClay, T.E. Moffitt, J. Mill, J. Martin, I.W. Craig, A. Taylor, R. Poulton, Role of genotype in the cycle of violence in maltreated children, *Science*, 297, 2002, 851–854.

Chen, C., D.G. Rainnie, R.W. Green, S. Tonegawa, Abnormal fear response and aggressive behavior in mutant mice deficient for alpha-calcium-calmodulin kinase II, *Science*, 266, 1994, 291–294.

Chiavegatto, S., V.L. Dawsons, L.A. Mamounas, V.E. Koliatsos, T.M. Dawson, R.J. Nelson, Brain serotonin dysfunction accounts for aggression in male mice lacking neuronal nitric oxide synthase, *Proc. Natl. Acad. Sci. U.S.A.*, 98, 2001, 1277–1281.

Coburn, C.A., Heredity of wildness and savageness in mice, *Behav. Monogr.*, 4, 1922, 1–71.

Gottesman, I.I. and H. Goldsmith, Developmental psychopathology of antisocial behavior: inserting genes into its ontogenesis and epigenesis, in C.A. Nelson, Ed., *Threats to Optimal Development: Integrating Biological, Psychological and Social Risk Factors*, Lawrence Erlbaum Associates, Hillsdale, N.J., 1994, pp. 69–104.

Grove, W.M., E.D. Eckert, L.L. Heston, T.J. Jr. Bouchard, N.L. Segal, D.T. Lykken, Heritability of substance abuse and antisocial behavior: a study of monozygotic twins raised apart, *Biol. Psychiatry*, 27, 1990, 293–304.

Guillot, P.-V., P.L. Roubertoux, W.E. Crusio, Hippocampal mossy fiber distributions and intermale aggression in seven inbred mouse strains, *Brain Res.*, 660, 1994, 167–169.

Keller, C.E., The association of the black, non-agouti, gene with behavior in the Norway rat, *J. Heredity*, 33, 1942, 371–384.

Kouri, E.M., S.E. Lukas, H.G. Pope Jr., P.S. Oliva, Increased aggressive responding in male volunteers following the administration of gradually increasing doses of testosterone cypionate, *Drug Alcohol Depend.*, 40, 1995, 73–79.

Lagerspetz, K., Genetic and social causes of aggressive behaviour in mice, *Scand. J. Psychol.*, 2, 1961, 167–173.

Lappalainen, J., J.C. Long, N. Ozaki, R.W. Robin, G.L. Brown, H. Naukkarinen, M. Virkkunen, M. Linnoila, D. Goldman, Linkage of antisocial alcoholism to the serotonin 5-HT1B receptor gene in 2 populations, *Arch. Gen. Psychiatry*, 55, 1998, 989–994.

Nielsen, D.A., D.V.M. Goldman, R. Tokola, R. Rawlings, M. Linnoila, Suicidality and 5-hydroxyindoeacetic acid concentration associated with a tryptophan hydroxylase polymorphism, *Arch. Gen. Psychiatry*, 51, 1994, 34–38.

Sandnabba, N.K., Predatory behaviour in females of two strains of mice selectively bred for isolation-induced intermale aggression, *Behav. Processes*, 34, 1994, 93–100.

Sandnabba, N.K. and S.R. Korpela, Effects of early exposure to mating on adult sexual behavior in male mice varying in their genetic disposition for aggressive behavior, *Aggress. Behav.*, 20, 1994, 429–439.

Sandnabba, N.K., K.M. Lagerspetz, E. Jensen, Effects of testosterone exposure and fighting experience on the aggressive behavior of female and male mice selectively bred for intermale aggression, *Horm. Behav.*, 28, 1994, 219–231.

Saudou, F., D.A. Amara, A. Dierich, M. Lemeur, S. Ramboz, L. Segu, M.C. Buhot, R. Hen, Enhanced aggressive behavior in mice lacking the 5ht1b receptor, *Science*, 265, 1994, 1875–1878.

Schenberg, L.C., C.A.L. Brandao, E.C. Vasquez, Role of periaqueductal gray matter in hypertension in spontaneously hypertensive rats, *Hypertension*, 26, 1995, 1125–1128.

J.P. Scott, Hereditary differences in social behavior, fighting of males, betwen two inbred strains of mice, *Anat. Rec. Suppl.*, 79, 1940, 103.

Sluyter, F., B.J. Meijeringh, G.A. Van Oortmerssen, J.M. Koolhaas, Studies on wild house mice, VIII: postnatal maternal influences on intermale aggression in reciprocal F1's, *Behav. Genet.*, 25, 1995, 367–370.

Sluyter, F., G.A. Van Oortmerssen, J.M. Koolhaas, Studies on wild house mice VI: Differential effects of the Y chromosome on intermale aggression, *Aggress. Behav.*, 20, 1994, 379–386.

Underwood, M.D. and J.J. Mann, The neurochemical genetics of serotonin in aggression, impulsivity, and suicide, in M.P. Mattson, Ed., *Neurobiology of Aggression: Understanding and Preventing Violence*, Humana Press, Totowa, N.J., 2003, pp. 65–72.

Utsurikawa, N., Temperamental differences between outbred and bred strains of the albino rat, *J. Anim. Behav.*, 7, 1917, 111–129.

Yerkes, R.M., The heredity of savageness and wildness in rats, *J. Anim. Behav.*, 7, 1913, 11–28.

11 Future Directions and Perspectives

A basic objective of this book is to provide both historical and up-to-date selective reviews of the major areas of research that comprise the neurobiology of aggression and rage in both animals and humans. After having reviewed the literature, it is appropriate at this time to step back and address the following questions: (1) Where should we go with our research at this time? Can we identify what would appear to be the most heuristic lines of research that should now be addressed? (2) Is it possible to use our knowledge of the neurobiology of aggression and rage in order to develop strategies by which we can control our impulsive aggressive tendencies in response to specific environmental events? The following discussion attempts to address these basic questions.

From our review and analysis of the literature, I would submit that the following lines of research represent fertile areas of investigation for the present and over the next decade. While I believe these areas of research to be important, the list is not intended in any sense to be exclusive.

As indicated in Chapter 3, a basic element essential to our understanding of the neurobiology of aggression and rage is knowledge of the underlying anatomical substrates over which various forms of aggression are mediated. While substantial progress along these lines has been made, there yet remain significant gaps in our knowledge of these pathways. For both defensive rage and predatory attack, we have little understanding of the lower brainstem circuitry mediating these processes, which include, in particular, the relationships of these pathways to cranial nerve nuclei and spinal cord motor and autonomic neurons that contribute to the expression of these behaviors. There are numerous contributions of this avenue of research. For one, it would enable a comparison of the pathways mediating similar forms of aggression in different species. By comparing the similarities as well as differences in the circuitry among various species, it would provide new insights into the organization and nature of the anatomic substrates for specific forms of aggression. In addition, such comparisons would provide a further basis by which knowledge of the receptor–neurotransmitter mechanisms situated at key synapses in the pathways that mediate rage and aggression. This kind of information would provide a better perspective of the commonality of neurotransmitters/receptors critical for the expression of these behaviors, the significance of which could have an important impact to future studies involving the neurochemistry of aggression and rage.

Related to future studies concerning neuroanatomical approaches are those that will seek to generate a better understanding of the neurochemical mechanisms integral to the expression and control of aggression and rage. Here, several lines of research would appear to be of considerable importance. One direction clearly in need of investigation is to explore in greater detail the role of neuropeptides that may act either as neurotransmitters or as neuromodulators and to identify their synaptic loci within neuronal cell groups where their effects are mediated. We have only begun to explore several of these peptides, including substance P, cholecystokinin, and opioids. There are many others, such as gut peptides in addition to cholecystokinin, and peptides normally associated with the regulation of pituitary function (i.e., releasing hormones). A second approach would be to explore how other peptides recently shown to be present in the central nervous system but which have just begun to be associated with the regulation of neurotransmitter functions. Examples of these include the role of cytokines, which was addressed briefly in Chapter 9, and leptins, which have been linked to the regulation of feeding behavior. In particular, these studies should be aimed, in part, at identifying the neurotransmitter–receptor mechanisms through which their effects are mediated. A third approach in need of exploration includes identification and analysis of the molecular mechanisms associated with the various neurotransmitters and neuromodulators that have already been shown to play important roles in the regulation of aggression and rage. This approach could have important implications for the development of "serenic" drugs for the treatment of violence and aggression. A fourth approach is our need to understand: (1) what neurotransmitters are released during the rage or aggressive process, (2) the synaptic regions where they are released, and (3) what effects blockade of their release at specific synaptic regions would have in regulating these behaviors.

Another area of investigation that deserves attention involves an aspect of the subject of hormones and aggression. Although considerable knowledge has been obtained concerning the roles of various sex hormones on aggression, we have very little understanding of the mechanisms underlying such hormonal modulation. In particular, we do not know where in the forebrain sex hormones may interact with neurotransmitter–receptors to modulate aggression and rage. Moreover, we also have no idea what these neurotransmitter–receptors are or how they may affect the local neuronal populations. Studies addressing these issues are essential for us to gain an understanding of how hormones such as testosterone and estrogen affect aggression and rage. Hopefully, in the next few years investigators will begin to focus on these problems.

An evolving area of research is the linkage between genetics and behavior. With the recent development of molecular methodologies, which have provided us with tools such as "knockout" mice, investigators are now in a position to be able to address in a different way the role of specific neurotransmitter–receptors in aggression and rage. In fact, this represents a very novel approach by which a wide variety of receptor systems could be studied for their effects on these processes. Indeed, a significant advantage of this methodology is that it can conceptually allow one to study receptor systems that are unavailable for analysis by traditional neurochemical or neuropharmacologic methods. Likewise, this methodology may also provide a better understanding of the genetic factors that contribute to aggression and rage, which is somewhat lacking in the literature. However, for this approach to be effective, several of the pitfalls and difficulties described in Chapter 10 have to be addressed and overcome.

Finally, there is also a need for further examination in a systematic manner to identify the relationship between brain regions displaying impaired function and aggressive and impulsive tendencies in humans. With the development of newer and more sophisticated methods such as f-MRIs and PET scans that can be applied to the study of human behavior, there is now a greater opportunity for investigators to be able to identify more precisely the neuronal cell groups and regions linked to the expression of aggression, rage, and impulsivity. In addition, there is also a need to develop better questionnaire and other testing techniques that can classify individuals more clearly in terms of the extent to which they fit into categories of predatory or defensive (affective) forms of aggression. Such efforts could then provide the basis for the design of experiments that could more clearly identify the relationship between transmitter levels and function and each of these forms of aggression.

Can our knowledge of the neurobiology of aggression and rage help us as individuals in reducing our tendencies to display aggressive, hostile, and impulsive behavior? As indicated in the beginning of Chapter 1, violence has been and continues to be a major concern in the United States and elsewhere. While a discussion of the subject of criminal activity and its control goes well beyond the scope of this book, the issue of our individual capacity to control and release ourselves from anger, impulsivity, and hostility relates very much to what has been discussed thus far. The question can thus be raised about whether the knowledge of the neurobiology of aggression and rage that we have acquired can now be applied to develop strategies to control these tendencies. In the following, I suggest how this objective may be accomplished.

The primary assumption on which the following arguments are based is that there is (are) a neural basis (or bases) for establishing these strategies. One such substrate, shown in Figure 11.1, focuses on the role of the prefrontal cortex. Several features unique to the prefrontal cortex make it a prime candidate in this context. One unique feature is that the prefrontal cortex receives inputs from all regions of the cerebral cortex, limbic structures, and the medial thalamus (Fuster, 1980). Such inputs allow for the mediation and integration of all types of sensory and emotional signals. Second, the prefrontal cortex projects both directly and indirectly to the hypothalamus, midbrain periaqueductal gray, and limbic structures, the most notable of which is the amygdala. Thus, from a neuroanatomical perspective, the prefrontal cortex is capable of receiving sensory information, integrating it, and providing a descending output to regions controlling emotional behavior in response to these signals. Third, the prefrontal cortex is also known as a region associated with higher cognitive functions as well as the regulation of emotional processes. Therefore, these properties of the prefrontal cortex endow it with the capacity to receive information from the external environment and translate that information into both a cognitive as well as emotional output. One other assumption concerning neurons in the prefrontal cortex should be indicated; namely, that these neurons are capable of conditioning. This assumption is quite reasonable since neurons in many areas of the cortex and other regions of the brain such as the hippocampal formation and cerebellum also have this property. If we can accept this analysis and the underlying assumptions as reasonable, we can now postulate how we can utilize the unique

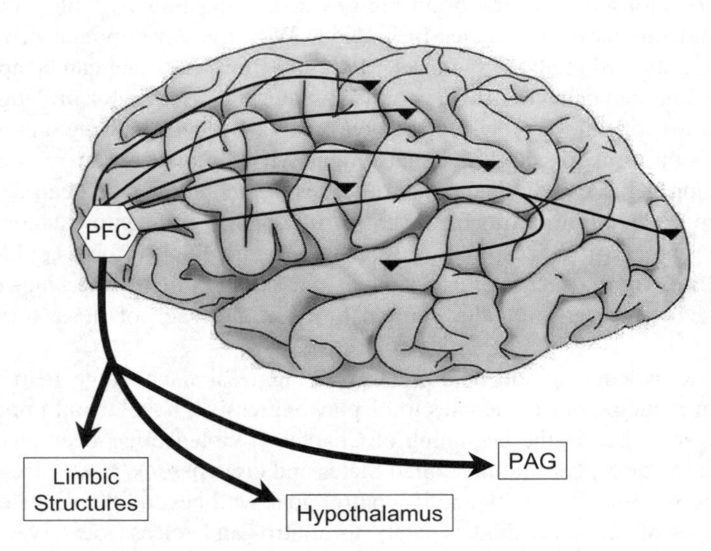

FIGURE 11.1 (1) Afferent sources of input to prefrontal cortex from other regions of cerebral cortex; and (2) outputs of prefrontal cortex to limbic structures, hypothalamus, and midbrain periaqueductal gray (PAG). The prefrontal cortex is unique in that it receives inputs from most other aspects of the cerebral cortex, including primary and secondary sensory regions. In addition, it has the capacity to influence structures of the forebrain that mediate or regulate aggression and rage behavior. These include the hypothalamus and PAG, from which these forms of behavior are expressed, and the limbic system, which powerfully modulates the hypothalamus and PAG and their associated behaviors.

properties of the prefrontal cortex to free ourselves from anger, hostility, and impulsive behavior. The following behavioral strategies are predicated on these conclusions concerning the prefrontal cortex, and make use of the principles of behavioral conditioning techniques, some of which have been described in a somewhat different context by Pliskin (1996).

In order for anger control strategies to be effective, it is necessary to first acknowledge the importance of achieving this goal. While not everyone would agree that such an objective is warranted or necessary, I would argue otherwise, and, in support of this position, the advantages of reducing feelings of anger are presented at this time. Examples of what anger reduction can achieve include

1. The positive feeling of experiencing self-improvement and control over one's feelings
2. The ability to be able to learn from others
3. A capacity to take responsibility for one's actions and not to blame others for actions taken
4. The positive effect it has on one's cardiovascular system by reducing the likelihood of elevating blood pressure

5. The avoidance of a behavioral response that is inherently aversive in nature. In fact, sages conveyed this position approximately 2000 years ago with the saying: "Who is strong? He who conquers his evil inclination, as it is said: 'Better is one slow to anger than a strong man, and one who rules over his spirit than a conqueror of a city'" (Ethics of the Fathers, Chapter 4). In other words, one's strength should not be measured in terms of physical prowess, but in terms of one's ability to control his passions and show restraint.

What then are the conditioning techniques that can be applied to control anger? An answer to this question basically requires a multidimensional approach involving the utilization of internal and external sensory signals, including those present in the environment. Concerning internal cues, one approach is to apply the principles of "feedback" that have been used with some success in controlling autonomic functions. Here, it is important to develop sensitivity to the internal stimuli associated at the time at which one's anger is beginning to develop. By sensing increased blood pressure and muscle tension in response to a given event, these sudden internal biological changes can be used by an individual as warning signals to back off from what he or she is about to do or say. In this way, such signals could serve as conditioning stimuli to induce a conditioned response (i.e., control of one's anger and impulsivity).

In addition to utilizing internal clues, much more can be said about external events that drive people to anger. In addressing this issue, a slight digression is necessary at this time. The question must first be asked: Is there a single factor more than any other that contributes to states of anger? It may be suggested that there is such a factor, and that factor is "arrogance" or "haughtiness," which is endemic to most of us. Obviously, the extent to which arrogance tends to affect response probabilities for anger, violence, and impulsive behavior varies among people. If our tendencies of arrogance or haughtiness are sources of anger, then the antidote for this behavior is "humility." The ability to develop humility is not simple, but is rather a life-long process. It requires that people realize they are not perfect and that they can make mistakes like everyone else. It requires that one appreciates the beauty, importance, and value of other individuals, and thus see others in a positive light. By viewing the world in this way, we are enabled to establish a different perception of how to interpret events that might impact on us. In other words, an external event that could lead an individual to express anger could have a more neutral or dampening effect on her or his emotionality. This kind of change basically involves a conditioning process requiring that an individual must constantly remind himself to take a positive view of others, in particular when an aversive or provocative situation arises. Several approaches can be applied to enhance a sense of humility. One strategy is to take a few moments at bedtime to evaluate one's conduct that day. If there was a particular event or response that was unpleasant or inappropriate, recapitulation of these events could serve to reduce the likelihood that such a response is repeated in the future. In essence, acknowledgment of the mistakes that we make facilitates, by virtue of a conditioning process, the ability to control impulsivity and anger. Through an introspective approach, people can develop a better sense of their own self-worth, which includes an accurate assessment of their strengths and weaknesses. This is also what humility is all about — knowing what we do well and what we do not do well. Conscious awareness of our strengths and weaknesses would also serve as important conditioning stimuli in allowing us to avoid inappropriate responses after experiencing a negative or aversive event.

A number of other conditioning procedures can be applied as well. One approach, presented here as an example, relates to a principle referred to as "reframing." This term is derived from the art world where a frame is a part of the picture. When the frame is changed, the appearance of the picture would also likely change as well. In order to better understand this concept, we must first ask the question: Why is it that two different people react in diametrically opposed ways (i.e., one becomes very angry and the other is nonresponsive) to a given event? The answer to this question relates to how each of these individuals perceives that event. It may be argued that the individual who reacts with anger to a specific event is using an inappropriate frame about the events. This reaction may be linked to feelings of insecurity or feelings that were hurt as result of a previous experienced. The trick here is for the individual to find ways of reframing such events. As noted above, one approach is to express humility, which is intrinsically bound up with modesty and a rational sense of self-worth. As a result, the individual may then look differently at a specific event, causing deflation of the expression of anger. Consider a situation that is probably familiar to all of us, namely, experiencing an insult. There are several ways in which an individual may react to the insult. One way of dealing with the insult is to elicit a verbal or physical assault on the person giving the insult, the effect of which would be to create a much more difficult and painful situation at a later time. Alternatively, the person receiving the insult may consider two possibilities: the first is that the insult is so distant from reality and the person delivering it is suffering from hallucinations or another major perceptual deficit. If this is the case, it would almost be silly to respond to someone who is suffering from such a perceptual disorder since a response becomes equally meaningless. The second possibility is that the person delivering the insult is correct in that the insult accurately reflects something negative in a statement or act by the individual at whom the insult is aimed. Friends are not likely to point out such personality or character flaws for fear of hurting the individual's feelings. Thus, it is left up to the one who may not care very much for the individual in question to express such an insult for the purpose of hurting his feelings or otherwise "putting him down." If this were the case, how then should the individual react? Again, he can express anger and more. Or, he can use a reframing technique. He may think as follows: "The person delivering the insult has done a favor for me. He, alone, has pointed out a personality flaw that nobody else would dare to do. Therefore, I have to be thankful for what he has done, even though his motives were far from righteous. By interpreting the insulting event in this manner, there is also no need to become angry or act impulsively since the individual delivering the insult has been very helpful to me." The key issue here is that the principle of conditioning is now linked to the mechanism of reframing. Specifically, if the individual now is cognizant of this argument, the next time that he experiences a situation where he is insulted, his thoughts of this argument will come to mind, producing a reframe of the event, and thus result in a deflation of any emotional response that might otherwise have been generated by the elicitation of the insulting remarks. The same arguments can now be applied to many other events where anger may be aroused, such as not acting out "road rage" in response to the poor driving of others, or reducing or eliminating the anger and irritation that can typically occur when stuck in the middle of a traffic jam.

The discussion presented above represents my views on how we can take the principles of neuroscience and behavioral conditioning linked to regional circuitry in the brain and use that information to develop ways of controlling anger, impulsivity, and violent behavior. Obviously, there are many other approaches that can be applied to the control of emotional behavior that have not been considered in this chapter, as such a discussion would have gone well beyond its scope and purpose. There are many excellent sources in the fields of psychology, sociology, and theology that focus on this issue, and such reading is certainly recommended.

REFERENCES

Ethics of the Fathers, Chapter 4, Mishna 1a.
Fuster, J.M., *The Prefrontal Cortex: Anatomy*, Raven Press, New York, 1980.
Pliskin, Z., *Anger! The Inner Teacher*, Artscroll, New York, 1996.

Index

A

B

E

F

G

H

I

K

L

M

N

O

P

R

S

T

V